DECISIONS AT THE WILDERNESS
AND SPOTSYLVANIA COURT HOUSE

OTHER BOOKS IN THE COMMAND DECISIONS
IN AMERICA'S CIVIL WAR SERIES

*Decisions at Stones River: The Sixteen Critical Decisions
That Defined the Battle*
Matt Spruill and Lee Spruill

*Decisions at Second Manassas: The Fourteen Critical Decisions
That Defined the Battle*
Matt Spruill III and Matt Spruill IV

*Decisions at Chickamauga: The Twenty-Four Critical Decisions
That Defined the Battle*
Dave Powell

*Decisions at Chattanooga: The Nineteen Critical Decisions
That Defined the Battle*
Larry Peterson

*Decisions of the Atlanta Campaign: The Twenty-One Critical Decisions
That Defined the Operation*
Larry Peterson

*Decisions of the 1862 Kentucky Campaign: The Twenty-Seven
Critical Decisions That Defined the Operation*
Larry Peterson

DECISIONS
AT THE WILDERNESS AND SPOTSYLVANIA COURT HOUSE

The Eighteen Critical Decisions
That Defined the Battles

Dave Townsend

Maps by Tim Kissel

COMMAND DECISIONS
IN AMERICA'S CIVIL WAR

The University of Tennessee Press / Knoxville

Copyright © 2020 by The University of Tennessee Press / Knoxville.
All Rights Reserved. Manufactured in the United States of America.
First Edition.

Library of Congress Cataloging-in-Publication Data

Names: Townsend, Dave, 1947– author.
Title: Decisions at the Wilderness and Spotsylvania Court House: the eighteen critical decisions that defined the battles / Dave Townsend ; maps by Tim Kissel.
Description: First edition. | Knoxville, TN: University of Tennessee Press, [2020] | Series: Command decisions in America's Civil War | Includes bibliographical references and index. |
Identifiers: LCCN 2019013798 (print) | LCCN 2019015938 (ebook) | ISBN 9781621905271 (Kindle) | ISBN 9781621905288 (pdf) | ISBN 9781621905264 (pbk.)
Subjects: LCSH: Overland Campaign, Va., 1864. | Wilderness, Battle of the, Va., 1864. | Spotsylvania Court House, Battle of, Va., 1864. | Tactics—Case studies. | Grant, Ulysses S. (Ulysses Simpson), 1822-1885—Military leadership. | Lee, Robert E. (Robert Edward), 1807-1870—Military leadership. | Virginia—History—Civil War, 1861-1865—Campaigns.
Classification: LCC E476.52 (ebook) | LCC E476.52 .T69 2020 (print) | DDC 973.7/36—dc23
LC record available at https://lccn.loc.gov/2019013798

CONTENTS

Preface	xi
Introduction	1
1. Before the Battles	9
2. The Battle of The Wilderness	43
3. Transition to Spotsylvania Court House	83
4. The Battle of Spotsylvania Court House	113
5. Conclusion	161
Appendix I. Battlefield Guide to the Critical Decisions at The Wilderness and Spotsylvania Court House	175
Appendix II. Union Order of Battle	303
Appendix III. Confederate Order of Battle	319
Notes	331
Bibliography	365
Index	375

ILLUSTRATIONS

Figures

Federal Wagon Park	4
Ulysses S. Grant, USA, and Robert E. Lee, CSA	5
Presidential Election Poster	11
Pres. Abraham Lincoln, USA	13
Lincoln's Nomination of Grant for Lieutenant General	16
Lieut. Gen. Ulysses S. Grant, USA	17
Maj Gen. George S. Meade, USA	21
Maj. Gen. Phil Sheridan, USA	25
Grant and Staff after the Battle of Spotsylvania Court House	30
Brig. Gen. Andrew A. Humphreys, USA	44
Maj. Gen. Gouverneur K. Warren, USA	47
Wilderness Tavern	48
Saunders Field, 1864	52
Saunders Field, Present Day	53
Terrain along the Orange Plank Road	57
Lieut. Gen. James Longstreet, CSA	59
Maj. Gen. Winfield S. Hancock, USA	63
Intersection of the Orange Plank Road and Brock Road	67
Confederate Entrenchments	68
Lieut. Gen. Richard S. Ewell, CSA	74

Maj. Gen. Jubal Early, CSA	74
Brig. Gen. John B. Gordon, CSA	75
Spotsylvania Court House	88
Todd's Tavern	96
Maj. Gen. Fitzhugh Lee, CSA	97
Maj. Gen. James E. B. Stuart, CSA	108
Laurel Hill, Present Day	117
Laurel Hill, 1864	121
Col. Emory Upton, USA	131
Upton's Attack along Dole's Salient	135
Apex of the Mule Shoe	139
Confederate Traverses	140
Hancock's View of the Salient and the "Bloody Angle"	149
Johnson's Defensive Works on the Salient	150
Gordon's View of the McCoull House	151
Remains of an Outbuilding of the Wilderness Tavern	186
View Looking West along the Original Orange Turnpike	187
View Down the Orange Turnpike in the 1860s	188
Battlefield Entrance and Grant's headquarters	191
View of Saunders Field Looking toward Ewell's Position in the 1860s	197
View of Saunders Field, Present-Day	197
Confederate Entrenchments at Saunders Field	213
Poague's Artillery Position across the Widow Tapp Field	225
Vermont Monument	244
Civil War–Era Photo of Todd's Tavern	247
Todd's Tavern Location, Present Day	247
View across Spindle Field to Laurel Hill	259
View of Spindle Field from Laurel Hill	260
Confederate Entrenchments at Spotsylvania Court House	265
Start of the "Wood Road"	267
View Looking South from Dole's Salient to Captain Smith's Confederate Battery	274
Scene of Capture of Johnson's Division	291
View Looking Northeast from the Bloody Angle	294
Harrison House	295

Maps

Grant's Options for Attacking Lee	35
Lee Moves to Attack Grant	45

Midday, May 5, 1864—Orange Turnpike	51
Midday, May 5, 1864—Orange Plank Road	55
Dawn, May 6, 1864—Hancock Attacks Hill's Corps	62
Afternoon, May 6, 1864—Lee Attacks	71
Evening, May 6, 1864—Gordon's Flank Attack	79
Grant's Plan for Moving South	89
Night, May 7–8, 1864—Anderson Pushes Through the Night	94
Midday, May 7, 1864—Lee's Delaying Action around Todd's Tavern	99
Night, May 7-8, 1864—Lee's Delaying Action around Todd's Tavern	100
Morning, May 8, 1864—Warren Attacks	118
May 9–20, 1864: Overview of Hancock's Movements	126
May 10, 1864, Evening—Upton's Attack	133
Evening, May 11, 1864—Confederate Artillery in the Mule Shoe	141
Dawn, May 12, 1864—Confederate Artillery Returning to the Mule Shoe	143
Dawn, May 12, 1864—Grant Commits a Full Corps	148
May 20–21, 1864—Grant Moves South, Lee Responds	159
Tour Stop Overview	184
Stop 1: Wilderness Tavern	185
Stop 2: Grant's/Meade's Headquarters	192
Stop 3: Saunders Field, Union Perspective	196
Stop 4: Saunders Field, Confederate Perspective	212
Stop 5: Chewning Farm	218
Stop 6: Widow Tapp Farm	224
Stop 7: Orange Plank and Brock Road Intersection	232
Stop 8: Todd's Tavern	246
Stop 9: Spindle Field/Laurel Hill	256
Stop 10: Upton's Attack on Dole's Salient	266
Stop 11: The Mule Shoe	280

Diagrams

Organization of Union Forces	38
Organization of the Army of Northern Virginia	38

PREFACE

Many years ago in high school, I read a book on the Civil War whose title I have long forgotten that included Robert E. Lee, and I was immediately captivated. Thus began a lifelong interest in the Civil War in general and a growing focus on Lee in particular. I read Douglas Southall Freeman's four-volume *R. E. Lee* and then his three-volume *Lee's Lieutenants* through college and my early engineering days. Over the next twenty years or so, I pursued this developing passion on my own, and with my wife graciously going to the battlefields with me (more for my sake than hers).

In the 1990s, I joined the Rocky Mountain Civil War Roundtable in the Denver area and found kindred spirits. The presentations at the meetings exposed me to constantly widening aspects of the Civil War. With study groups focusing on a particular battle or campaign followed by field trips back east to visit the relevant location, I found a whole new dimension to learning.

It was enjoyable and interesting to study a subject in-depth and make presentations, so when fellow roundtable members and authors Matt Spruill and Larry Peterson approached me about a book project on the Civil War, I was both flattered and intrigued. The challenge of writing a book on the critical decisions at the Battles of The Wilderness and Spotsylvania Court House has deepened my appreciation for the effort of previous historical authors. My decision to cover both battles in one book resulted from of the strong coupling of the fighting at the two locations. I felt that Grant's May 11, 1864, letter to his chief of staff Maj. Gen. Henry Halleck validated this decision. Therein, Grant referred to the conflict as "the sixth day of very heavy fighting." Unlike the orders and correspondence contained in official orders issued during the action, officers' formal reports were sometimes written well after the fighting. Officers also divided the flow of the fighting into individual "epochs" and addressed them as more discrete occurrences.[1]

The first challenge for me and now you, the reader, is that this work takes a different perspective and approach to the fighting. Its fundamental premise is that the reader is somewhat familiar with what happened during the battles, and is interested as to why various major events occurred. A methodology examining the critical decisions made at different times provides insight into why the fighting at The Wilderness and Spotsylvania Court House occurred and unfolded as it did. Once this approach is understood, it can be applied to any battle, campaign, or historical event. This book is part of the Command Decisions in America's Civil War series utilizing this methodology. The criterion for a critical decision is that it shaped not only the events immediately following, but also the rest of the campaign or battle. If a particular critical decision had not been made or a different decision had been made, the subsequent events of Grant's Overland Campaign of 1864 would have been quite different.

The chart below shows the decisions hierarchy. At the bottom are the many and various decisions, above that are a lesser number of important decisions, and at the top are a very few critical decisions.

I developed an initial set of decisions by studying primary and secondary sources and walking the battlefields themselves. As the manuscript evolved,

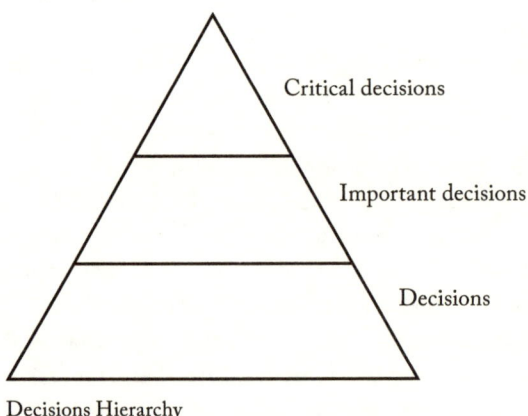

Decisions Hierarchy

I refined these decisions to produce the final set. Many important choices flow from a critical decision. For example, when Lee attacked Grant as the Army of the Potomac moved through The Wilderness, he caused Union Maj. Gen. Gouverneur K. Warren to pull troops from an excellent position at the Chewning Farm between two Confederate corps on the first day of fighting. Warren had to reinforce his threatened front at Saunders Field.

When possible, I have endeavored to use source documentation such as the *Official Records*, letters, lectures, and memoirs of the battles' participants. As essential as source documents are, they still carry authors' personal biases, and their statements need to be tempered with this understanding. The starting point for Civil War source documentation is the 128-volume *The War of the Rebellion: A Compilation of the Official Records of the Union and Confederate Armies*, typically referred to simply as the *Official Records* (OR). During the fighting in northern Virginia in May 1864, Congress introduced and passed a joint resolution to print the Union armies' official records and preserve the documents of actual participants in the war. These eyewitness reports proved the timeliest accounts of events, as one of the editorial goals for the *OR* was including documents written during the war. The submitted reports were not evaluated for accuracy, and once submitted, they could not be altered.

Unfortunately, after the 1864 Overland Campaign started with fighting at The Wilderness and Spotsylvania Court House, the Confederates made fewer reports that could be included in the *Official Records*. This reduced documentation limited insight into some of the actual communications and circumstances of their operations. At times, reports were written months after the fact, and they could be self-serving for the authors. For instance, Sheridan did not write his *OR* report of the April–August 1864 Union cavalry operations until May 1866, and his accounting of the action was much more positive than the facts indicated. Finally, the heavy casualties from intense fighting ravaged the officer corps of both armies. Reports that were written by junior officer replacements who may not have had the all the information their predecessors had. The opposite situation occurred when losses in the junior ranks made some information unavailable for the commanders' reports. Brig. Gen. Alexander Webb, of Hancock's Corps, opened his report with "This report is necessarily incomplete, from my inability to obtain any reports from regimental commanders, most, if not all, of them being killed." Even with these limitations, the *Official Records* reflect the best information available when decisions were being made.[2]

By their very nature, individual memoirs were written after the war and with inevitable biases. Grant wrote his memoirs twenty years after the war. He had had a controversial presidency, and southern publications had disparaged his generalship by citing his *OR* statement that he intended to defeat Lee "by mere attrition." Grant composed the portion of his memoirs about the Overland Campaign and the fighting at The Wilderness and Spotsylvania Court House as he battled the final stages of cancer and rushed to complete his manuscript. In discussing the initial contact between Lee's army and Warren's Corps on May 5, 1864, Grant stated, "Warren was, therefore ordered to attack as soon as he could prepare for it." This is at odds with the *Official Records* which documents Grant's orders to Meade to pitch into the enemy "without giving time for dispositions."[3]

Grant's pain and medications undoubtedly affected his perception and recollection of the campaign. He ignored published accounts at odds with Sheridan's narrative of the cavalry's actions. In addition, he incorrectly depicted Lee as having never come out of his entrenchments when, in fact, Lee had attacked the Federals more consistently than Grant's later writings suggested. Maj. Gen. Phil Sheridan wrote his memoirs after Maj. Gen. George Meade's death. Therefore, only Sheridan's account of his contentious meeting with Meade that resulted in Grant's critical decision is available. Maj. Gen. John Gordon's reminiscences appeared in 1905 after his postwar political career as Georgia governor and senator, and no doubt exaggerated his accomplishments.[4]

One aspect of the critical decision-making process that cannot be effectively quantified is the available time frame to make the decsionand how it affects the decision maker's thought process. Lincoln had the luxury of time to get the necessary information to evaluate his options before nominating Grant for lieutenant general. The rapid pace of battle radically reduces the time window and environment for making a decision. In the midst of the fighting on the evening of May 11, 1864, Gen. Robert E. Lee worried about Grant's movements. Lee had access to confusing information, and he had only limited time to make a decision. Even so, the situation demanded an immediate resolution. Conversely, while Grant evaluated his options for addressing the Meade/Sheridan dispute on May 8, he not only had time but was in a relatively stable camp headquarters as his troops moved south. Grant chose not to use the available time, instead making a quick decision, seemingly without thought as to the implications for his command structure or the tactical situation. At The Wilderness, Confederate Lieut. Gen. Richard Ewell squandered his limited wibdow of time to authorize an effective flank attack. He made his decision too late. In all cases, the stress of time constraints illuminates an aspect of the decision-maker's character; Lincoln was painstaking and thoughtful, Lee decisive, Grant impulsive, and Ewell hesitant.

A final consideration remains: men in the ranks had to implement most of these decisions, and then as now, how orders were carried out affected the results of those decisions. Just after winning his second term in office, Lincoln referred to the necessity of a presidential election during the rebellion and stated, "Human-nature will not change." I would argue as a corollary to this statement that the troops actually doing the fighting do not change. Bob Moulder of the Rocky Mountains Civil War Roundtable has given a talk entitled "Dear Mother and Friends at Home" based on letters written by John Howell Phillips, a captain in the Union army. If the talk had not been given at a Civil War Roundtable, the audience would not have been able to determine whether Phillips was writing about the Civil War, World War I or II, Vietnam, or Afghanistan. The "human-nature" of the people who make and implement the hard critical decisions endures—regardless of when they live, people are still people. Union and Confederate troops who carried out these decisions exhibited incredible courage and valor.[5]

This book groups critical decisions into four time frames: before the battle, The Wilderness, the transition, and Spotsylvania Court House.

> Before the Battles
> > Lincoln Decides Grant Will Lead the Union Armies
> > Grant Decides on the Key Commanders of the
> > > Army of the Potomac
> >
> > Grant Decides to Locate with the Army of the Potomac
> > Grant Decides On an Offensive Operation Against Lee
>
> The Battle Of The Wilderness
> > Lee Decides to Take the Initiative
> > Lee Decides to Leave His Troops in Place
> > Lee Decides to Keep the Initiative
> > Ewell Decides to Delay a Critical Flank Attack
>
> Transition To Spotsylvania Court House
> > Grant Decides to Move South
> > Anderson Decides to Push On
> > Fitzhugh Lee Decides to Stay and Fight
> > Grant Decides to Let Sheridan Pursue Stuart
>
> The Battle Of Spotsylvania Court House
> > Warren Decides to Redeem His Reputation
> > Grant Decides How to Use His Corps
> > Upton Leads a New Type of Attack

> Lee Decides to Withdraw the Artillery from the Salient
> Grant Decides to Commit a Full Corps
> Grant Decides to Move South from Spotsylvania Court House

The format for each decision is as follows:

> Situation—This background information sets the stage for a decision that had to be made.
> Options—A summary of the realistic options available presents the pros and cons of each.
> The decision
> Results/impact—This section addresses what happened based on the decision that was made.
> Alternate decision and scenario—Discussion of the possible results of a different decision is included when appropriate. The intent here is not to present an alternative history, but rather a different perspective on the results/impact of the actual decision.

As you read you will notice that the Union and the Confederacy used similar, but often different, methods to identify units. Therefore, some explanatory comments are appropriate. Both sides used the same method to identify units at the company, battalion, and regimental level. Companies were identified by a letter—e.g., A Company. Regiments and battalions were usually identified by a number—e.g., Eight (or 8th) Maryland, Forty-fourth (or 44th) Alabama. Above regimental level, both sides diverged in how units were identified

The official designations of Union brigades, divisions, and corps were numeric and begin with a capital letter. Examples include First Brigade, First Division, or Col. Emory Upton's Second Brigade, Brig. Gen. David Russell's First Division, Maj. Gen. John Sedgwick's Sixth Corps. When referring to a brigade or division belonging to or commanded by an individual, lowercase letters are used. For example, Upton's brigade, Russel's division, or Sedgwick's corps.

Early in the war the Confederacy used a numbering and a name system for unit designations. As the war progressed, the numbering system was used less and the name system was most commonly used. The official designations of Confederate brigades, divisions, and corps were the commanders' names followed by Brigades, Divisions, or Corps. For example, Gordon's Brigade, Early's Division, of Ewell's Corps.

The Confederate system can sometimes be confusing. The unit officially designated Kershaw's Brigade at The Wilderness had earlier been commanded by Brig. Gen Joseph Kershaw—hence its official designation. At The Wilderness it was commanded by Col. John Henagan, but it was not called Henagan's Brigade as the designation was officially Kershaw's Brigade. Lowercase letters are used when referring to brigade, division, or corps belonging to or commanded by an individual. Examples include Brig. Gen. John Gordon's brigade, Maj. Gen. Jubal Early's division, or Lieut. Gen. Richard Ewell's corps. As with anything pertaining to the Civil War, there are always exceptions.

No amount of study is a substitute for physically walking a battlefield. Therefore, where feasible, a detailed battlefield guide in appendix 1 links actual locations to the critical decisions. The critical decisions approach to studying the Battles of The Wilderness and Spotsylvania Court House presents a challenge: What balance is appropriate for the level of detail about the fighting presented in the main text and that included in the battlefield tour in appendix 1? Can an appropriate balance minimize duplication? Hopefully, I have managed this dilemma. The main text includes enough detail about the battles to provide understanding of the circumstances and results of a critical decision. Also, the main text encourages the reader to delve into the appendix, which includes additional details such as the actual orders issued during the fighting, officers' reports submitted after the fighting, and period photographs. Although there is nothing quite like walking an actual battlefield, reading appendix 1 provides a deeper appreciation of events that relate to the critical decisions.

Most of the stops are within the National Park Service boundaries. Walking the terrain where many of the decisions were made leads to a better appreciation of the tactical environment at the time of the battles and the obstacles the troops faced. The time of year when the reader takes the tour will provide different perspectives on the challenges the armies encountered. When the foliage is absent, the topography is more apparent, and when the dense vegetation is present, the impact on troop movements is more obvious. Unlike the maps in the text, the tour maps represent current topography.

INTRODUCTION

By the spring of 1864, Americans' early fantasies that the Civil War would end quickly had been dispelled. Previously unknown towns and places—like Shiloh, Chancellorsville, and Gettysburg, to name a few—were now forever etched in the country's history. Sentiment was divided in the North; the press had diverging views of the fighting, and criticism of how the war was being prosecuted was wide spread. The national debt was growing at a rate of $3 million to $4 million a day. Volunteering was lagging, draft riots had erupted in New York, and the mercenaries replacing enlistees were less effective as soldiers. A new dimension to the struggle was the fact that 1864 was a presidential election year. Without Union success on the battlefield, a new administration would likely be elected, and the South might be able to gain its independence politically rather than militarily.[1]

The North had made substantial gains. With the fall of Vicksburg the previous July, Union forces had taken control of the Mississippi River and greatly reduced the effectiveness of Confederate states west of the river to help supply the Confederacy. In addition, Tennessee was predominantly under Union control. Troops in the Military Division of the Mississippi under Maj. Gen. William T. Sherman now maintained a foothold south of Chattanooga in northern Georgia; this base threatened the Confederacy's heartland and Gen. Joe Johnston's Army of the Tennessee. In the West, the Confederacy still controlled Arkansas, Texas, and most of Louisiana.[2]

In the East, where the fighting had been subjected to more scrutiny from Washington and its politicians, the situation was more disappointing for the

Union. The Confederate capital in Richmond was no longer under immediate threat from Union troops. Gen. Robert E. Lee's Army of Northern Virginia had made its winter quarters south of the Rapidan River, only forty miles away from where the initial fighting at First Manassas had taken place in July 1861. Maj. Gen. George Gordon Meade's Army of the Potomac had stayed north of the river and was preparing for a spring movement south. Maj. Gen. Benjamin Butler's Army of the James was located at Ft. Monroe at the mouth of the James River, and the refurbished Ninth Corps under Maj. Gen. Ambrose Burnside was in Annapolis, Maryland.

Recently promoted to the rank of lieutenant general and now overseeing all Union armies, Ulysses Grant described the situation of the opposing forces in the East as "substantially in the same relationships towards each other as three years before, or when the war began." Further, Grant felt that the Northern newspapers magnified Confederate victories and minimized Union ones, asserting that the press "expressed dissatisfaction with whatever victories were gained because they were not more complete."[3]

By 1864 the influence of the US Military Academy at West Point was becoming more obvious. Although political generals still led troops, as for example Butler with the Army of the James, most commanders fighting at The Wilderness and Spotsylvania Court House were West Point graduates. Grant was a graduate, as were *all* Union and Confederate army and corps commanders and their successors. If division commanders and their replacements following casualties are included, two-thirds were West Point graduates. These officers who were involved with or influenced many of the decisions this book will discuss had a similar military education that shaped their thinking. Today, the graduating classes at West Point number roughly a thousand each year, but from Lee's class of 1829 to the class of 1860, which included division commanders Brig. Gens. Wesley Merritt and James Wilson, the average graduation class numbered only about forty-three. The two graduating classes in 1861 (May and June) were similar in size and included brigade commanders Col. Emory Upton and Brig. Gen. George Custer. Thus, not only were many of these men classmates, but the significantly smaller graduating classes at that time would have put them in much closer contact with each other, thereby providing better insight as to how their comrades and adversaries would act.

Note: Appendixes II and III provides the combined Union and Confederate Orders of Battle respectively for the fighting at The Wilderness and Spotsylvania Court House with some additional information. Where appropriate, the class year for graduation from the US Military Academy (USMA) or Virginia Military Institute (VMI) is provided. Replacement officers, necessitated by casualties, are shown below the initial serving officer and his casualty. The type of casualty, Killed, Wounded, Mortally Wounded, or Captured is shown first followed by the location, Wilderness or Spotsylvania Court House. Thus, Maj. Gen. John Sedgwick, commander of the Sixth Corps who was killed at Spotsylvania Court House and his replacement, Brig. Gen. Horatio Wright are shown in the Order of Battle as follows:

> Maj. Gen. John Sedgwick (K-S) (USMA 1837)
> (Brig. Gen. Horatio G. Wright) (USMA 1841)

Now, after three years of war, the fundamentals of the struggle were changing—in the North's favor. Federal military operations so far had not been effectively coordinated between the East and West and therefore had not brought consistent pressure on the Confederacy. Grant stated, "Before this time these various [Union] armies had acted separately and independently of each other, giving the enemy an opportunity often of depleting one command, not pressed, to reinforce another more actively engaged. I determined to stop this." As a result, Grant implemented an overall coordinated strategy with the objective of destroying the Confederacy's military capacity. This strategy would eliminate the South's previous utilization of interior lines that had allowed them to shift troops to threatened areas.[4]

The industrial disparity between the North and the South grew more significant. While Northern troops were well fed, Lee was continually struggling to supply his men. In a letter to his wife in January 1864, Lee stated "[I] had to disperse the cavalry as much as possible, to obtain forage for their horses . . . Provisions for the men are very scarce." He also cited a report from one division asserting that "over four hundred men were barefooted and over a thousand without blankets." Lee constantly warned Davis that the lack of supplies for his men was making it increasingly difficult to maintain the army's effectiveness. Confederate troops would expose themselves to severe enemy fire in attempts to get food from the haversacks of fallen Federal soldiers.[5]

The eastern portion of Grant's overall strategy against the Confederacy was called the Overland Campaign, of which the Battles of The Wilderness and Spotsylvania Court House were the opening actions. While Lee struggled to supply his army, Grant would cross the Rapidan River with 4,300 wagons, 835 ambulances, and over 56,000 mules and horses to support the Army of the Potomac and Burnside's Ninth Corps.[6]

Key to the upcoming campaign would be the two iconic leaders of the Civil War, Lee and Grant, who would face each other for the first time. Both were aggressive in their fighting philosophy and had demonstrated military capability, but they commanded in different ways. Partially out of concern for

Federal wagon park near Brandy Station, 1864. Library of Congress.

his corps commanders, Lee was typically in the field. On several occasions he was tempted to personally lead Confederate assaults. Although Grant would go to the front, he characteristically managed the army's efforts from headquarters. On the first day of fighting at the Battle of The Wilderness, he calmly sat on a stump at headquarters and whittled sticks, and his new leather gloves, to pieces.

The Battles of The Wilderness and Spotsylvania Court House initiated the first fighting in Grant's 1864 Overland Campaign, and they also laid the foundations of the postwar debates and critiques concerning Lee's and Grant's overall strategic vision and generalship. The discussion itself is outside the scope of this book. However, some perspective on the commanders' fundamental differences in political and command responsibility is relevant. Grant commanded all US Army forces and reported to Lincoln, who, as president, was tasked with defending the US Constitution. The Constitution's preamble states that the document was established in part "in order to form a more perfect Union." Grant and Lincoln were engaged in the war to preserve the whole Union. Emancipation, abolishing slavery, impacted the war, but it did not become part of the US Constitution until 1865.

Lee at this time commanded only the Army of Northern Virginia. In the military hierarchy, he would have been equivalent to Meade and focused only on this combat theater. Lee served under Confederate president Jefferson Davis, who was defending a constitution tailored after that of the United States but with a critical difference, a preamble that added the phrase "each state acting in its sovereign and independent charter." Stating that he could "take no part in an invasion" of the South, Lee declined an offer to command Union forces presented to him by Francis Blair on Lincoln's behalf. Lee resigned his commission from the US Army immediately after Virginia seceded, and in his resignation letter to commanding Gen. Winfield Scott he stated, "Save in defense of my native state, I desire never to again draw my sword." In 1868 he stated that his duty had dictated his resignation: "I could have taken no other course without dishonor." Both Lee and Grant were aware of the overall military situation in the divided nation, but only Grant had a constitutional foundation, a presidential mandate, and a military position for putting forth and implementing an overall strategy.[7]

After the wounding of Gen. Joseph Johnston outside of Richmond, Lee had been put in "command of the armies in Eastern Virginia and North Carolina" on June 1, 1862. That same day, Special Orders No. 22 referred to this command as the Army of Northern Virginia, and Lee would lead it until the surrender at Appomattox Court House on April 9, 1865. He had earned his men's respect and devotion. He was the unquestioned leader of the Army of Northern Virginia and a symbol of the Confederacy in the North and the South. Indeed, Lee's army had not lost a major battle in Virginia since he had taken command. As devoted to his men as they were to him, Lee was typically near the front in the upcoming battles. Only the emotional and physical intervention of his men prevented him from personally leading counterattacks on several occasions. The intervening troops effectively stopped the enemy each time.[8] Finally, Lee had a good working relationship with Confederate president Jefferson Davis, and the two maintained communication through correspondence and personal meetings. Lee meticulously informed Davis of his plans and solicited his opinions.

The adversaries—Ulysses S. Grant and Robert E. Lee. Library of Congress.

In contrast, Grant was new to the Army of the Potomac and the fighting in Virginia. He came east with an impressive record of victories, his latest being "the battle above the clouds" at Chattanooga that had lifted the siege of the city and forced the Confederates back into Georgia. Grant's persistent and aggressive style differed dramatically from what had previously been employed in the East. Sometimes his impulsive actions would have unintended negative results. As the general-in-chief, he would be able to implement his overall strategy on a much broader scale than ever before. Although he would personally make reconnaissance of the fighting, Grant would typically remain at his headquarters issuing orders and receiving reports from his subordinates.

Unlike his predecessors, Grant enjoyed the trust and confidence of Lincoln. The men's relationship was similar to that between Lee and Davis. Both of them came from humble backgrounds and had to overcome adversity to get to their present positions. Both had good common sense and "shrank from posing for effect, or indulging in mock heroics." Grant had demonstrated his military competency in his past performances. He avoided lecturing the president (some of McClellan's correspondence had been condescending, bordering on insubordinate), understood the linkage of his military strategy and Lincoln's political strategy, and shared his overall thoughts with Lincoln. In return, Lincoln allowed Grant unprecedented discretion in employing and managing the Union armies.[9]

Another dynamic of the overall situation in the Overland Campaign would be troop replacement. Lincoln tapped the North's larger population base in February and March 1864 by ordering the recruitment of seven hundred thousand troops from the states, a level the South could never approach. In the Confederacy, troops' enlistments were for the duration of the war, and replacements were integrated into existing units with veteran experience. Union replacements were done differently. Some of the veteran units in the Army of the Potomac were nearing the end of their three-year enlistments and would need to be replaced along with the casualties from fighting. The North's political expediency of allowing the states to raise new units and thereby appoint new colonels would result in whole units going into combat with no fighting experience. Meanwhile, veteran units such as Col. Emory Upton's brigade had been reduced to regimental size. The presence of a veteran or an untried unit at a critical place and time could dramatically affect the results of a combat decision.[10]

The fighting would reach new levels of intensity and brutality, with May 1864 producing eighty thousand casualties, the most in a single month in our nation's entire history of war. To put this in perspective, the sixteen thousand casualties Meade's chief of staff Maj. Gen. Andrew Humphreys reported on May 12, 1864, all incurred at the Mule Shoe at Spotsylvania Court House, were over three times the number reported at First Bull Run in July 1861. Moreover, there were twice as many casualties at the Mule Shoe as there were American casualties on D-Day; modern estimates put the total casualties from the former battle at seventeen thousand. Amid the carnage at The Wilderness and Spotsylvania Court House, the adversaries engaged in incredible acts of compassion and camaraderie—what Lincoln referred to as "the better angels of our nature." When the Union's Brig. Gen. James S. Wadsworth was mortally wounded at The Wilderness, his body fell into Confederate hands. Lee wrote Meade and promised that Wadsworth's body would be returned to the Army of the Potomac. The men exchanged cordial correspondence, and when Wadsworth's remains were returned to Union lines, Meade sent Lee "sincere thanks for [his] kind consideration."[11]

On the evening of May 10, 1864, Col. Emory Upton's Union troops briefly breached the Confederate lines in an attack at Dole's Salient. That night, after the lines had been restored, the bands of the two adversaries played competing sectional tunes, ending with "Home Sweet Home." In what must have been a most poignant moment, the troops loudly cheered the performance.[12] On May 13, one day after the bloody fighting at the Mule Shoe, Grant entered a farmhouse to write some dispatches and found a badly wounded Confederate soldier sitting on the only chair in the room. The Confederate offered the chair to Grant, and, following some good-natured bantering, Grant graciously declined. The general ordered a surgeon to do all he could for the soldier, who survived the war as a result.[13]

Field fortifications would play an ever-increasing role in the fighting of 1864. Meade's chief of staff, Brig. Gen. Andrew Humphreys, would describe the use of entrenchment in the fighting at The Wilderness and Spotsylvania Court House as being "in a manner unknown to European warfare, and, indeed, in a manner new to warfare in this country." The Confederates were forced into becoming particularly adept at rapidly making fieldworks for protection, although they were often forced to use "bayonets & tin cups" for their construction. The Federals were fortunate to have as many as 2,500 engineering troops engaged in the building and improvement of their entrenchments.[14]

The utilization and effectiveness of sharpshooters would increase on both sides during the campaign. Consequently, the interval between fighting grew almost as dangerous as the fighting itself, and courier duty carrying information between commands became increasingly hazardous. The Wilderness would be the first battle where Confederate sharpshooters fought as a separate organization. When the enemy was initially engaged, these sharpshooters would be detached to help protect the flanks. Artillerymen were prime targets. One Confederate sharpshooter noted, "Artillerymen could stand anything else better than they could sharpshooting." At Spotsylvania Court House, a Rebel sharpshooter "hit every man who stepped into the breach of the gun" for an hour until he was in turn killed by a Union sharpshooter. On May 9, a Confederate sharpshooter would kill Union Maj. Gen. John Sedgwick, and every officer that had shown himself in that area had been hit. Confederate sharpshooters also made Federals' reconnaissance of their defenses at the Mule Shoe more difficult. Union sharpshooters forced the Confederate defenders to make traverses every few yards along the Mule Shoe's interior defensive line for protection of the troops.[15]

In the upcoming fighting, the Federals felt that they were hindered in their operations not only by the challenge of the topography and the lack of roads in the area, but also by the poor maps they used when making operational decisions. While Lee's army was more familiar with the terrain and the inhabitants were eager to provide information on the enemy's whereabouts, the Federals continually had problems finding reliable guides. Union troops relied on maps that were often inaccurate and were printed on paper that wore out after only a few days of being carried in pockets. Sometimes the maps recorded distances that varied more than two miles from actual measurements, as well as showing erroneous directions. In Hancock's movement on the morning of May 5, he noted in his report to Meade's headquarters, "It must be a greater distance to Todd's Tavern by the road I came than appears on the map."

The common use of family names for farms and places added to the Federals' confusion and at times caused severe errors. On May 9, 1864, Burnside was ordered to use his infantry to try and identify the roads leading to Spotsylvania Court House from the northeast. Grant referred to the "Gate" house as shown on the Federal maps. Burnside subsequently reported that the troops were at Gayle's house, that there was no such place as Gate, and that there was "no way of connecting with General Hancock" as Grant had intended. Burnside would be operating in the important area east of Spotsylvania Court House along the Fredericksburg Road. Maps of this location showed the distance from the Ni River crossing the road to the town as two miles farther out than its actual location.[16]

The fighting at The Wilderness and Spotsylvania Court House would be driven by eighteen critical decisions made by a few men and implemented by countless thousands of soldiers who would write a new chapter in American history. This book focuses on those decisions.

CHAPTER 1

BEFORE THE BATTLES

If you have not read the preface, please return there and read the definition of a critical decision to better understand this different approach to the Battles of The Wilderness and Spotsylvania Court House.

Some appreciation of the conditions in the North helps contextualize the four critical decisions made before the fighting. The Union experienced renewed optimism with the coming of spring in 1864 and the improved weather that allowed military operations to begin. Newly promoted to lieutenant general and heading all Union armies, Ulysses S. Grant faced high expectations. On March 18, the *New York Herald* stated, "If, with General Grant at the head and struggle as it now is, we cannot put the rebellion down in the coming summer, we can never put it down." The front page of that same paper also addressed the reality of the coming year, the presidential campaign, and the "placing of General George B. McClellan's name for the next President."[1]

Northern public opinion would focus on the East, with Gen. Robert E. Lee as the Confederacy's foremost symbol of success. Despite achievements in the "Western Theater," Grant believed that Federal forces in the East were not positioned much differently than they had been in 1862 after Maj. Gen. George B. McClellan's failed Peninsula Campaign. With the larger population in the East (more people lived in New York, Massachusetts, and Pennsylvania than in all the Union's western states combined) and the influential New York papers heavily covering Virginia, the North's opinion would be heavily shaped by the fighting in the region. In 1862, Lincoln noted the disparity between successes in the West and failures in the East (McClellan's Peninsula Campaign): "Yet it seems unreasonable that a series of success, extending through half-a-year and clearing more than a hundred thousand square miles of country, should help us so little, while a single half-defeat should hurt us so much."[2]

The logistical situation for the Army of Northern Virginia forced Lee into a waiting mode. He had repeatedly warned President Davis that the lack of supplies severely impeded his ability to operate, and he had been forced to winter his three army infantry corps in different locations. Stating, "[A] disaster to the railroad would render it impossible for me to keep the army together," Lee stationed Longstreet's recently returned divisions near the railroad junction at Gordonsville to facilitate any needed protection.[3]

The journey to Appomattox Court House, a still-unknown hamlet in rural Virginia, began with a clash of epic and bloody proportions at The Wilderness and Spotsylvania Court House. The continuous fighting between Grant and Lee's armies in the upcoming struggles would be like nothing the country had ever seen, and the iconic faces of the two men would forever be associated with the Civil War.

Lincoln Decides Grant Will Lead the Union Armies

Situation

The winter of 1863–64 was a challenging time for President Abraham Lincoln. The Union army boasted more men, material, and resources than the Confederacy. But after two and a half years of fighting since the Battle of First Bull Run / Manassas on July 16, 1861, the results were still disappointing. The Union's notions of a quick war—moving on to Richmond and fighting one more battle (also the South's opinion)—had been forced into submission.

The complexion of the war had changed dramatically. After the firing on Fort Sumter in April 1861, Lincoln issued a proclamation calling for seventy-five thousand militiamen for ninety days (implicit in this time frame is the assumption that the troops would not be needed for any longer). The response in the North was overwhelmingly positive. Many states asked to send more than their quota of troops, and a quarter of a million people turned out for a New York City Union rally. The Southern response was Virginia, Arkansas, North Carolina, and Tennessee seceding from the Union and joining the seven states that had previously seceded to form the Confederate States of America.

Following the Battle of Antietam in September 1862, Lincoln issued the Emancipation Proclamation, which took effect the following January. The

Presidential election poster from 1864 attacking Lincoln. Library of Congress.

initial objective of the war to reunite the nation was now expanded to include abolishing slavery. But in July 1863, two weeks after the victory at Gettysburg, the new military draft lottery was enacted, prompting antidraft riots in New York City. In four days of rioting, over one hundred people were killed and millions of dollars in damages incurred. The stakes for reunification were at their highest, as 1864 was a presidential election year. Moreover, the Republican Party was far from unified on many matters; even Lincoln's renomination for the presidency was uncertain. The election would be a referendum not only on war, but also on emancipation. If the Democrats won, peace negotiations would likely quickly follow, and emancipation could be abandoned. If the Republicans won, the war would continue until the South surrendered and emancipation was implemented.[4]

By the spring of 1864, the main Union army in the East, the Army of the Potomac, lacked aggressiveness and was plagued by political intrigue. In contrast, its counterpart, Robert E. Lee's Army of Northern Virginia, had demonstrated aggressive military actions and had been very successful in Virginia. Lee was the undisputed commander of this Southern army and had good rapport with Confederate president Jefferson Davis. In July 1862, Lincoln acknowledged the disadvantages of trying to be his own general-in-chief and called Maj. Gen. William H. Halleck from the West to Washington to assume this role. Maj. Gen. George B. McClellan, who had just finished the unsuccessful Peninsula Campaign, considered Halleck his inferior and did nothing to improve the climate in the command structure. Following the Battle of Antietam in September 1862, McClellan received orders to pursue the Confederates. "The President directs that you cross the Potomac and give battle to the enemy or drive him south . . . You will immediately report what line you adopt and when you intend to cross the river." But McClellan made excuses and delayed, and in the end, nothing was done. In 1862, after only a few months in his new assignment in Washington, Halleck lamented privately ,"I am sick, tired, and disgusted with the condition of military affairs here in the East . . . There is an immobility here that exceeds all that any man can conceive of."[5]

McClellan was subsequently replaced by Maj. Gen. Ambrose E. Burnside, who mounted one of the few winter campaigns of the Civil War. In December 1862, Burnside ordered the Army of the Potomac to attack a well-defended Confederate position at Fredericksburg, Virginia, and was soundly defeated. The morale in the Army of the Potomac plunged to new lows. Four generals of the Sixth Corps went outside the chain of command and complained directly to President Lincoln about Burnside's lack of leadership. Maj. Gen. "Fighting Joe" Hooker assumed command next, but following his defeat at Chancellorsville, Lincoln worried that Hooker did not want to fight Lee again. Foreshadowing Grant's orders to Meade in a prescient moment of Grant's orders to Meade the following year, Lincoln wired Hooker, "I think *Lee's* Army, not *Richmond*, is your true objective point." When Hooker was slow to pursue Lee north on June 28, 1863, Lincoln relieved Hooker of command and named Maj. Gen. George Meade in his place.[6]

Meade was commanding the Army of the Potomac at Gettysburg when it defeated Lee and his Army of Northern Virginia. But his slow pursuit allowed the retreating Confederates to slip across the Potomac River, disappointing President Lincoln in what might have been able to be accomplished. On July 4, 1863, the day after Lee's defeat at Gettysburg, Meade issued General Order Number 68 on behalf of the country to thank the Army of the Potomac for its victory, and to emphasize the need to "drive [the invader] from our soil." Speaking to his secretary John Hay, Lincoln expressed his frustration that Meade, like McClellan, did not grasp the full context of the war: "Will our generals never get that idea out of their heads? The whole country is our soil."

Later in the year, Meade tried to maneuver around Lee in what became known as the Mine Run Campaign, but the Confederates' defensive position was much too strong, and no major attack was made. Although it was the right tactical decision for the Union, it had negative political implications. Early in March 1864, members of the powerful Joint Committee on the Conduct of the War felt "the incompetency of the general in command of the Army [of the

Pres. Abraham Lincoln.
Library of Congress.

Potomac]" and deemed it "therefore necessary to present to the president their demand to remove Meade and replace him with someone more competent to command." When asked what general they would recommend, "they said that for themselves they would be content with General Hooker, believing him to be competent." If the president considered another general more competent, the committee members declared, "Then let him be appointed."[7]

As Lincoln needed to find a commander that "could grasp the whole field and get out of the Army what he knew was in it," the command structure stakes were raised. Late in 1863, Rep. Elihu B. Washburn from Illinois (Grant's district) introduced legislation to re-create the rank of lieutenant general, and thus the officer who would command all Union armies. Although Winfield Scott held this rank as a brevet, only George Washington had previously held this rank permanently.[8]

Options

Lincoln had a limited field to choose from, for the bill stipulated that only existing major generals would qualify for the new rank. Although previous Army of the Potomac commanders were West Point graduates, their performance had been disappointing. Political generals at lower command levels had generally fared worse. In an April 1864 letter to Major General Sherman, Major General Halleck stated, "It seems but little better than murder to give important commands to such men as Banks, Butler, McClernand, Sigel and

Lew Wallace, and yet it seems impossible to prevent it." Elevating previous commanders of the Army of the Potomac did not look promising. Though popular with the army's troops, McClellan was now being considered as a Democratic presidential candidate to run against Lincoln in the fall and would not have been a realistic option. Based on Hooker's past performance, it is unlikely that Lincoln considered the committee's suggestion, and he could not ignore the committee's views on Meade. Lincoln thus had three viable options. Since he had to nominate someone for the position, he could decide not to recommend anyone for the rank and continue to use the power of his office to direct the war effort. Halleck was a senior officer who had been in Washington since 1862. He had a good overall view of the army's operations, and he understood the city and its influence on waging the war. Finally, the rising star in the West, Maj. Gen. Ulysses Grant, was getting a lot of attention in both the Northern papers and Congress.[9]

Option 1

Lincoln had not been actively involved with the bill re-creating the rank of lieutenant general, so no political investment compelled him to nominate anyone. He was undergoing the painful experience of gaining expertise as a wartime president. He had expanded the fighting's objective to include emancipation, thereby greatly reducing the likelihood of England or France aiding the Confederacy. But if Lincoln did not make a change, neither the looming political challenges nor the military ones would be solved. The disappointing results of the Union's efforts would likely continue.[10]

Option 2

With his role in Washington as Lincoln's general-in-chief, Halleck had a good view of the overall military situation and would be a viable option for lieutenant general. He knew how things worked in the capital, he had been a reasonable administrator, and Lincoln considered him a friend. Understandably, some thought he might "be the lucky man." But Halleck viewed himself as an advisor and a scholar. He was not particularly popular with Congress due to his slowness and reluctance to take responsibility for his actions. Some congressmen who supported the bill restoring the rank of lieutenant general did so to get rid of him. A proposed amendment to the bill required the new lieutenant general to "supersede Gen. Halleck."[11]

Option 3

Grant's successful military performance in the West was the best of any Union commander. Indeed, his early victories there had brought him to the attention of the government in the East. The capture of Vicksburg the previous year had been a major accomplishment. After delivering his Gettysburg Address on November 19, 1863, Lincoln received a telegram from Secretary of War Edmond Stanton stating that Grant was preparing to attack—just the type of action the president wanted. The result was Grant's remarkable victory at Chattanooga in late 1863 that solidified his status as the North's most accomplished military commander. But Grant wanted to remain in his current post. While his victory at Vicksburg had made him a candidate to lead the Army of the Potomac, Grant was grateful that Halleck and Dana had used their influence to help him maintain command in the West.[12]

Also, 1864 was an election year, and major political factors at play needed to be considered in addition to military importance. An astute politician, Lincoln knew that the eight previous presidents had served only a single term; a sitting president had not been reelected since Andrew Jackson in his second term in March 1837. At the same time, the Republican Party disagreed as to who should run. Radical Republicans considered Lincoln's former political appointee Maj. Gen. John C. Fremont as a protest candidate, and Lincoln's former secretary of the Treasury Salmon Chase had also been proposed for the nomination. In addition, McClellan was making a push to be the Democratic nominee, seeking the military path to the White House that had been used many times before. Finally, Grant, believed to be a Democrat, had fielded private inquiries about being a candidate. Grant was not interested, but Lincoln would not have been aware of these "private" communications. However, the president would have been aware that the bill for the revised rank had at one time had expressly recommended Grant. The possibility of a Lieutenant General Grant running against and defeating Lincoln for the presidency was unacceptable.[13]

Decision

Lincoln had never met Grant, and he was anxious to understand the general's possible political ambitions and evaluate whether he might have "that presidential grub." In a conversation with Congressman Washburn, Grant's political sponsor and Lincoln's confidant, the president noted that all he knew about Grant the person he had learned from Washburn. Lincoln asked, "Who besides you [Washburn] knows anything about Grant?" The congressman recommended Joseph Russell Jones, the Northern Illinois District US marshal, to Lincoln. As a close friend of Grant's, Jones would have been very knowledgeable about any of Grant's political ambitions. Lincoln summoned Jones to Washington, but the details of the meeting are unclear. During the winter of 1863, Grant carefully composed his correspondence to deter potential political supporters. Jones showed Lincoln a letter that Grant had sent him just before

Lincoln's nomination of Grant for the rank of lieutenant general. Library of Congress.

his trip to Washington. Grant told his friend, "Nothing would induce me to think of being a presidential candidate, particularly so long as there is a possibility of having Mr. Lincoln re-elected." Upon reading the letter, Lincoln told Jones, "You will never know how gratifying that is to me." This was enough for the president. When the bill reviving the rank of lieutenant general was passed, Lincoln signed it into law, sent a brief nomination memo to the Senate, and summoned Grant to Washington. The above brief note initiated the first of the eighteen critical decisions.[14]

Results/Impact

Just as President Lincoln had concerns about Grant's political ambitions, Grant was anxious about Lincoln as commander in chief. Lincoln had placed the politically appointed Maj. Gen. John McClernand over Maj. Gen. William T. Sherman, one of Grant's most trusted commanders. Before his victory at Vicksburg, Grant had had McClernand removed from his post. McClernand had protested to Lincoln, and in February 1864 the president basically overrode Grant and restored McClernand to command of the Thirteenth Corps, then in Louisiana. Grant also felt that the president had given McClellan too much advice and "destroyed his military career."[15]

The rank of lieutenant general came with powerful implications. It was established by a formal act of Congress, and it carried more prestige and authority than a presidential appointment. "In effect the government was staking everything on the bet that Grant was going to win the war." Grant would now be in command of all the Union armies, at that time the largest military force in the world. In contrast to his previous experiences with his generals, Lincoln quickly developed a good working relationship with his new general-in-chief. An "unshakable understanding and trust [emerged] between the President" and Grant. Shortly after Grant's promotion, President Lincoln, responding to a question from his assistant secretary William Stoddard, said, "Stoddard, Grant is the first general I've had! He's a general! . . . I'm glad to find a man who can go ahead without me." It seemed to Stoddard that the president felt "someone competent" had taken over the responsibility of the army.[16]

Grant would quickly implement an overall strategy that coordinated the various Northern armies' efforts to constantly pressure all possible Confederate positions. This was the first time in the war such a plan had been instituted. The Confederacy would no longer be able to shift troops to strengthen a particular area, as it had done in sending Longstreet's Corps to Chickamauga the previous year.

Lincoln and Grant's partnership started on a very positive note. On March 10 Lincoln presented Grant's commission to him at the White House in front of his cabinet members and a small group of individuals. For his part, Lincoln expressed his and America's appreciation for what Grant's leadership had accomplished and reliance on the new lieutenant general to do "what

Lieut. Gen. Ulysses S. Grant, commander of US Army forces (USMA class of 1843). Library of Congress.

remains to be done in the existing struggle." In response, Grant stated that he understood the responsibility of his position. He demonstrated his commitment and understanding by leaving the next day to meet with Meade in the field and to begin carrying out his new responsibilities.[17]

While Grant's headquarters were at Culpeper from late March until the start of the campaign, he would regularly visit the president and the secretary of war, promoting his working relationship with Lincoln. The correspondence between Lincoln and Grant written just before the Federal army moved south provides insight into the developing trust between them. When Lincoln wrote to Grant, he stated that he did not know or care to know the "particulars" of his plans. He also informed his newcommander, that "If there is anything wanting which is within my power to give, do not fail to let me know it." Grant responded to this correspondence with his appreciation for the President's support. In what must have been music to Lincoln's ears, Grant closed with the statement that if the success achieved was less than hoped for, "the fault is not with you."[18]

Although difficult times were ahead, Lincoln never regretted his decision of promoting Grant. The impact of Grant's command of all the Union armies would extend well beyond his Overland Campaign in Virginia to essentially every major battle for the rest of the war.

Grant Decides on the Key Commanders of the Army of the Potomac

Situation

On March 10, 1864, the day after his official appointment, newly promoted Lieut. Gen. Ulysses S. Grant traveled from Washington to Meade's headquarters of the Army of the Potomac. The encampment at Brandy Station in Virginia was the largest one of the Civil War. Grant faced the formidable task of deciding how to reshape the character and fighting effectiveness of the Union's largest field army. This decision featured two interlocking components: who should command the Army of the Potomac, and who should command its corps. Together, these two personnel issues would determine the degree to which this army would realize its potential.[19]

This Army of the Potomac posed a unique set of factors. It had originally been formed under and had its character shaped by Maj. Gen. George B. McClellan, who was now pursuing a presidential bid against Lincoln in 1864. Many of the army's officers, including the then Brig. Gen. George G. Meade, now the current commander of this army as a major general, had served under

McClellan in 1862, and they still exhibited McClellan's cautious approach to waging war. The officers' caution stemmed from the considerable respect they had gained for the capabilities of Lee and the Army of Northern Virginia. Wanting a more aggressive approach to waging the war, the Lincoln administration had selected Grant for the job. The Army of the Potomac's theater of operation was close to Washington, and the ease of access to congressmen and senators made political intrigue and influential military postings a constant factor in its operations.[20]

Situation—Army Commander

Meade had replaced the previous commander of the Army of the Potomac, "Fighting Joe" Hooker, just days before the Battle of Gettysburg in July 1863. Although the battle had been a Union victory, Meade had not been able to aggressively pursue the Confederates or to mount any offensive operations since then. Prior to his visit to Brandy Station, Grant felt that Meade's removal was necessary to change the army's mind-set. In a December 1863 letter to Grant, Assistant Secretary of War Charles Dana expressed the White House's and War Department's feelings about the Army of the Potomac: "From that army nothing is to be hoped under its present commander." Meade expected to be replaced. In a letter to his wife the previous day, he had stated, "I understand he [Grant] is indoctrinated with the notion of the superiority of western armies and that the failure of the Army of the Potomac to accomplish anything is due to their commanders."[21]

Options—Army Commander

When members of the Committee on the Conduct of the War had informed President Lincoln that Meade should be removed, they had suggested Major General Hooker to once again command the Army of the Potomac. Hooker had been maneuvering for the position. After his defeat at the Battle of Chancellorsville in May 1863, he had been removed as the army's commander and sent west, where he commanded the Eleventh and Twelfth Corps at the Battle of Chattanooga. Thus, Grant was aware of Hooker's ability, but he was likely also aware that Hooker was trying to restore his reputation.[22]

After Grant, the most recognizable commander in the West was Maj. Gen. William T. Sherman. When he met with Grant, Meade specifically mentioned Sherman as his possible replacement. Like Grant, Sherman was unfamiliar with the Army of the Potomac's personnel. It was doubtful that his brashness and informality would mesh well with this army's more reserved character. However, the major reason Sherman was unlikely to head the Army of the Potomac was that Grant thought Sherman "could not be spared in the West." Grant and Sherman both felt that the war could be won in the West, and Sherman's aggressive style and familiarity with the troops was key.[23]

Another possible candidate was Maj. Gen. William "Baldy" Smith. He had served with the Army of the Potomac early in the war, and he had been involved with the army's leadership controversy following the Battle of Fredericksburg. Smith lost the command of his corps, and the Senate did not approve his promotion to the rank of major general at that time. His strong ties to McClellan and his brash, open criticism of his superiors had contributed to his fate. He was then transferred west in October 1863. Smith had just been promoted to the rank of major general due in large part to Grant's support. Grant had been impressed with Smith's performance at Chattanooga, when he had presented Maj. Gen. William Rosecrans and then Grant, Rosecrans's successor, with a plan to open a desperately needed new supply line.[24]

Referring to the meeting at Brandy Station, Grant's aide-de-camp Lieutenant Colonel Comstock wrote in his diary on March 10, 1864, "Came up (Meade with us) this morning. Hooker et. Al. are trying to get Meade removed & Hooker or another in his place. Grant who at Chattanooga thought Baldy Smith should have it [command of the Army of the Potomac]." "Baldy Smith, whose ambitions were no secret, came to Brandy Station with Grant. However, Smith did not stay to dinner."[25]

A final option was to retain Meade as head of the Army of the Potomac. Meade was thoroughly familiar with the army's men and officers, and with Lee and the Army of Northern Virginia's capabilities. But the political environment was decidedly against him. Washington had widely criticized Meade's inability to effectively pursue Lee's retreating army after Gettysburg. In his correspondence with Meade on July 14, Halleck stated, "I need hardly say to you that the escape of Lee's army without another battle has created great dissatisfaction in the mind of the President." Meade's decision not to attack Lee at Mine Run the past winter was viewed as further evidence of his overly cautious leadership. Unlike Washington political figures, the troops who were in the field appreciated Meade's sound judgment and respected his refusal to bow to political pressures to attack the strong Confederate defenses at Mine Run.[26]

Decision—Army Commander

At their first meeting on March 10, Meade's opening remarks immediately determined Grant's decision. Meade stated that the task before the army was so great that personal feelings should not stand in the way of selecting the "right men." Therefore, he would serve to the best of his ability wherever he was placed. Meade had suggested Sherman as a possible replacement, but Grant wanted Sherman to stay in the West and so informed Meade that he had no intention of replacing him. Grant felt that, before he arrived in Washington, he had been misled by "persistent" newspaper statements about the administration's desire to remove Meade. When he did arrive, he found that the there was no reason to replace the Army of the Potomac's commander. In his memoirs, written over twenty years later, Grant provided insight into this seemingly surprising decision. Meade's willingness to serve wherever he should be placed impressed the new general-in-chief even more than his victory at Gettysburg the year before. "It is men who wait to be selected, and not those who seek, from whom we may always expect the most efficient service," Grant stated. Even so, as late as March 17, Comstock wrote in his diary of Meade's concern that Baldy Smith might replace him.[27]

In correspondence with his wife, Meade revealed that he was "very much pleased with" Grant, saying, "[He demonstrated] much more capacity and character than I expected." In a truly prophetic closing, Meade, who would retain his position as commander of the army, wrote, "You may look now for the Army of the Potomac putting laurels on the brows of another rather than your husband."[28]

Maj. Gen. George G. Meade (USMA class of 1835), commander of the Army of the Potomac. Library of Congress.

Results/Impact—Army Commander

The biggest advantage to selecting Meade was that he was thoroughly familiar with the personnel of this army and the area where the fighting would take place. In addition, he had a healthy respect for and insight into the Army of Northern Virginia. Meade himself thought that he could be a more effective commander under Grant than he had been under the limitations of Halleck, the previous general-in-chief. This decision would have some awkward consequences, and Grant was aware of the "embarrassing" situation it put the two men in. The new general-in-chief would endeavor to make Meade feel that his position was the same as if Grant were in Washington. While Grant's quiet and seemingly unemotional disposition contrasted with Meade's fiery temperament, the men's approach to fighting differed as well. Grant favored aggressive tactics, while Meade was inclined to choose a more defensive or conservative approach. Initially, Grant would issue suggestions to Meade, such as when ordering the movements to Spotsylvania. But later he would make direct orders. For example, Grant ordered Sheridan to go after Stuart, seemingly undermining Meade's authority.[29]

Grant wanted to use Maj. Gen. William "Baldy" Smith's military ability and thus assigned him as a division commander for Maj. Gen. Benjamin Butler, the politically connected but militarily inept leader of the Army of the James. Grant "was not long in finding out that the objections to Smith's promotion were well founded." Smith's typical open criticism of his commanders subsequently resulted in his removal. Grant's opinion of the officer would likely have diminished wherever he was assigned, ending in similar results.[30]

Alternate Decision—Army Commander

As the Committee on the Conduct of the War had suggested, Hooker would have been an intriguing alternative for command. Had Hooker been reinstated as commander of the Army of the Potomac, his more aggressive style of fighting might have complemented Grant's better than Meade's. Under Grant's steady leadership, Hooker might not have repeated his lapse of resolve at Chancellorsville. In 1863, his actions as the army's previous commander had restored the soldiers' confidence. Moreover, Hooker's effort to improve the cavalry's effectiveness by consolidating its various units into a single corps could have enhanced a working relationship with new cavalry leadership.

Grant had made it clear that going outside the chain of command would not be tolerated, and Hooker's political maneuvering might have caused some issues. Although Grant described Hooker's movement at Chattanooga as "brilliant," he nonetheless characterized him "as a dangerous man" and faulted him for not being "subordinate to his superiors." In stark contrast to his rationale for selecting Meade, Grant felt that Hooker "was ambitious to the extent of caring nothing for the rights of others." When engaged in battle, Hooker tended to get separated from the main body of the army and then try to operate as an independent command.[31]

President Lincoln had lost confidence in Hooker, but whomever Grant wanted for the position, he would get. Had Grant not made the seemingly snap decision about Meade and instead gone with the committee's suggestion, the conflict between Grant and Hooker might have rivaled that between the Army of Northern Virginia and the Army of the Potomac.

Situation—Corps Commanders

Once the decision had been made to retain Meade as commander of the Army of the Potomac, Grant could address the options for the corps commanders. Before Grant took over, Meade had requested permission to reorganize the five infantry corps into three. Relying on Meade's familiarity with the Army of the Potomac's personnel, Grant implemented this reorganization. Brig. Gen. John Newton was removed from command, and his First Corps was consolidated with the Fifth Corps. Maj. Gen. George Sykes of the Fifth was also removed, and Maj. Gen. Gouverneur Warren took charge of the expanded corps. Finally, Maj. Gen. William French was removed from the Third Corps, two divisions were reassigned to the Second Corps, and one division was reassigned to the Sixth Corps.

Following his meeting with Major General Meade at army headquarters in Culpepper, Virginia, Grant returned to Washington and then immediately went to Nashville, Tennessee, to reconnect and discuss overall strategy with Generals Sherman, McPherson, Dodge, Logan, and Sheridan. Details of the meeting are not available. But the discussions might have planted seeds for a possible solution to Grant's concern about the cavalry leadership.

Earlier in the year, Brig. Gen. Judson Kilpatrick, commander of the Cavalry Corps's Third Division, had proposed a daring raid against Richmond including specific actions against Confederate president Jefferson Davis. Maj. Gen. Alfred Pleasonton, Cavalry Corps commander at the time, had approved the plan. Unfortunately, Pleasonton was a self-promoter who seemed to routinely claim greater results than the facts would support. The raid, often called the Dahlgren Affair after the fallen Col. Ulric Dahlgren. When his body was searched, orders were found concerning efforts to assassinate Confederate President Davis. The operation was a military failure and a political disaster. Kilpatrick had previously been criticized for ordering a disastrous cavalry charge against Confederate infantry on the third day at Gettysburg. The fiasco resulted in Pleasonton's and Kilpatrick's transfer to the West.[32]

After his return to Washington on March 23, 1864, Grant met with President Lincoln and Maj. Gen. Henry Halleck and indicated his "dissatisfaction" with the cavalry's lack of performance. Grant felt more could be accomplished "under a thorough leader."[33]

Options—Cavalry Corps Commander

Two possible corps commanders were Brig. Gens. Wesley Merritt and David Gregg, both longtime Army of the Potomac cavalry commanders. Merritt had served under Brig. Gen. John Buford as a brigade commander and had commanded the First Division after Buford's death in December 1863. Gregg had been a division commander since February 1863, and he had performed well at Gettysburg (East Cavalry Field). When Pleasonton was on leave in early 1864, Gregg had been acting corps commander by virtue of his seniority. Given that the Army of the Potomac's commander and the infantry corps officers all came from this army, and given that Grant was concerned about the cavalry's leadership, looking outside the unit was an intriguing possibility. If Grant took this route, the question would be whether cavalry experience or leadership ability was the more important skill set for the corps commander.[34]

Decision—Cavalry Corps Commanders

Although changes in the cavalry command were not a surprise, the officer that Grant picked was. In the meeting previously noted, Grant stated that he wanted the best man in the army for the position, and Halleck suggested Sheridan. Grant replied, "The very man I want." Grant's decision to pick Sheridan seems to have been as hasty as his choice to select Meade. Sheridan was ordered east that day. Although Grant liked Sheridan's aggressive fighting spirit and was very impressed with his performance at Chattanooga, this apparent snap decision on a vital assignment would negatively impact the upcoming campaign.[35]

Grant now commanded all Union army forces and had just chosen the corps commander for the Army of the Potomac's cavalry. Ironically, he had at one time stated that he would have given anything to command a brigade of cavalry in this same army. He felt he could have done "some good" in that position.[36]

Results/Impact—Corps Commander

Sheridan was "staggered" when he got orders for his new assignment. His only previous experience as a cavalry officer was in 1862, when he had commanded the Second Michigan for four months. He only knew a few of the

Maj. Gen. Phil Sheridan (USMA class of 1853), commander of the Army of the Potomac cavalry corps. Library of Congress.

officers in the Army of the Potomac, and his fiery temperament would soon put him at odds with Meade, his new immediate commander. Completing the cavalry reorganization, Brig. Gen. James Wilson, Grant's former aide, was assigned to command the Third Division (previously commanded by Brigadier General Kilpatrick). Brig. Gen. Alfred Torbert was appointed to command the First Division (previously commanded by Brig. Gen. Wesley Merritt). It is unclear whether Wilson's assignment was Grant's or Sheridan's idea, but Wilson was known to both. He had served in the Army of the Potomac as a member of McClellan's staff in 1862 before being ordered west and serving under Grant. Wilson's cavalry experience consisted of a few months of administrative duties in the Cavalry Bureau, but no line assignment. Torbert had served with the army since the Peninsula Campaign in 1862, but only as an infantry officer. Thus, the Army of the Potomac would start the Overland Campaign with three of its four top cavalry commanders having only four months of line cavalry experience.[37]

This choice of cavalry leadership would significantly impact the Battles of The Wilderness and Spotsylvania Court House. Wilson's failure to provide an accurate location for Lee's army on May 4, 1864, led Meade to believe Lee was still in his Mine Run defenses. Wilson "had neglected to take precautions that would have been second nature to a more seasoned commander." When Lee launched his two-army corps attack along the Orange Turnpike and Plank Road, Union forces were caught totally unaware. "They [Sheridan's

Cavalry] had failed miserably in the first function [scouting enemy positions] permitting Lee's entire army to approach undetected." If Meade had had the correct information, he most likely would have put strong Union forces across both routes and defeated the Rebel advance.[38]

Army commander Meade and Cavalry Corps commander Sheridan's personality conflict and differing opinion of how cavalry should be utilized would come to a head in the initial movements to Spotsylvania Court House on May 8, 1864.

Alternate Decision and Scenario

Had Grant put a more experienced cavalry commander, such as Brig. Gen. David Gregg, in charge of the cavalry, it is unlikely that the unqualified Wilson would have been given the command of a cavalry division. The failures to warn the Federal army of the Confederate threat on both the Orange Turnpike and the Orange Plank Road might have been avoided, and the communications between the infantry and the cavalry could only have been better. The personality clash between Meade and Sheridan would not have happened. Furthermore, Grant might have had the benefit of a seasoned cavalry commander and his men to provide him better information as to Lee and his troop dispositions.

Grant Decides to Locate with the Army of the Potomac

Situation

With all the dramatic changes to the Army of the Potomac resulting from Grant's becoming the new commanding general, the importance of where Grant would establish his headquarters is often overlooked. When he was summoned to Washington to receive his promotion, Grant expressed his desire to command from the Western Theater, where he was familiar with the army, the officers and troops, and the terrain where the upcoming western campaign would be waged. On March 4, 1864, before leaving for Washington, Grant wrote to his close friend and most trusted subordinate Sherman, stating that he would not accept the position of lieutenant general if it required him to make his headquarters in Washington. It is highly unlikely that Grant would have followed through on this assertion. Grant possessed a strong sense of duty and believed that a soldier should do his best where assigned. In August 30, 1863, correspondence to Congressman Washburn, Grant confessed reluctance to command the Army of the Potomac and expressed strong distaste for Washington. At the same time, he declared he "would not positively disobey an order." Those convictions would apply now. Sherman responded to Grant, "Take to yourself the whole Mississippi Valley" and finish the work that had been started. In a private message in early March 1864, Sherman had urged Grant not to stay in Washington because Maj. Gen Henry Halleck was more qualified to deal with the "intrigue and policy."[39]

Arriving in Washington with the intention of staying in the West, Grant then realized that the city's political influence over military matters would not allow him to pursue his overall strategy from afar. In short, he needed to be in the East. The powerful and influential Joint Committee on the Conduct of the War was a constant thorn in the side of military operations. Created in December 1861 to investigate the Union defeat at Ball's Bluff, the group's current focus was the Army of the Potomac and its Gettysburg and Mine Run Campaigns. The committee's investigations into western operations glossed over situations. In contrast, Union officers in the East regularly appeared before the committee for investigation of their actions. The committee, dominated by Radical Republicans, was often at odds with the administration's military decisions and the war's conduct. Grant witnessed its intrusive character firsthand when Meade accompanied him to Washington. While Grant was to head west to confer with the theater's generals on overall strategy, Meade was once again to meet with the committee "to answer Dan Butterfield's falsehoods."[40]

Meade's chief of staff Maj. Gen. Andrew Humphreys had a unique perspective on Washington's influence on the Army of the Potomac's operation. Humphrey had served continuously with this army since McClellan was its initial commander, and he had participated in all the army's battles. He felt that the capital city had "thwarted some generals, and interfered with all who had commanded the Army of the Potomac since the beginning of the war." Humphreys believed Grant would not be able to manage the operations of all the armies "unless he was near the capital."[41]

Maj. Gen. Henry Halleck became part of the solution to Grant's dilemma of needing to be in the East but not wanting to be in Washington. Although lacking skills as a field commander, Halleck was an able military administrator. His limited view of his previous role had led Lincoln to refer to him as a first-rate clerk. His relationship with Grant while in the West and in Washington was convoluted. When Grant, the new general-in-chief, arrived in Washington, Halleck, the previous general-in-chief, requested to be relieved. Grant astutely appointed Halleck as chief of staff—a newly created position. Halleck would remain in Washington to coordinate orders and, to a degree, shield Grant from Washington politics. On March 12, 1864, the War Department issued General Order Number 98, stating, "The headquarters of the Army will be in Washington [Halleck], and also with Lieutenant-General Grant in the field."[42]

The emerging technology of the "magnetic telegraph" became the other aspect of the solution for Grant's location dilemma. The telegraph would allow Grant to quickly communicate with his commanders and Washington regardless of his headquarters location. Over the course of the war, over fifteen thousand miles of telegraph wire would be laid for the Union armies. While in Nashville, Tennessee, in early 1864 before moving east, Grant had a telegraph installed at his headquarters and used it to discuss both sides' troop disposition with commanders and Washington officials. Sherman stated that the importance of the telegraph "was illustrated by the perfect concert of action between the armies in Virginia and Georgia during 1864."[43]

Options

As Sherman's leadership in the western command would continue the Federals' aggressive tactics, Grant was now free to decide where "in the field" back east he would establish his headquarters. Grant's innovative approach of commanding from the field, departed radically from Halleck's methods; the previous general-in-chief viewed his role as more of a "military adviser of the Secretary of War and the President." Having decided against making his headquarters in Washington, Grant would make his base of operations in the East by establishing it with an army in the field. Maj. Gen. Ambrose Burnside had previously been brought back east for recruiting purposes, and he once again commanded the Ninth Corps of over twenty thousand men currently located in Annapolis, Maryland. Regarded "as a reinforcement to the Army of the Potomac," this corps was not a viable option for Grant's headquarters. The remaining options included establishing a base with Major General Butler and his Army of the James in lower Virginia or with Major General Meade and his Army of the Potomac in northern Virginia. After assuming his new rank, Grant visited Butler and Meade at their respective headquarters in the field.[44]

Option 1

On January 19, 1864, at the request of then General-in-Chief Halleck, Grant had suggested an eastern strategy of attacking the Confederacy through the Carolinas with sixty thousand men, thereby depriving Lee's army of supplies. By establishing his headquarters with Butler, Grant would physically be in a position to effectively implement this strategy. While meeting with Butler at Fort Monroe, Grant sought his ideas for the upcoming campaign. Butler's thoughts were in line with Grant's thinking, and orders were issued to that effect. Butler had been one of Lincoln's earliest political general appointments, but he had no formal military training. Comstock, who was also at the meeting, described Butler as "sharp, shrewd, able, without conscience or modesty—overbearing." His military performance to date was mixed at best. Grant felt that the Army of the James would benefit from a professionally trained soldier, which his own presence would certainly provide.[45]

Option 2

The Army of the Potomac was the largest Union army in the field, and the one closest to Washington. This latter fact had always inhibited the army's movements. When Grant visited Sherman and other officers in Nashville, he described the Army of the Potomac as the "finest army he had ever seen." But at the same time, it had not lived up to its potential. Maj. Gen. George B. McClellan had created the Army of the Potomac, and many of its current officers (including Meade) had served under him. The army still exhibited much of McClellan's cautious approach to waging war on the South, as well as an inclination to withdraw after battle to replenish personnel and refurbish materiel rather than continue pressing the enemy for an advantage.[46]

Decision

Grant decided to located his base of operations with Meade. When he visited Meade on March 10, 1864, Grant indicated that, while he was in the northern Virginia area, his "intention" was to remain with this army. In his memoirs, Grant rationalized the conflicting circumstances of being headquartered with Meade while having Meade feel that he was in command of the Army of the Potomac: "To avoid the necessity of having to give orders direct, I established my headquarters near his, unless there were reasons for locating them elsewhere." While visiting Nashville to confer with Sherman and the other western commanders on overall strategy, Grant informed them of his decision, praised the Army of the Potomac, and stated that Sherman would command the armies in the West. In a March 22, 1864, letter to his wife, Meade stated that Grant's executive skills dictated that he needed to be in the field and that part of his objective was to avoid Washington and "its entourage." To provide the needed professional military perspective within the Army of the James, Maj. Gen. Baldy Smith was assigned to Major General Butler's army to command the right wing of his forces.[47]

Results/Impact

The impact of this decision was both immediate and enduring. Grant was sensitive to the command situation it put Meade in to have their headquarters in the same location, and he endeavored to make all orders for the Army of the Potomac's movements to Meade. Grant typically framed these orders as suggestions. The orders to Meade directing the Army of the Potomac's May 7,

Grant and staff after the Battle of Spotsylvania Court House. (Grant is in front of the tree with his legs crossed; Meade, with the blurred face, is on the far end of the bench to Grant's left.) Library of Congress.

1864, movement to Spotsylvania Court House was written as follows: "I think it would be advisable in making this change." Burnside's Ninth Army Corps movement orders referenced Meade's orders, which included the nonsuggestion "Burnside will move on the plank road." For his part, Meade understood the pressure and high expectations Grant faced from the administration, Congress, and public opinion. He felt that Grant acted on his suggestions quickly, and the relationship between the two was generally amicable.[48]

Grant was very diplomatic when referring to Meade's roles and responsibilities, stating that Meade had immediate command of the Army of the Potomac, while he (Grant) had "supervision" of all the armies. Although he stated that his "instructions" to Meade were general in nature, with the details and the execution left to Meade, Grant's actual orders could be detailed and specific. Although his orders for the movement to Spotsylvania on May 7 were couched in a tone of advice, Grant specified what troops were to relocate and in what order. He would also set specific times for attacks, often in an effort to produce an integrated movement with all parts of the Army of the Potomac and Burnside's Ninth Corps.[49]

By locating his headquarters with Meade, Grant quickly came to appreciate him. As confirmation of his regard for the commander's ability, Grant recommended Meade for promotion to the rank of major general in the regular army on May 13, 1864, the day after the assault on the Mule Shoe at Spotsylvania Court House. More significant is the fact that Grant simultaneously recommended Sherman for the same rank. Grant had worked with Sherman for years in the West and with Meade for only a few months in the East, yet he recommended their advancement at the same time. "I would not like to see one of these promotions at this time without seeing the other," the lieutenant general asserted.[50]

By colocating with Meade, Grant avoided any issues or possible conflicts of seniority. Meade's November 1862 commission to the rank of major general of volunteers placed him junior to Major General Burnside, who commanded the Ninth Army. The Ninth would need to coordinate extensively with the Army of the Potomac. Others senior to Meade who were operating in the East were Major General Butler, commanding the Army of the James, and Major General Sigel, commanding troops in the Shenandoah Valley.

The most dramatic and lasting impact of Grant's locating his headquarters with Meade would be the redefinition of the character of the Army of the Potomac. Now there would be a strong correlation between what Grant wanted to do and the speed with which his preferences would be implemented. What Meade called Grant's "indomitable energy and great tenacity of purpose" would be directly incorporated into the Army of the Potomac's character in the coming campaign—due in large part to Grant's personal presence with the army.[51]

Just as McClellan's personality had defined the previous character of the Army of the Potomac, Grant's headquarters being located with the Army of the Potomac would now reshape it into a relentless fighting force. After the battle of Shiloh in 1862 and the subsequent Confederate efforts, Grant "gave up all idea of saving the Union except by complete conquest." This sentiment was uncannily similar to Lincoln's assertion after the Battle of Antietam later the same year that the war would end not because of "strategy" but because of "hard fighting." Grant's current rank and physical location with the Army of the Potomac would now allow this "conquest" that would require the "hard fighting" to begin. The two most iconic generals of the Civil War, Lee and Grant, would now meet head on in battle.[52]

Grant Decides On an Offensive Operation Against Lee

Situation

As the war entered its fourth year, the North saw reason for both hope and concern: hope that Grant would bring his achievements in the West to the

Eastern Theater, and concern that the three previous years had brought little success against Lee in Virginia. Grant wanted to develop an overall integrated strategy that would reduce wasted manpower and effectively utilize all Union forces to attack the Confederacy simultaneously on multiple fronts. Not since 1861, when Maj. Gen. Winfield Scott had devised the Anaconda Plan, had there been an overall unified approach. To negate the potential advantage the Confederacy's interior lines afforded, Grant would "take the initiative in the Spring Campaign to work all parts of the Army together, and somewhat towards a common center." After his confirmation as lieutenant general, his initial step was to meet personally with Major Generals Meade, Sherman, and Butler.[53]

When Grant assumed his new position, he saw the Union armies as organized into nineteen departments, of which four in the West had been combined into a single military division. Grant would stop the previous practice of these units acting independently and instead implement an overall coordinated offensive strategy. To this end, he regarded the Army of the Potomac as the center wing and Butler's Army of the James as the left wing of his operations.[54]

Grant met with Maj. Gen. Meade at Meade's headquarters near Brandy Station, Virginia, on March 10 as the initial step in planning his overall campaign to win the war. Grant returned to Washington the following day and then traveled west, where he met with Sherman and other officers that had served under him, then sketched out an overall campaign. Sherman would later define the scheme: "He [Grant] was to go for Lee and I was to go for Joe Johnston. That was his plan." Returning east, Grant next met Maj. Gen. Benjamin Butler on April 1 at Fort Monroe (Old Point Comfort) and discussed strategy and shared views. The meeting went well, and Grant returned to his headquarters at Culpepper to solidify and begin implementing his overall strategy.[55]

Grant used two guiding principles in developing his general strategy: use the greatest number of troops practicable, and hammer the enemy continuously.[56] His plan to simultaneously meet the Confederates on multiple fronts and limit their ability to reinforce each other shaped his decision as to how to attack Lee. Grant was aware of the impact that the movement of Lieut. Gen. James Longstreet's corps to Chickamauga had had on the Confederate victory. The underlying problem was that the Union's lack of an overall coordinated strategy for the war allowed the Confederates to shift their limited troops from one threatened area to another. Grant's overarching plan would ensure that the Rebels could not do so again.

To apply pressure on all fronts, Grant issued the following orders:[57]

Maj. Gen. Nathan Banks would abandon Texas and move on Mobile.[58]

Maj. Gen. William T. Sherman was to advance against the Confederate Army of Tennessee in the West and render it ineffective.

Maj. Gen. James Butler was dispatched up the James River to capture City Point and threaten Richmond.

Maj. Gen. Franz Sigel was to advance up (south) the Shenandoah Valley to destroy this Confederate "breadbasket."

Maj. Gen. George Meade's Army of the Potomac would move against Lee and the Army of Northern Virginia.

Maj. Gen. Ambrose E. Burnside was charged with protecting the Army of the Potomac's supply lines. Grant stated, "I will give him [Burnside] the defense of the road from Bull Run as far south as we wish to hold it." (Note: Burnside's twenty thousand men numbered more than any single corps in the Army of Northern Virginia.)[59]

Of these movements, only those of Major Generals Sherman and Meade would have success. In both cases, Grant directed their efforts against the Confederate armies rather than a fixed objective, as had previously been the goal. To Sherman, he wrote, "You I propose to move against Johnston's Army, to break it up and to get into the interior of the enemy's country as far as you can, inflicting all the damage you can against their War resources." Ironically, on April 9, 1864, exactly one year before Lee's surrender at Appomattox Court House, Grant wrote Meade, "Lee's Army will be your objective point. Wherever Lee goes there you will go also." The similarity of Lincoln's June 1863 suggestion to Hooker and Grant's order to Meade that Lee, not Richmond, was the army's objective underscores the men's shared view of how the war should be pursued. This common approach was a fundamental element in their growing trust and confidence in one another. Unlike the situation in the West, where Sherman would develop his own specific approach to "get at Johnston," in the East it was Grant, not Meade, who would decide how to attack Lee and his Army of Northern Virginia.[60]

Options

Grant had won at Vicksburg by utilizing a siege, and he had won at Chattanooga by breaking one. Thirteen years after the war, Grant said he had preferred this approach and "considered the plan with great care," but he had decided it entailed too many logistical problems and risks for the country. Sieges take time to implement and show progress, something the Northern civilian population in a presidential election year might not have viewed too favorably. Ironically, a siege of Richmond is what worried Lee the most. Neither Grant's 1865 official report nor other articles mention this option.[61]

The lessons of the aborted November 1863 Mine Run Campaign against Lee's entrenched forces and the rugged terrain south of the Union army's camps rendered a frontal assault unrealistic. Thus, Grant had three approaches to "get at Lee."[62] He could use the Army of the Potomac to effectively pin Lee's army in place while attacking the Confederates' supply lines from the south with Butler's Army of the James. Another option was using the vastly superior Union naval power to support an approach from the east up Virginia waterways, as Maj. Gen. McClellan had tried. Finally, Grant could initiate battle and move against one of Lee's flanks.

Option 1

In January, before his appointment as lieutenant general, Grant had encouraged Halleck to abandon the previous routes of the advance on Richmond, and to instead move troops up through North Carolina and threaten Lee. Grant estimated that a moving force of sixty thousand troops could implement this operation and start from Suffolk, Virginia (twenty-five miles from Butler's current location at Fort Monroe). Federals would find little military opposition in this area. Additionally, the interior railway that was so vital for supplying the Army of Northern Virginia would be so vulnerable that Lee would have to commit a significant portion of his army to guard it, thereby forcing the evacuation of Virginia. Halleck, the then general-in-chief, responded that there was no place to get the sixty thousand troops, that the perceived strength of the defenses around Washington was overstated, and that any weakening of Meade's forces might "uncover" the capital and result in Lee's army once again moving north. The sentiment in the North would force all of the troops deployed for the operation to be recalled. Finally, this option would be inconsistent with Grant's aggressive style and his intention to use the Army of the Potomac to go after Lee and his army.[63]

Option 2

Grant could take advantage of the US Navy's control of the major waterways of the Chesapeake and mount a joint offensive operation against Lee. Grant had effectively utilized naval forces to provide logistical support and firepower in the West, and it might be possible to duplicate that success here. Although Lincoln professed that he did not want to know Grant's plans, he did have some ideas of his own. In a meeting with Grant, Lincoln proposed using waterways to move troops and to protect the army's flanks. Stanton and Halleck

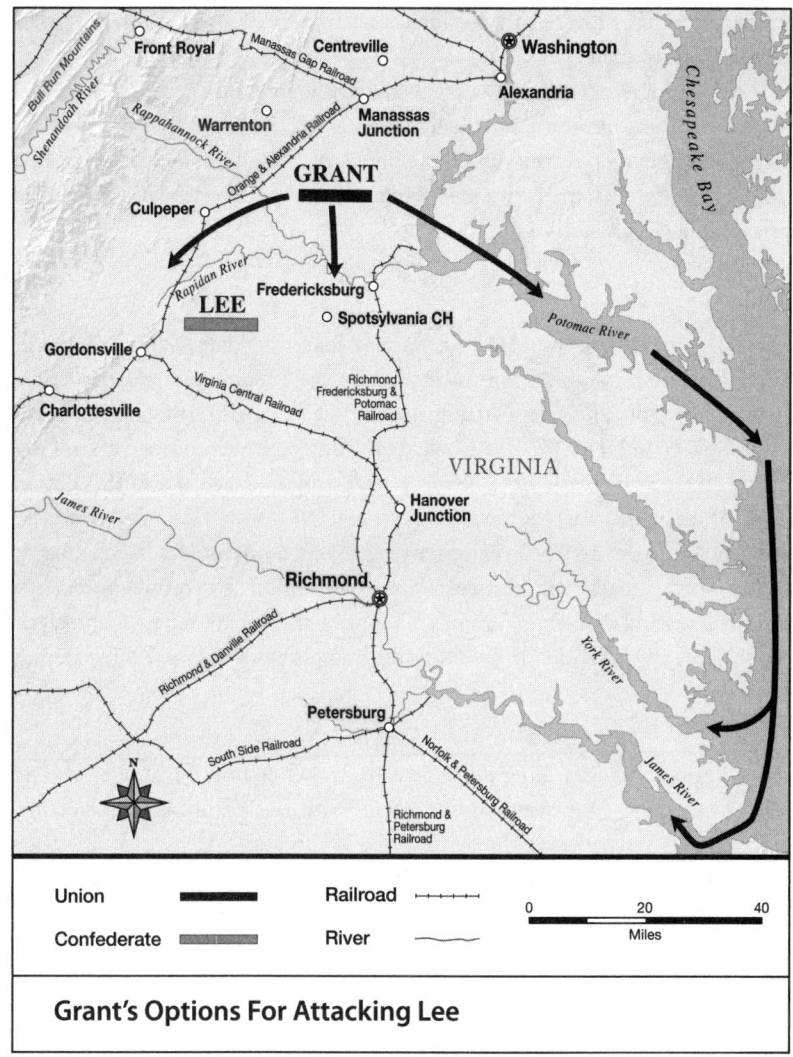

Grant's Options For Attacking Lee

had cautioned Grant that the president was too kindhearted to refuse anyone asking about plans, so Grant did not communicate his to Lincoln. Grant realized, but did not tell Lincoln, that these same streams would also shield Lee. Furthermore, Grant reasoned that attacking from the east and forcing Lee into the strong defenses of Richmond was not to the Union's advantage. Grant considered the Richmond defenses so good that "one man inside to defend was more than equal to five outside besieging or assaulting." The fall of Richmond would naturally follow the capture of Lee's army. Grant felt that it was better to fight Lee's army outside these strong Richmond defenses rather than in front of them.[64]

Lee would be aware of any such naval movement, and, "judging from" his previous actions, he would establish positions well in advance of the existing Richmond defenses. Enveloping these defenses "would not have been practicable." Also, the Federals' extended flanks would be subject to attacks "under unfavorable circumstances."[65]

Option 3

A flanking movement would force Lee out of his strong defensive position near Mine Run and into more open country where the Union's numerical superiority would be more effective. Moving against Lee's army in the field so far from Richmond would give Grant's whole army ample opportunities for flanking attacks, thereby increasing the chances of destroying Lee's army. This option would also ensure Washington would be shielded from a Confederate attack. The problem was that the previous attempts of this nature had ended badly for the North. In addition, the sheer size of the Union army would pose a significant logistical challenge. Grant's Vicksburg and Chattanooga Campaigns enforced his appreciation for good logistical support as a necessity for success.[66]

Decision

Grant decided on a flanking movement to draw Lee out of his defenses and then engage him. One question remained: With Lee's army as the objective, which flank would present the best chance of success?[67]

If Union troops crossed the Rapidan River above Lee's left flank and moved west, Lee would be unable to ignore the threat to Richmond. Nor could he attempt an attack on Washington. This advance would move through more open country, affording Federal troops greater maneuverability and better use of their superior artillery, and allowing them to concentrate forces. However, the Army of the Potomac's ability to direct and coordinate Butler's movements with its own would be greatly impaired. Another challenge would be to protect the Alexandria Railroad and its depot. Wagon trains could provide only fifteen days' supplies before requiring replenishment from the railroad. Any advance would require more and more troops to protect the lengthening rail line and the wagon trains. The criticality and vulnerability of this railroad became apparent when Maj. Gen. John Pope tried this route in 1862. The resulting Battle of Second Manassas almost destroyed the Union Army of Virginia. Abandoning the railroad supply line for the navigable waterways to the east would require a strong covering force.[68]

If they moved south below Lee's right flank, Federal troops could use Brandy Station as a base of supplies until another was secured on the York or James River. The Alexandria Railroad supply line would not be needed, and a shorter link to the navigable water connecting to Washington could be utilized. Humphreys felt that by moving to the Union's left flank (Confederate right), "no protecting force would be necessary to cover these short land routes." The challenge with this flanking route was the ten or fifteen miles of rough terrain south of the Rapidan that Union troops would have to traverse before reaching more open and maneuverable terrain. It would be a nightmare to move, control, and communicate with soldiers in The Wilderness, a tangled second-growth forest with an almost impenetrable undergrowth. Artillery could be effectively used only along the few main roads and the scattered small farms with open fields.[69]

Appreciating the importance of logistics, Grant decided to move south on Lee's right flank and cover ground that had proved costly in the past. In a conversation with his staff officer Lieutenant Colonel Porter the night before the campaign began, Grant stated that he had "weighted very carefully the advantages and disadvantages of moving against Lee's left or right." He had determined that, although a movement to the left "promised more decisive results," the communications and logistics "presented too many serious difficulties." The key to this advance would be how rapidly the Union army could traverse The Wilderness and get to the more open country where the Union advantage in men and materiel could be brought to bear.[70]

Results/Impact

The campaign started with Union troops moving across the fords on the Rapidan unopposed on May 4, 1864. The Army of the Potomac would make this movement with over 4,300 wagons, 835 ambulances, and 120,000 men. To appreciate the logistical magnitude of this wagon train, it is estimated that if the wagons had been placed end to end, they would have stretched 60 miles—the approximate distance from the Rapidan to Richmond. Concern for the protection of the supply trains halted the first day of the infantry's march earlier than originally planned. The slow advance of this huge wagon train (the wagons had not finished crossing Ely's Ford until two o'clock on the afternoon of May 5) resulted in the Battle of The Wilderness. Humphreys' plan called for moving through this area on the first day, but the strategy had to be modified and the advance halted.[71]

That night, Grant and Meade pitched their headquarters tents near Wilderness Tavern along the Orange Turnpike. A newspaperman whom Grant was acquainted with approached and asked how long it would take to get to

Richmond. Grant replied, "I will agree to be there in about four days, that is if General Lee becomes a party to the agreement; but if he objects, the trip will undoubtedly be prolonged." As it turned out, Lee would have very strenuous objections resulting in 36,400 Union and 23,400 Confederate casualties.[72]

Opposing Forces' Structure at Beginning of the Campaign

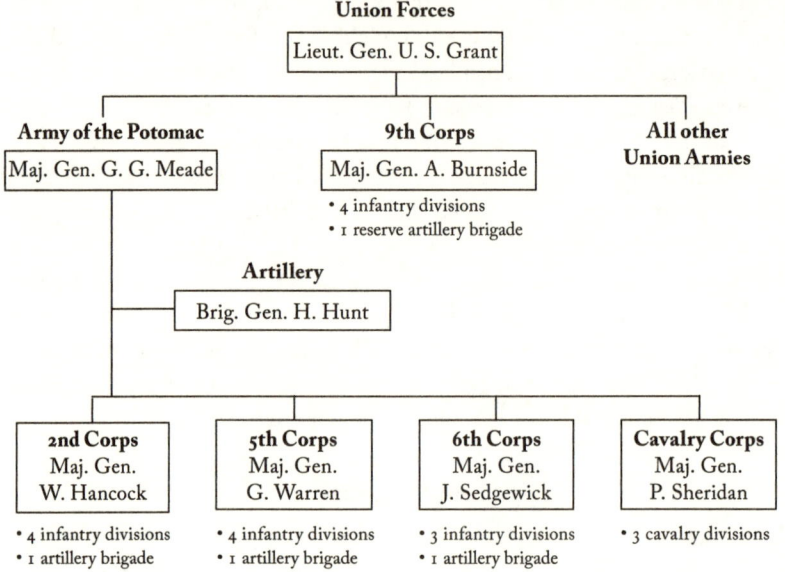

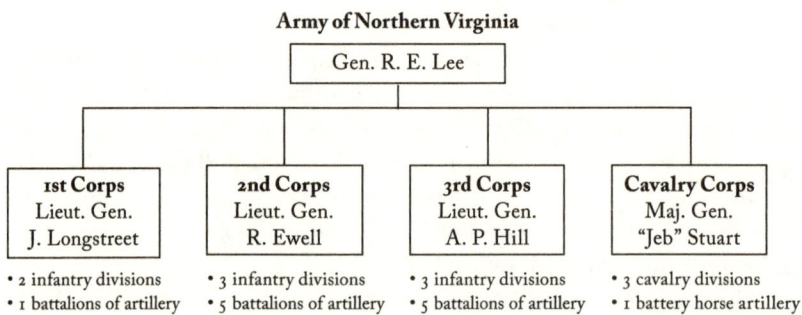

Alternate Decision and Scenario

In an interview in 1878, Grant discussed the alternate flanking movement to Lee's left. As he often did, Grant cited his unfamiliarity with the officers, men, and capabilities of the eastern armies. If he could have Sherman, who he had previously stated was too valuable in the West, and Sheridan, who was already with him, he " would not have hesitated for a moment." In Grant's view, Lee would have been forced to move to counter such a flanking movement, but unlike The Wilderness which provided good cover and two roads to attack Grant's flank, the Union would have been on more open ground.

But Grant's 1878 comments neglect the issue of supplying his army that had so concerned him in 1864. Had the Union army moved to Lee's left, it would have passed through countryside "so exhausted of all food or forage" that soldiers would have had to carry their own supplies and ammunition. The supply line would have been vulnerable to Confederate attacks, thus limiting either the troop mobility or strength required for the maneuver. Longstreet had remained near Gordonsville to be in a position to counter any such move, and the fighting that took place at The Wilderness and Spotsylvania Court House would have occurred elsewhere. Only the unlikely occurrence of a catastrophic defeat of the Union forces in Virginia might have tested Grant's resolve to go after Lee and his army.[73]

Timeline Summary

The Wilderness

<u>May 4, 1864</u>

> Union forces start night crossing of the Rapidan River at Germanna and Ely's Fords.
> Lieut. Gen. Richard Ewell's and Lieut. Gen. Ambrose (A. P.) Hill's corps leave their winter encampments.

<u>May 5</u>

> Ewell and Maj. Gen. Gouverneur Warren clash at Saunders Field along the Orange Turnpike in the first fighting of the campaign.
> Hill and Maj. Gen. Winfield Hancock clash along the Orange Plank Road near the Brock Road intersection.

<u>May 6</u>

> Hancock initiates a dawn assault on Hill's disorganized troops and drives them back.

Lieut. Gen. James Longstreet arrives just in time to prevent Hancock's assault from destroying Hill's Corps.

Longstreet executes a successful flanking attack on Hancock, who then takes a defensive position at the Brock Road.

Longstreet is wounded, and the Confederate movement stalls.

Gen. Robert E. Lee makes a late attack on Hancock that fails.

Finally given permission, Brig. Gen. John Gordon makes a flank attack but is too late in the day to achieve more than limited success.

May 7

Cavalry fights near Todd's Tavern.

Maj. Gen. Fitzhugh Lee's cavalry fights a delaying action along Brock Road.

Grant issues orders for a night move to Spotsylvania Court House.

Warren and Maj. Gen. Richard Anderson move toward Spotsylvania Court House in a race for position advantage.

Spotsylvania Court House

May 8

Anderson's Confederates arrive at Laurel Hill just minutes before Warren's advance troops.

Warren initiates the first of several attacks against Laurel Hill, all of them unsuccessful.

Union and Confederate troops continue to deploy.

May 9

The Mule Shoe defensive line is complete.

Maj. Gen. John Sedgwick, Union commander of the Sixth Corps, is killed by sharpshooter fire.

Maj. Gen. Phil Sheridan and his cavalry depart the Army of the Potomac to pursue part of Maj. Gen. James "Jeb" Stuart's Confederate cavalry.

Maj. Gen. Ambrose Burnside's Ninth Corps arrives on the Union left flank east of Spotsylvania Court House.

Hancock probes the Confederate left flank.

Confederates strengthen their left flank near the Po River.

May 10

Hancock tests the Confederates' left flank along the Po River and finds it too strong.

Before the Battles

Warren again unsuccessfully attacks Laurel Hill.
Col. Emory Upton leads a late-day attack at Dole's Salient with limited success.

May 11

Hancock's Corps makes a night march from the Union right flank to the center for a dawn attack on Mule Shoe.
Lee pulls most of the artillery from the Mule Shoe.

May 12

Supported by Brig. Gen. Horatio Wright's corps, Hancock attacks the Mule Shoe at dawn.
After initial success, the Union attack stalls.
After fierce Confederate counterattacks, the line is temporarily restored.
Twenty hours of intense fighting follow.
Confederates construct a secondary defensive line at the base of the Mule Shoe.

May 13

At 3:00 a.m., Confederates pull out of the salient to their new line.
Federals assess the carnage.
Hancock's Corps is pulled back to recover.

May 13–17

Grant maneuvers for opportunities, Lee effectively counters

May 18

Grant orders Hancock/Wright to attack the Confederates' new defensive line at the base of the Mule Shoe—the attack fails.

May 19

Grant prepares to maneuver south.
Ewell makes a reconnaissance in force on the Union right flank and barely escapes disaster.
Grant delays for a day his initial move south.

May 20

Hancock's Corps moves south from Spotsylvania Court House, starting the next phase of the Overland Campaign.

CHAPTER 2

THE BATTLE OF THE WILDERNESS

With the arrival of spring in 1864, the roads to the south began to dry, allowing military operations to begin. Grant would initiate the part of his overall strategy that focused on Lee and his Army of Northern Virginia by crossing the Rapidan River. History would designate this as Grant's Overland Campaign, and the war's top two generals would meet in battle for the first time. This initial encounter between Lee and Grant would take place in the thick, tangled foliage just west of Chancellorsville Battlefield, where the two armies had clashed one year before. The area would limit the use of artillery, cause problems for the cavalry, and make infantry fighting extremely difficult. Four critical decisions shaped this opening encounter in what would simply be called "The Wilderness."

Lee Decides to Take the Initiative

Situation

As dawn broke on May 4, 1864, lead elements of the largest army ever assembled in the country were advancing toward the Rapidan, thereby initiating Grant's decision to move against Lee's right flank.[1] Meade's chief of staff Maj. Gen. A. A. Humphreys developed the details of the Federal movement with information gained the previous November, when Meade had tried to maneuver against Lee in what would be called the Mine Run Campaign.

Brig. Gen. Andrew A. Humphreys (USMA class of 1831), Meade's chief of staff. Library of Congress.

The Federal movement started in the morning, but Confederates observed it as soon as the fog lifted. Although the Confederates had early warning and were able to thwart the advance, Humphreys felt that Lee's response had been slow to the Federal advance at Mine Run.[2]

Based on that experience, Humphreys's plan for initiating the current campaign would set the Union army in motion at midnight to "pass out of the Wilderness" and into more open ground before a general engagement. Maj. Gen. Gouverneur Warren's Fifth Corps was ordered to leave at midnight on May 3, while Maj. Gen. John Sedgwick's Sixth Corps would start toward Germanna Ford at 4:00 a.m. To the east, Maj. Gen. Winfield Hancock's Second Corps would move toward Ely's Ford at midnight. The challenge would be for the supply trains to keep up with the troop movements.[3]

Unlike Grant, whose forces were concentrated and well provisioned, Lee had had to disperse his army to feed and supply it during the winter. In January 1864 letters to Confederate secretary of war James Seddon, Lee expressed concern about the negative impact of insufficient provisions: "Unless there is a change, I fear the army cannot be kept effective." Lee wrote to Confederate president Jefferson Davis of the criticality of his supply line, stating, any disruption "would render it impossible for me to keep the army together." Without that supply route, the Confederates might even have to retreat to North Carolina. Consequently, this situation would have a major bearing on Lee's

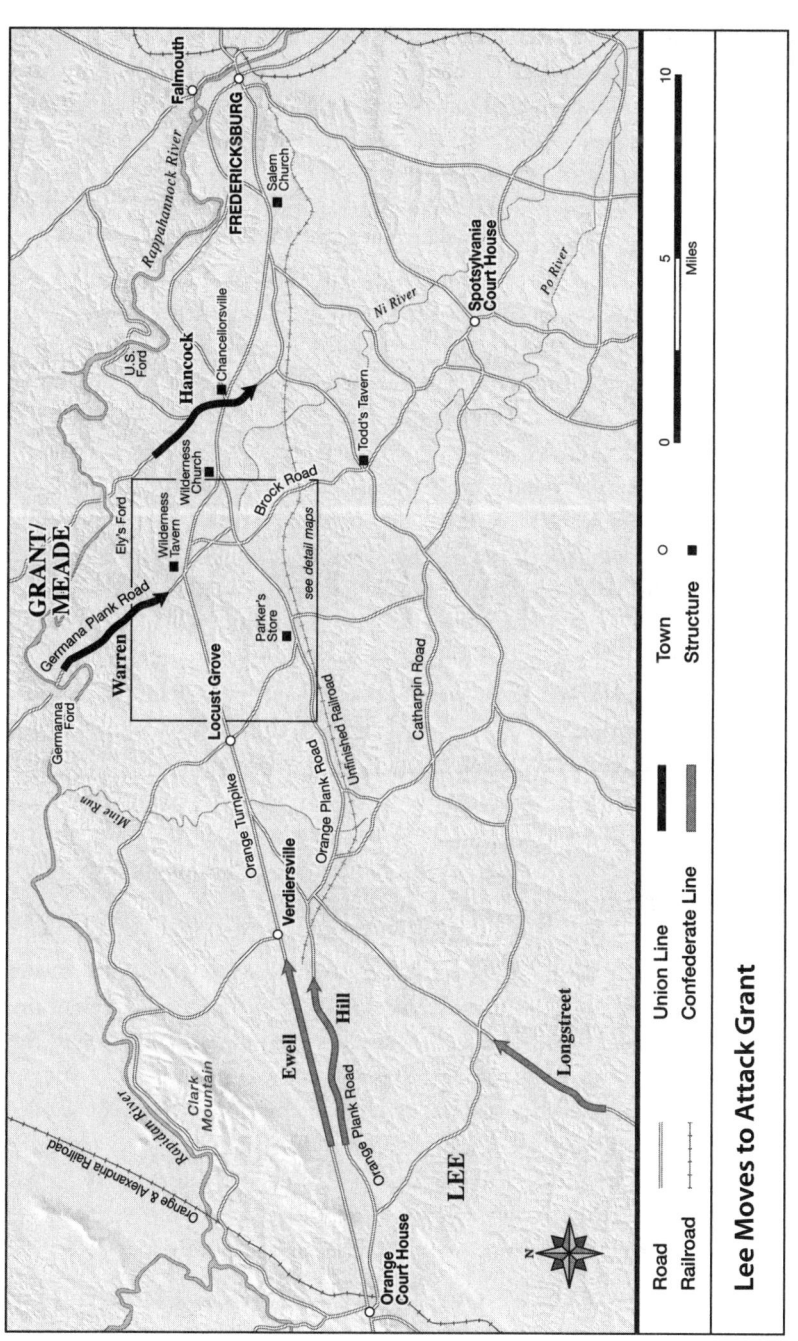

Lee Moves to Attack Grant

decision-making. In March correspondence with Longstreet, who was still in Tennessee, Lee shared his concern over the "scarcity of supplies," which he called "the controlling element to which everything must yield." Lee felt that if Grant were allowed to freely maneuver south of the Rapidan, a siege would eventually result, and that would be the end.[4]

To alleviate the problem of provisioning his army as much as possible, Lee had had to disperse the corps. Lieutenant General Ewell's Second Corps was camped east of Orange Court House, and Lieut. Gen. Ambrose Powell (A. P.) Hill's Third Corps was south of the town. Lee's headquarters was located near Orange Court House. Longstreet's First Corps, recently returned from Tennessee, was twenty miles away at Gordonsville. Concerned that Grant might try to move west on his left flank, Lee kept Longstreet near Gordonsville. This location also provided Longstreet and his men quick access to the railroad should they have to move quickly to defend Richmond from Grant's integrated plan. Despite the privations, the Army of Northern Virginia's troops would face their numerically larger enemy with "undaunted spirit."[5]

In 1864, the high ground of Clark's Mountain, about eight miles northeast of Orange Court House, offered Lee and his officers a good view of the vast Union army. The Rebels could clearly see Federal troop movements that May 4 morning; Lee was thus aware of the advance to the Rapidan. Grant's move across the river was basically unopposed, and both commanders expressed satisfaction at this development. Grant, who had been apprehensive about a contested river crossing, considered the maneuver "a great success." Lee, who felt Grant was making Hooker's mistake of the previous year by moving into the dense terrain, "hoped the result would be even more disastrous to Grant" than it had been for Hooker.[6]

Pursuant to orders, Warren's Corps moved to Old Wilderness Tavern, and by 4:00 p.m., his troops were making camp near the intersection of the Orange Turnpike and Germanna Plank Road. "It was in consideration of the fact that it was not practicable in this region to move great trains along the projected flank of the army simultaneously with the troops that led to fixing the halting-places of the heads of the infantry columns." Brig. Gen. Alexander Webb of Hancock's Second Corps estimated the length of the supply-train wagons at sixty-five miles. This halt was in contrast to Humphreys's original timetable to quickly move through The Wilderness and was the "first misfortune of the campaign." Fortune had intervened on the Confederates' side.[7]

Options

Although Lee was an audacious commander, the condition of his army and the presence of such a large Federal force in northern Virginia made an at-

Maj. Gen. Gouverneur K. Warren (USMA class of 1850), commander of the Fifth Corps, Army of the Potomac. Library of Congress.

tempt at Washington or any other Northern city unrealistic. Lee had three viable options to meet the Union army's advance. He could consolidate his three dispersed corps and meet Grant south of The Wilderness from a fortified position, possibly along the North Anna River. Or he could launch a full attack with Hill's and Ewell's Corps and press whatever opportunities might develop. Finally, Lee could use a variation of the second option in which Hill and Ewell would attack and halt the Union army's advance through The Wilderness. Longstreet would then come up so that, when combined, the full force of the Army of Northern Virginia could be utilized.

Option 1

By unifying his three corps, Lee could meet Grant with his full force and, given the Southern forces' demonstrated ability to move more rapidly, at a place of his own choosing. But the time required to get the Army of Northern Virginia to a chosen position would allow the Federals to move unhindered through The Wilderness and be on open ground that would greatly facilitate Grant's maneuverability, take advantage of his superior artillery, and allow him to dictate how the battle would be fought. Such a Federal turning movement would also make the strong defenses at Mine Run useless. Grant would be closer to Richmond and have a more direct path to the Confederate capital than Lee would.[8]

Wilderness Tavern, site of the Fifth Corps encampment, May 4, 1864. Library of Congress.

Option 2

By initiating a flanking attack with Hill and Ewell, Lee would be able to take advantage of the terrain. The Wilderness was a second-growth forest with limited roads on which to move troops and few open areas in which to maneuver and bring the power of the Union artillery to bear. To a degree, then, it would negate Grant's numerical advantages. By keeping Longstreet near Gordonsville, Confederates could quickly counter any Union movement toward their left flank. The Mine Run defenses that the Confederates had just left could be used as a fallback defensive position should the Federals counterattack in force and start to gain a considerable advantage over Hill and Ewell.[9]

The biggest issue would be the disparity in the opposing forces. Not only would the Rebels be attacking, but they would also be greatly outnumbered. The tactical advantage of a flanking movement would partially offset this drawback, but the Confederates could not exploit any advantage they might gain since they lacked ready reserves. More likely, the Federals' greater manpower would quickly turn any temporary Rebel advantage into a disaster.

The past performance of Ewell and Hill, the two corps commanders who would have to execute the attack, had raised questions about their leadership ability at the corps level. After the death of "Stonewall" Jackson the previous year only a few miles from The Wilderness Battlefield, Lee had been forced to reorganize the Army of Northern Virginia. A. P. Hill and Richard Ewell had both been promoted from division to corps commanders. This would be the first major engagement since Gettysburg where the performance of both generals had been disappointing. The two men had health problems, with Hill suffering from chronic stomach issues and Ewell having lost a leg at Second Manassas. Ewell had also gained a wife, thereby displeasing his men, who felt her influence had an undesirable effect on his military performance.[10]

Finally, a major drawback to this plan was that a gap would form between the two corps as they moved along the Orange Turnpike and Orange Plank Roads. This distance, coupled with the terrain, would make it extremely difficult for troops of either corps to support the other should the Confederates make a breakthrough, or should disaster arise and the Federals attack one corps.

When planning the Union's advance, Humphreys considered these first two options the most likely to be executed by the Confederates during the Overland Campaign.[11]

Option 3

By using Ewell and Hill to attack the flank of the Union advance, Lee could hold the enemy in check until Longstreet could cover the distance from Gordonsville to the battle. This option would use the tangled terrain of The Wilderness to the Confederacy's best advantage. In 1864, the area featured few established roads. The Orange Plank and Orange Turnpike Roads ran roughly west to east from Hill's and Ewell's encampment location and intersected the vital north–south Brock Road that Union troops would have to use to pass through The Wilderness. The dense undergrowth and thick foliage made it difficult to coordinate the movement of large troop units and thus limited the effectiveness of the Union's size advantage in a fight. Artillery could be used only in a few areas, again diminishing the Federals' numerical advantage. The knowledge of local roads in the Battle of Chancellorsville that had been fought nearby the year before had allowed the Confederates to deliver a crushing though costly flank attack. In The Wilderness, the Southern army would again have an advantage in the support of locals for information on the limited roads available for maneuver. In contrast, Grant was completely unfamiliar with the unique challenges the area posed. The Union army was "well acquainted with the chief roads passing through that region known as the Wilderness," but not with the smaller "numerous wood-roads" in the area.[12]

The concerns with Hill and Ewell stated in the previous option would still apply here. Added problems were that Longstreet was habitually slow and would have the farthest to travel to the intended battlefield. While this option would also result in a gap between the two corps, the fact that Longstreet would be moving up and could be used where most needed—if he arrived in time—would somewhat mitigate that concern.

Decision

Lee decided his best option would be a flanking attack on the Union army as it moved through The Wilderness. This course of action would hold the Federal advance while Longstreet rushed to the fighting to deliver a decisive blow on the Union southern flank. Attempting to maintain the ability to dictate events rather than surrendering the initiative to Grant appears to have been a major factor in Lee's choice. This movement would require Lee's forty thousand men of the Second and Third Corps to hold the enemy for a day and Longstreet's forces to arrive at the planned time.[13]

Results/Impact

To implement this strategy, Lee ordered Ewell and Hill to move by noon on May 4. Ewell would camp near Robertson's Tavern, only three miles from Warren's stopping place, and Hill moved near the hamlet of New Verdiersville along the Orange Turnpike and Orange Plank Road respectively. Ewell had the shorter distance to travel. Longstreet was ordered to march from Gordonsville to Todd's Tavern on the Brock Road starting about 4:00 p.m. on May 4. Circumstances dictated that Longstreet would be needed at a different location, and at 1:00 a.m. on May 5, new orders were issued that changed the route to travel via the Orange Plank Road to Parker's Store. Lee's thirty-hour delay in November to counter Meade's Mine Run movement factored into the planned Union advance through The Wilderness. But this time Lee expected a Union flanking movement, and he "moved more promptly toward the Army of the Potomac than he had during the preceding November."[14]

As was typical of Lee's orders in the past, his orders to Ewell and Hill were somewhat discretionary and cautioned not to bring on a general engagement until Longstreet could be brought up. "He sent word to General Ewell not to advance too fast, for fear of getting entangled with the enemy while still in advance and out of reach of Hill. . . . Above all, Gen. E. was not to get his troops entangled, so as to be unable to disengage them, in case the enemy was in force."[15]

The two roads used for the Confederate advance diverged as they went east and formed a dangerous gap between the two corps. Ewell had to travel a shorter distance along the northern Orange Turnpike from his encampment

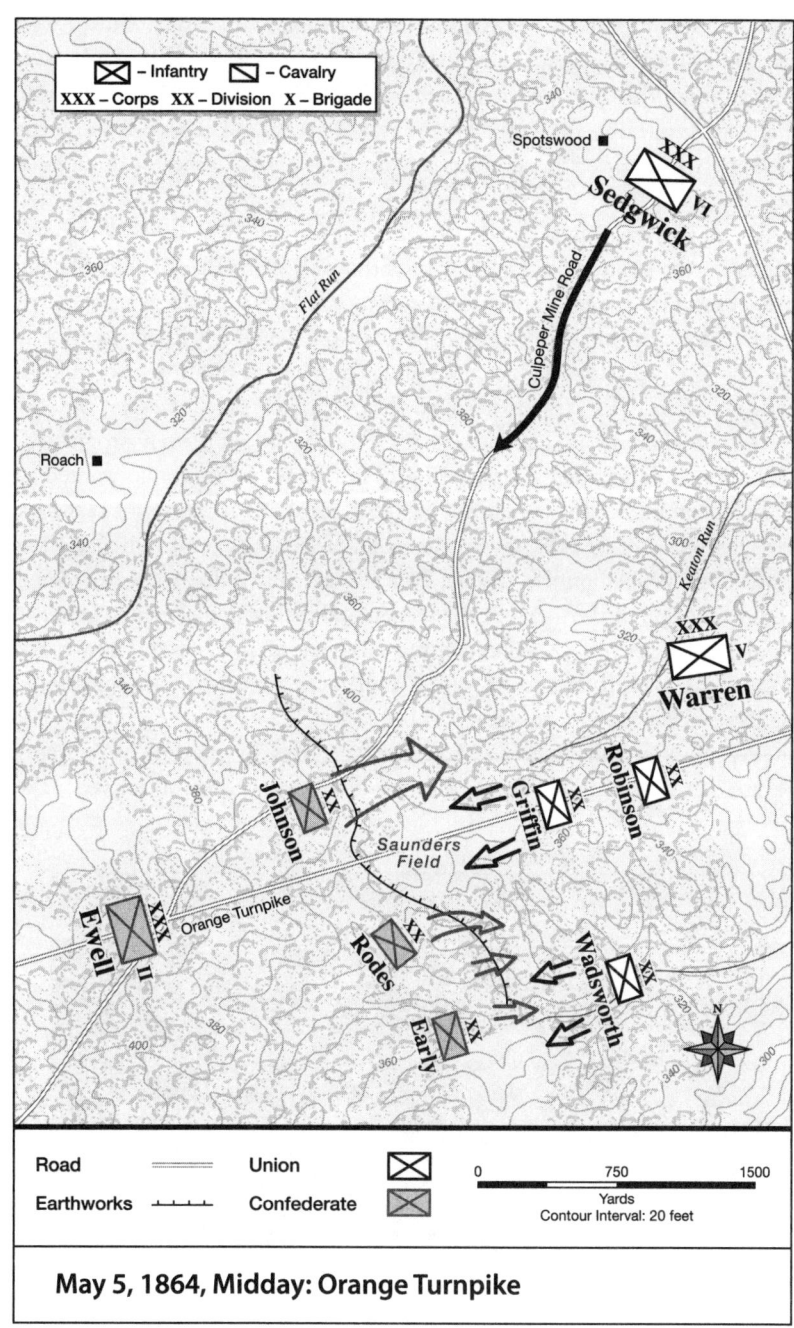

May 5, 1864, Midday: Orange Turnpike

than Hill had to travel along the southern Orange Plank Road. Thus, Ewell was initially farther along toward the Union troops than Hill. Both sides were aware of this gap and the vulnerability it represented. Coupled with the challenging terrain, this gap made it difficult for the two Confederate corps to support each other if necessary in the face of the numerically superior Federals.

By midday on May 5, Ewell's troops engaged those of Warren's Corps located near the Wilderness Tavern west of the Germanna Plank Road. Shortly thereafter, Hill's men encountered advanced units of Sedgwick's Corps coming down from the Wilderness Tavern on the Brock Road, as well as Hancock's Corps (Hancock commanding) moving up the Brock Road from Todd's Tavern. Hill's troops engaged the enemy forces just west of the important Brock and Orange Plank Roads intersection. As the day progressed, Union troop deployments exposed Hill's left flank. Lee initially wanted to divert upcoming troops to plug the gap between his two corps, but Hancock put such pressure on Hill on the afternoon of May 5 that all of his arriving troops were needed on the Orange Plank Road. Burnside had been ordered in the direction of the gap, but the terrain and his natural slowness kept the Ninth Corps from effectively exploiting this opening. Lee's decision to attack along these two roads and his quick implementation of the troops' movements set the stage for the rest of the battle.[16]

Warren's lead divisions moved out of the area by the Old Wilderness Tavern on the morning of May 5 with Parker's Store as their objective for the

Saunders Field in 1864, view across the Orange Turnpike toward the Confederate position. Library of Congress.

day. Brig. Gen. Charles Griffin's First Division was the last division to leave, and the soldiers were preparing to move out when they discovered Ewell's Confederates on the Orange Turnpike. Grant sent the following orders when informed of the situation: "If any opportunity presents itself of pitching into any part of Lee's army, do so without giving time for dispositions." For his part, Meade felt that Lee had left a division to fool the Federals and was trying to concentrate his forces toward North Anna. Warren, fearing the Confederate formation overlapped his flanks, ordered his Third and Fourth Divisions back to correct the situation at 7:30 a.m. The terrain greatly retarded the timing of the desired linkup, and Warren wanted to delay the attack until Sedgwick would be up. At 1:00 p.m. an impatient Meade ordered the attack to commence without Sedgwick. This five-and-one-half-hour delay in the attack—Meade had wanted the assault to start at 7:30 a.m.—allowed Ewell's men to improve on a good defensive position and consolidate their troops.[17]

Griffin's men attacked west across an open area known as Saunders Field and met heavy resistance. This initial assault achieved some success south of the Orange Turnpike, but as the Union forces pushed through, their flanks were exposed, forcing them to pull back. Wadsworth's attack through the woods farther south of the Orange Turnpike was checked and then pushed back.

While the initial contact was taking place along the Orange Turnpike, the Union troops encountered another surprise. Grant, impatient to get things

Saunders Field, view toward the Confederate position (similar to the 1864 photo). Courtesy of the author.

moving, had joined Meade at his Wilderness headquarters at about 10:00 a.m. At 10:15 a.m. Meade received a message from Brig. Gen. Samuel Crawford, Warren's Third Division commander: "There is brisk skirmishing at the Store [Parker's] between our own and the enemy's cavalry." Located at the Chewning Farm between the two advancing Confederate corps, Crawford had an excellent view of the Orange Plank Road. The aforementioned Union cavalry troops were Lieut. Col. John Hammond's Fifth New York of the inexperienced Brig. Gen. James Wilson's Third Cavalry Division, and they were desperately trying to stem the Confederate advance. This skirmishing was the precursor to Hill's Confederate infantry moving on the Orange Plank Road toward the vital intersection with the Brock Road—the tenuous link between Hancock and Warren's Corps.

Hancock, near Todd's Tavern about four miles from the Brock and Orange Plank Roads junction, was ordered to move to counter the threat. But with Hill only about a mile away from the prize, Hancock needed time to arrive. Therefore, Brig. Gen. George Getty of Sedgwick's Sixth Army Corps, who at the time was near headquarters and only about two miles from the junction, was sent to the threatened position. Getty and his staff arrived at the junction just as Confederate soldiers were coming into view. The lead unit (Brig. Gen. Frank Wheaton) was immediately ordered up and was able to drive Hill's men back. Getty later reported that Confederates had gotten so close that "the rebel dead and wounded were found within 30 yards of the cross-roads." Crawford had been west of the junction at Chewning Farm and in a position to attack Hill's left flank when Warren ordered him to move back toward Warren's left flank south of the Orange Turnpike. There, Crawford would support Wadsworth in the attack against Ewell on the Orange Turnpike. Lee's gamble to hold the Union in The Wilderness until Longstreet could come up was working.[18]

In the past, the Army of Northern Virginia had typically been outnumbered. Now, however, the disparity in men and war materiel and the tenaciousness of the new Union leadership made the right decisions even more imperative for Lee. Longstreet, Grant's prewar friend and best man at his wedding, provided insight into Grant. While conferring with Lee and his staff before the fighting occured, Longstreet stated, "That man will fight us every day and every hour till the end of the war." As Grant's Overland Campaign unfolded, Longstreet's prediction would be borne out by some of the most brutal fighting yet seen, and an unprecedented number of casualties.[19]

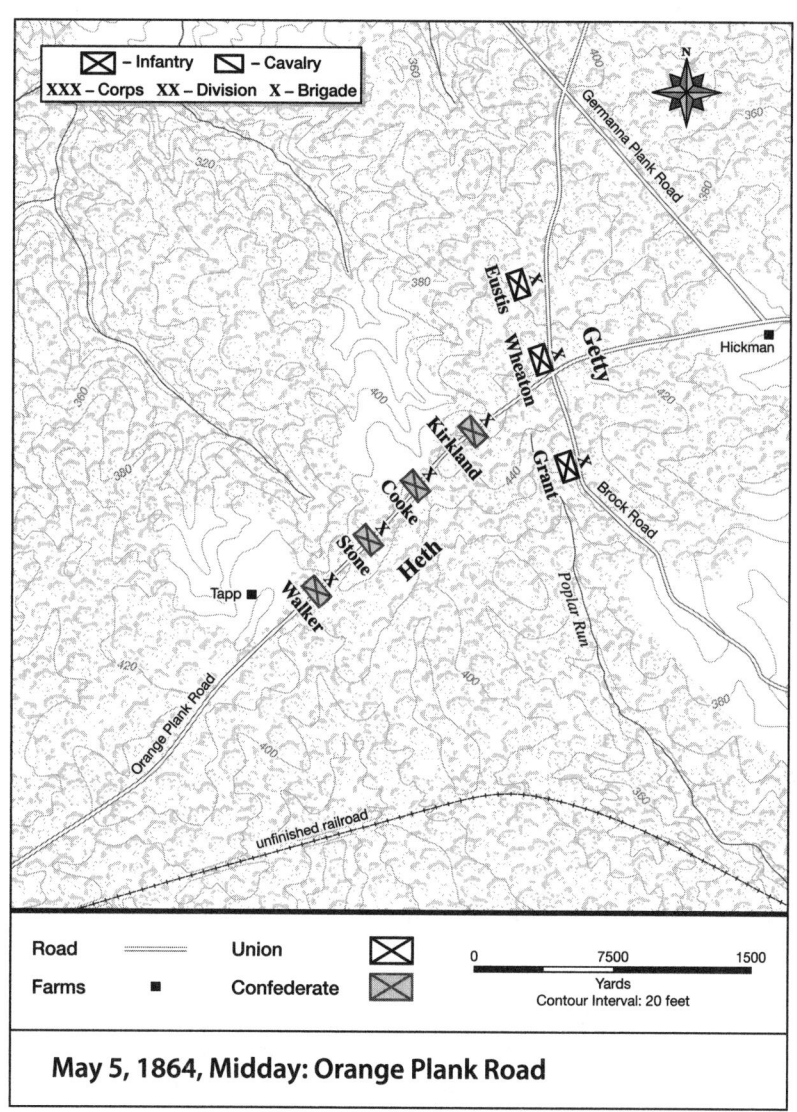

May 5, 1864, Midday: Orange Plank Road

Alternate Decision and Scenario

Had Lee tried to fully engage Grant's troops as they moved south rather than holding and delaying them, the obstacles that kept the Federals from advancing through the dense foliage and taking Confederate positions would have worked against him. Given Grant's numerical advantage and as yet untested aggressive nature in the East, this approach would likely have made Lee much more susceptible to massive enemy counterattacks and the resulting casualties. With Ewell lacking reinforcements on the Orange Turnpike and Longstreet too far away to quickly support Hill on the Orange Plank Road, the fighting would likely have forced Lee to a defensive position along the North Anna River. Meade thought Lee would initially move to this location which supported his assumption that the Confederate forces that Warren encountered were only a delaying force. The dramatic race to Spotsylvania Court House would not have happened, and the casualties that occurred there would have been replaced by those at another previously unknown location.

Lee Decides to Leave His Troops in Place

Situation

When darkness on the evening of May 5 forced the fighting to stop in the vicinity of the vital Orange Plank and Brock Road intersection, Hill's men "had marched and fought all day." Although it had been hotly contested, the junction remained in Union control. As Lee had hoped, his attacks had stopped the Union advance from moving beyond the tangled terrain of The Wilderness. Meade, who had initially thought that Lee was still near the strong Mine Run defenses, was glad to fight anywhere else but there. As night fell, the day's fighting along the Orange Plank Road was ending in a stalemate.[20]

As with the morning attacks by Warren's Fifth Corps before all troops were in place on the Orange Turnpike, the Union high command repeated its impatience here. Uncoordinated piecemeal attacks took the place of an effective massed assault along the Orange Plank Road. First, Brig. Gen. George Getty's division of Sedgwick's Corps had attacked Brig. Gen. Henry Heth's strong defensive works without any support and failed. When Hancock's Corps arrived, rather than sending a massed assault, he fed in his divisions of Maj. Gen. David Birney, then Brig. Gen. Gershom Mott, then Brig. John Gibbon, and finally Brig. Gen. Francis Barlow one at a time. With the arrival of Barlow's troops, Hancock outnumbered Heth's lone division by a five-to-

Typical terrain of fighting along the Orange Plank Road. Library of Congress.

one margin. The Union assaults had come close to destroying Heth's division, but the dense terrain masked how close they were to success. At 5:50 p.m., Lieutenant Colonel Lyman, Meade's aide-de-camp who was with Hancock at his field headquarters, wrote to Meade that Hancock was barely holding his own, and that reinforcements "would be most advisable." By evening, a coordinated pincer movement by Brig. Gen. James Wadsworth of Warren's Fifth Corps from the north and Barlow from the south was planned. Fortunately for Heth's men, darkness limited the attack before it could have any real effect. Humphreys felt that had there been "an hour more of daylight . . . he [Hill] would have been driven from the field."[21]

Credit was due to the stout Confederate defense. Lee had taken an active role in the battle's progress, ordering Brig. Gen. Cadmus Wilcox of Hill's Corps into and then from the critical gap position to support Heth's left flank. Hill performed well on this day as a corps commander. At various times, the Confederates made desperate counterattacks to try and maintain their positions. When Wadsworth was moving against Hill's left flank late in the day, the only available troops were the 150 men of the Fifth Alabama Battalion of Brig. Gen. Henry Walker's brigade. With the thick undergrowth basically making them invisible to the Union troops, the screaming Alabamians attacked and were able to halt this Union advance.[22]

At about 5:00 p.m. on May 5, the intense fighting along the Orange Plank Road forced Lee to revise his plan for Longstreet's First Corps. Rather than having these troops move along the Catharpin Road running south of the Orange Plank Road, he would have them turn north and come up the Orange Plank Road itself, thus bringing Longstreet up for direct support of Hill. Lee sent Lieut. Col. Charles S. Venable from his staff to find Longstreet at Richard's Shop and give him the new orders. Longstreet's men had marched twenty-eight miles that day to get to their current location at Richard's Shop. At 7:00 p.m. on May 5, Lee expected Longstreet and Anderson's division of Hill's Third Corps to be "up early." The fighting on Hill's front was still "raging," and Federal troops were reported to be "massing" in his front. With the exposed gap between Hill and Ewell, Lee wanted Ewell to be "ready to act" first thing in the morning, presumably to keep Warren and Sedgwick stationary at a minimum.[23]

As darkness closed in and the fighting stopped, the combatants were literally yards apart in some places. One brigade was located in a vulnerable position "with its naked flank perpendicular to the enemy's line." Both sides dug earthworks as best they could where they were. Hill's troops had been driven into disjointed clusters along their front. Wilcox's line was "very irregular" and "required to be re-arranged." It had no battle formation, and the men sought shelter behind whatever they could find. Anticipating being relieved by morning, ammunition was not ordered up. One Confederate described the haphazard disposition of the men's line in this manner: "[It] lay in the shape of a semi circle, exhausted and bleeding, with but little order or distinctive organization." As was typical for the Rebels, a soldier said, "We were without rations and had little water during the night." Even getting water with the jumbled lines could be dangerous. During the night when the fighting had stopped, Colonel Baldwin of the First Massachusetts crept a short distance to get water from a stream and was captured. Colonel Davidson of the Seventh North Carolina was captured while trying to get water near the same place. In some locations, Confederates were so close to the Federal positions "that they were all night within hearing of the voices of Hancock's men."[24]

As is so often the case, there are conflicting accounts of what occurred next. According to Col. William H. Palmer, Hill's chief of staff, Generals Heth and Wilcox met with Hill on the night of May 5 and informed him, "Their lines in the woods were like a worm fence, at every angle, and when they had undertaken to straighten them the enemy had captured our men and we captured theirs." Referring to his men's weak position, Heth told Hill, "A skirmish line could drive both my divisions and Wilcox's, situated as we are now. We shall certainly be attacked early in the morning." Later that evening,

Wilcox went to Lee's tent by himself "intending to report the condition of his front."[25]

Heth was correct in his conclusion that the Federals would attack in the morning. But once again, they would not do so with as much force as possible. Based on information obtained from Confederate prisoners, Grant erroneously felt that Pickett's division of Longstreet's Corps had arrived (it was actually in Richmond), and only Field's and Kershaw's divisions had yet been encountered. Barlow's First Division was withdrawn from the Federal left and posted south along the Brock Road to defend against Pickett's phantom division's possible movements. This move not only reduced the forces initially involved in the planned Federal assault the next morning, but also placed soldiers in an excellent location for part of the attack. Barlow's movement on the Federal left flank would have struck the Confederates on their exposed right flank and put troops in a position to inflict serious damage.[26]

Options

With Longstreet expected up before more fighting occurred, should Hill's men be withdrawn and lines re-formed in a better defensive position, or should they be left in place?

Lieut. Gen. James Longstreet, CSA (USMA class of 1842), First Army Corps Commander. Library of Congress.

Option 1

The Confederates were expecting the fighting to resume in the morning. Pulling back the exposed troops and re-forming them would provide a better defensive position. While Union troops had been sent into battle in a rather piecemeal fashion on May 5, almost all the Union troops in this sector would now be in place for a massive coordinated assault. The Confederate lines were so tangled that one section could not provide support for another if necessary. From stronger defensive works farther back, Hill's men would be better able to withstand the expected attack. A skirmish line could be left at the position currently occupied, the rest of the troops withdrawn, and the line rectified. The Orange Plank Road was the only road the Federals could use for the attack against Hill, and by taking strong positions on either side, they would force the enemy to advance mostly through the difficult terrain. Heth's men could be positioned on the south side and Wilcox's on the north of the Plank Road, respectively.[27]

By moving back, the Confederates would give up hard-won ground and be farther from the critical road junction. The dense foliage and terrain would make even the simplest movements challenging, but withdrawing exhausted troops in the face of a vigilant enemy in the dark (civil twilight ended at 7:34 p.m.) would be almost impossible, and any noise would result in firing in that direction.

Option 2

The Rebels could also remain in place at their current position along the line. Hill's men had started marching at daylight and gone directly into intense fighting as they neared the Brock Road intersection in the early afternoon. By nightfall, they were exhausted. The terrain and the Confederates' limited defensive works offered limited protection, and Hill's soldiers had been able to withstand repeated Federal assaults during the day from where they were. But there was a risk to leaving the Confederates in place. The critical factor was that Longstreet's Corps was supposed to be up before dawn to relieve Hill's men, and his troops were not expected to participate in the next day's anticipated fighting. But Longstreet had a history of being slow. The closeness of the Union forces would preclude Hill's men from re-forming and strengthening their lines in place. The success or failure of this option depended on when Longstreet's Corps would be on the field.

Decision

Knowing the troops were exhausted, and expecting Longstreet's Corps to be up before dawn, Lee chose to leave the soldiers in their vulnerable position.

He notified Hill and Wilcox of his decision after midnight. In Heth's recollection, Hill learned of the extremely vulnerable situation the line was in, noted that Longstreet would be up and Heth's division would not be doing any fighting the next day and "I [Hill] do not wish them disturbed." With Lee and Hill camped so close together, a written order to Hill would have been unnecessary; Lee could simply have communicated his wishes directly to Hill. Wilcox, still concerned about the situation of his line, went to see Lee (Hill was ill) at about 9:00 p.m. Lee initiated the meeting by telling Wilcox that Hill's third division commanded by Brig. Gen Richard Anderson "and Longstreet will be up, and the two divisions that have been so actively engaged will be relieved before day." Wilcox returned to his lines with no further action taken.

What Lee did not know was that, when Venable and Longstreet met, Venable had failed to communicate the sense of urgency for Longstreet to move. When Venable returned before dark, Lee was "anxious." When Lee discussed his concern with Stuart, Stuart promised Lee that he would inform Longstreet and then dispatched his chief of staff Maj. Henry B. McClellan. Shortly after the meeting with Wilcox, McClellan returned, and Lee learned that Longstreet's lead division (Field's) would not start from Richard's Shop until one in the morning. Compounding the situation, Longstreet got lost and had to retrace some of his march. His troops fell farther behind, and they would be unable to assume the position Lee so desperately needed them in at dawn.[28]

Results/Impact

Given the pounding Hill's Corps had taken the previous day, Grant felt that it was about to break. He planned a major assault for the morning of May 6 on the Orange Plank Road using Hancock's Corps and Getty's three brigades of Sedgwick's Corps temporarily under Hancock. The Confederate left on the Orange Turnpike, manned by Ewell's Corps, would be held in place by Warren's Fifth and Sedgwick's Sixth Corps. Burnside's Ninth Corps would push through the gap between the two Confederate corps, then turn south and strike Hill's flank and rear.[29]

Grant wanted to start the attack at 4:30 a.m., just before civil daylight began. After conferring with his generals, Meade wanted to postpone the attack until 6:00 a.m. Grant only agreed to wait until 5:00 a.m., and the Confederates considered this delay "fortunate." At five o'clock the Union assault began. Grant sent seven divisions against Hill's two, and "serious disaster seemed imminent." With the poor defensive position the Confederates occupied by staying in place, Hill's Third Corps was shattered, and it retreated in panic. Hancock's assault lost some momentum as it moved west through confusing and difficult terrain but continued to push on. Once again, the

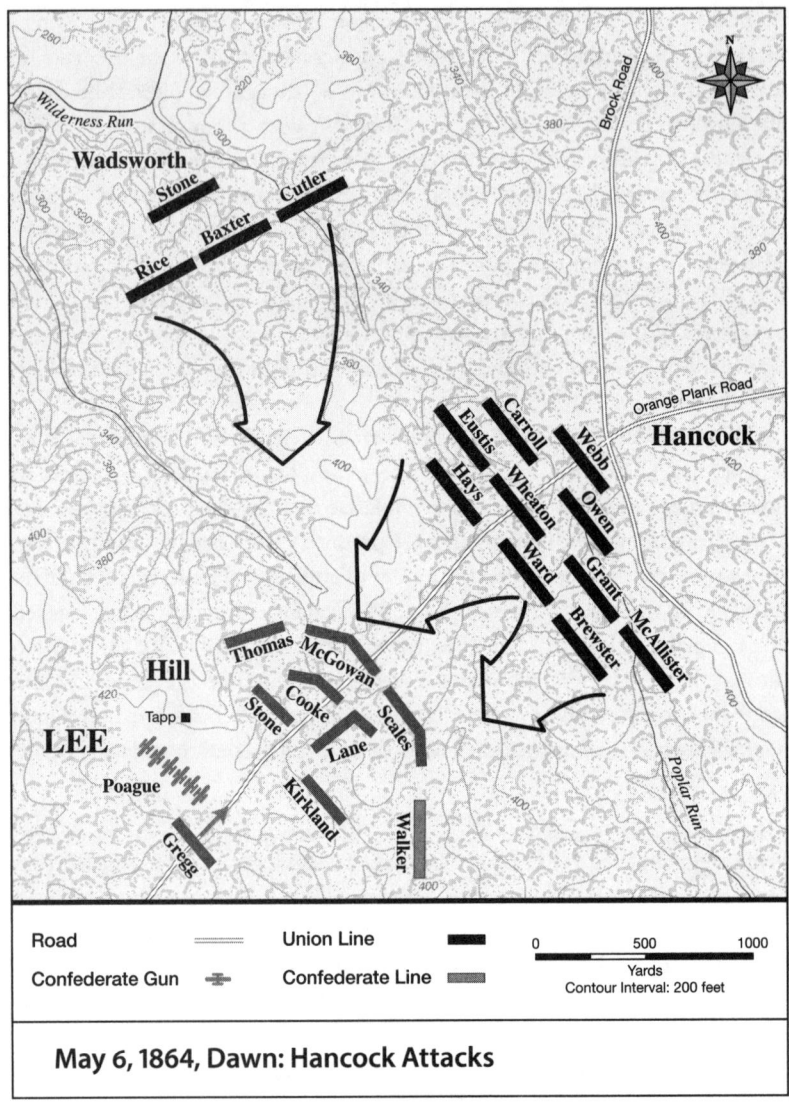

May 6, 1864, Dawn: Hancock Attacks

dense landscape helped Lee's troops. Grant felt that if the country had been more open and Hancock had seen the confusion in the Rebel lines, Hancock would have effectively pressed his advantage and pushed Lee to the defenses of Richmond.[30]

At Lee's headquarters along the Orange Plank Road and the Widow Tapp Farm (fewer than two miles west of the Orange Plank and Brock Road junctions), Hill's troops came streaming back. Poague's artillery (Hill's Corps) was positioned on the west side of the open field, and the artillerist began firing to try and stem the Union onslaught. Hancock's men filled the road as they pushed westward. The artillery fire "was unexpected, as no artillery had been used the day before," and it was momentarily effective at slowing the advance. It would not be able to hold off the weight of Hancock's full corps for long, however, as the assault started to flank the Confederate position. The danger was so great that Lee sent his adjutant, Lieut. Col. Walter Taylor, "back to Parker's store, to get the trains ready for a movement to the rear."[31]

Miraculously, at about 6:30 a.m. lead units of Longstreet's First Corps began to arrive. It was Brig. Gen. John Gregg's "Texas" brigade of Field's division, Longstreet's First Corps moving up the Orange Plank Road. In the excitement, Lee was ready to personally lead these troops in a countercharge, but Gregg's troops refused to go forward until their beloved commander retreated to safety. To Lee's shouted "Texans always move them," Gregg's troops delivered a crushing counterattack to blunt the Union assault on the north side of the Orange Plank Road. It was a costly attack for Gregg's men; of the eight hundred that made this attack, they "lost half their number."[32]

As more Confederates came up, Longstreet took charge. Putting Brig. Gen. Joseph Kershaw on the right and Maj. Gen. Charles Field on the left

Maj. Gen. Winfield S. Hancock (USMA class of 1844), Second Corps commander (seated), and division commanders Brigadier General Barlow (USMA class of 1861), Major General Birney, and Brigadier General Gibbon (USMA class of 1847). Library of Congress.

of the Orange Plank Road, it took a few hours to stabilize the shaken Confederate line. Longstreet's last-second arrival saved Hill's Corps from total destruction. Burnside's troops had once again fallen several hours behind. Burnside's flanking movement on Hill's left, intended to be coordinate with Hancock's frontal assault down the Orange Plank Road, would have been devastating. Burnside's inability to hit the Confederate flank combined with the last-moment arrival of Longstreet and the effectiveness of Poague's artillery in stopping Hancock's frontal assault saved Hill's corps. Though costly, Lee's gamble of leaving Hill's troops in place worked.[33]

Alternate Decision and Scenario

The delayed arrival of Longstreet's Corps had a major impact on the fighting along the Orange Plank Road. But delay resulted from issues concerning the execution of the decision to move them, not the decision itself.

Had Hill's troops moved back and re-formed, they most likely would have been in a much stronger position to receive Hancock's attack. The typical deployment of pickets could have given earlier warning of Hancock's movements. As Heth asserted, "Had Wilcox and I been in line of battle, we could easily have repulsed the advance of the Federals." Poague's Confederate battery located on the western side of the Widow Tapp Farm effectively stemmed the Federal assault, and the unit would have been even more efficient with infantry support to keep its men from being flanked. Longstreet's troops arrived just in time but lacked an adequate defensive position in the morning. As a result, the troops had to initiate costly countercharges that pushed the Federals back from where the Confederates' defensive line would have been. Lee knew he had lost a great opportunity, and Heth declared, "I think General Lee never forgave Wilcox or me for this awful blunder." In their defense, which Lee did not accept, Heth and Wilcox had followed the explicit orders of their commanding officer, Hill.[34]

The next day, Longstreet would launch a surprise flank attack from a sunken road on the Federal left flank. His men successfully pushed Hancock's troops back to the Brock Road and the critical intersection. If Hill had moved his troops back, established their defensive line in the earthworks on the west side of the Widow Tapp field, and set up even a semblance of defensive works on the south side of the Orange Plank Road, the Federal assault could have been stopped at a location much more favorable to the Confederates. The actual costly countercharges that pushed the Federals back would not have been necessary.

With a defensive position closer to the Tapp Farm, Longstreet's flank attack the next day would have hit Hancock's troops more in the rear. More importantly, Longstreet's troops would have been closer to the vital Brock and Orange Plank Road intersection than many of Hancock's own men. Had the Confederates controlled the intersection, and had a large part of Hancock's Corps been between two Confederate corps, the draw resulting from the actual two days of fighting might have been dramatically different. The best fighting corps in the Army of the Potomac might well have been crippled. Moreover, the Brock Road leading to Spotsylvania might have been blocked by infantry, not a small force of Confederate cavalry that could only delay a Union movement south.

In an 1868 interview, Lee stated that Longstreet was often slow, declaring that, had Longstreet been in time, he (Lee) "would have struck the enemy on the flank while they were engaged in front." Longstreet's flank attack was very effective at rolling up Hancock's left flank later in the day, even without rest for the soldiers. By arriving before dawn, getting a little rest and rations, then hitting Hancock's troops while they were fighting Hill's entrenched troops, Longstreet's flank attack in this scenario might have proved even more effective and captured the important intersection. The situation that resulted in Longstreet's wounding the next day would not have existed. Finally, the tactical situation for the next critical decision would likely have been dramatically different.[35]

Lee Decides to Keep the Initiative

Situation

The morning of May 6, 1864, had started with a crushing blow to the Confederates along the Orange Plank Road. But by the afternoon of May 6, the fortunes of battle had reversed—the Union had suffered a major assault. Could the Confederates capitalize on the change of momentum?

Fighting for the critical Orange Plank and Brock Road intersection had been raging since early morning. At 11:00 a.m. the Confederates executed a devastating flanking attack that threatened the capture of the Union-held crossroads. Early on the morning of May 6, Grant's main offensive along the Orange Plank Road unfolded successfully but then stalled. Starting at 5:00 a.m., Union troops on the Orange Plank Road had completely pushed Confederate troops from their ill-defined defensive works west of the Brock Road. With the crucial timing of the arrival of Longstreet's Corps, the initial success of the Union assault had been stopped. On the northern front of the battle, Major General Ewell's efforts on the Orange Turnpike had blunted Grant's plan of operations. For now, Grant had lost the initiative.[36]

When Longstreet had arrived in the morning, he had taken over operational command of the Confederate forces along the Orange Plank Road. Scouts of Brigadier General Wofford (Kershaw's division, Longstreet's Corps) discovered that their lines overlapped the Federal left flank and wanted to attack. Hoping to find an opportunity, Lee sent the army's newly arrived chief engineer, Maj. Gen. Martin Smith, to Longstreet to probe Hancock's line for weaknesses. Smith had arrived in Virginia only the previous month from the Western Theater, where he had designed the defenses at Vicksburg. Although little-known to the Army of Northern Virginia's officers, he was about to make a big impression. Smith's reconnaissance had revealed an unfinished railroad bed (grading work had stopped in 1861, and no track had been laid down) running roughly parallel to and about three-quarters of a mile south of the Orange Plank Road. This railroad cut should not have surprised either side. Both armies had moved through this area during the Mine Run operations in November and December 1863. Parts of Hancock's Corps had actually camped near here and even marched across the intersection of the railroad cut and Brock Road.[37]

Whether Smith or Brig. Gen. William T. Wofford (commander of the Confederate brigade on the right flank) suggested the opportunity for a flanking attack is not clear. But after Lee and Longstreet conferred, Lieut. Col. Moxley Sorrel (Longstreet's chief of staff), who had been with Smith on his reconnaissance, was assigned to coordinate the planned flanking movement. Unlike the uncoordinated Union attacks, Longstreet planned to launch a frontal assault to support and increase the effectiveness of the flanking maneuver. He advised Lieutenant Colonel Sorrell, "Don't start until you have everything ready. I [Longstreet] shall be waiting for your gunfire, and be on hand with fresh troops for further advance." The flanking attack was composed of four brigades (selected because of their proximity to the planned attack location) from four different divisions—Brig. Gen. William Wofford of Kershaw's division, Brig. Gen. "Tige" Anderson of Field's division (both of Longstreet's Corps), Brig. Gen. "Billy" Mahone of Anderson's division, and Col. John Stone of Heth's division (both of Hill's Corps). It is uncertain which officer was actually in charge of the flank attack. Mahone was the senior officer of the four, but Longstreet's subsequent reports contain conflicting statements, causing confusion.[38]

The flanking attack was extremely effective, pushing all Federal units in the area back toward the Brock Road defenses. As the Confederates approached the Orange Plank Road, Brig. Gen. James Wadsworth tried to secure the Union left flank from total disaster by attacking west along the Orange Plank Road. His men met with intense musketry fire and failed to stop the collapse. Wadsworth was mortally wounded in this effort.[39]

The surprise and results of Wofford's and Mahone's eleven regiments had demonstrated the effectiveness of a well-executed flanking movement. Although fighting had reduced the size of these regiments, they were able to drive the equivalent of a Federal corps back into its defensive works. As in the previous year at the Battle of Chancellorsville, after a successful flanking attack, the Gods of War once again turned their hands against the Confederacy. On the extreme right of the flanking movement, the Twelfth Virginia of Mahone's brigade had advanced north of the Orange Plank Road and had become separated from the other units. To regroup, the troops turned around and headed south toward the road as the remainder of the brigade was moving north. At the same time, Longstreet was heading east on the Orange Plank Road to ascertain the best way to press the success of the flank attack. Accompanying Longstreet were Brig. Gen. Joseph B. Kershaw (division commander) and Brig. Gen. Micah Jenkins (brigade commander) and his brigade. Adding to the irony, Jenkins's men wore new dark-gray uniforms that could be mistaken for Union uniforms.[40]

Thinking the men advancing eastward on the Orange Plank Road were enemy troops, some of Mahone's men opened fire. The firing was quickly stopped, but not before Longstreet had been critically wounded in the neck and Jenkins had been killed. Confederate Brig. Gen. E. Porter Alexander, Longstreet's chief of artillery, wrote after the war, "Longstreet's fall seemed actually to paralyze our whole corps," adding, "From the accounts of those who

Critical intersection of the Orange Plank Road (lower left to right) and Brock Road (upper left to right). Courtesy of the author.

The Battle of The Wilderness

Confederate entrenchments near the Orange Plank Road. Note how difficult it would have been to move troops through this area. Library of Congress.

participated in his attack . . . I have always believed that, but for Longstreet's fall," Grant would have been forced to retreat back across the Rapidan. As with the year before, when "Stonewall" Jackson's own men had brought him down after a successful flank attack, Longstreet was brought down in what was most likely his finest hour. Unlike the year before, when Jackson had been struck down in the dark, Longstreet was felled at midday, when plenty of daylight was left for fighting.[41]

Options

A critical decision needed to be made immediately. Could the Confederates' dramatic reversals in fortune from the morning and this attack's momentum be exploited to gain a decisive victory? Or should the gains to this point be consolidated and a strong defensive position be prepared to meet the next assault by the numerically superior Union troops? Until the Confederates could take the Brock and Orange Plank Roads intersection, the recent success would be incomplete."[42]

Option 1

Confederate troops could maintain the momentum and press the advantage. Lee's fighting spirit was up, and he gave no indication of relinquishing the initiative. He had tried to lead an attack that morning at the Widow Tapp Farm, and he had only reluctantly demurred to his troops' wishes that he not personally lead the countercharge. The flanking attack had been very successful. Hancock later stated that the Confederates had "rolled me up like a wet blanket," and it took him several hours to get reorganized. Lee had expressed to Jefferson Davis his concern that once Grant crossed the Rapidan it would only be a matter of time before Richmond itself was threatened. With Ewell securely holding the Orange Turnpike, an opportunity presented itself on this front to deliver a crushing blow that could alleviate that concern.[43]

In all the past encounters with the Army of the Potomac following a hard-fought battle in Virginia, the Union had pulled back. Lee was painfully aware of his troops' starvation and the continual stress maneuvering would entail. Recent medical studies have validated Lee's concern, stating that long periods of poor diet and hard fighting and maneuvering are "associated with significantly reduced physical performance." His troops and the enemy were in place. What better opportunity might present itself? Through Sorrel, Longstreet had urged Lee to continue building on what had been accomplished. Lee knew the overall plan of the assault, but only Longstreet had knowledge of its details.[44]

Many of the troops that would be called on had been fighting and/or marching for hours. Command and control would be difficult since the flank attack had involved troops intermingled from various commands. Delivering the necessary striking power for a successful attack would take time to organize the units for an effective coordinated assault. The tangled and confusing terrain of The Wilderness—the same challenge the Union had faced in its offensive attempts—would be a major challenge to implementing such a movement. Confederate Major General Field, who had temporarily succeeded Longstreet after his wounding, felt nothing could be done until the troops could be re-formed.[45]

Option 2

Alternatively, the Confederates could pull back and re-form the lines, thereby encouraging the aggressive Grant to attack his outnumbered enemy in a defensive position. A major reason the battle was being fought in The Wilderness was that the terrain diminished the Union's superiority in men and artillery and provided the Confederates an advantage. By consolidating and organizing their forces, the Rebels would be in a better position to once again exploit this unique terrain. Lee remained mindful of the gap between Ewell's Corps along the Orange Turnpike and the remainder of the Army of Northern Virginia along the Orange Plank Road. To guard against Grant's attempts to exploit this gap and expose his frontal assault to a devastating flanking movement, Lee would have to direct some of the men of Heth's and Anderson's divisions against Burnside. The force in front of Ewell and the distance he would have to travel made support from this quarter unrealistic. Re-forming the lines would provide a better position along the Orange Plank Road and allow troops to be deployed to possibly exploit the gap for another dramatic movement. Only Longstreet's timely arrival in the morning had averted disaster when the disorganized Confederate defenses had collapsed. The Confederates were once again in a disorganized deployment. Furthermore, a vital commander had been lost, and very few fresh troops were available.

The problem was that the Federals still controlled the vital intersection that connected their troops and provided a good route south to threaten Richmond, thereby forcing Lee to countermove. If Lee pulled back, all of the gains and momentum obtained so far would be for naught.

Decision

Lee chose to press the attack, and it would prove to be one of the Army of Northern Virginia's last major assaults of the war. Never again would Lee have the ability to mount an offensive of this magnitude against the forces Grant continually confronted him with. In addition to the Federals, Lee would be fighting the clock. It would take time to regroup and get the disorganized and intermingled units into position to make a unified force with sufficient strength to mount an attack. Maj. Gen. Charles Field, temporarily commanding the First Corps after Longstreet's wounding, felt no advance was possible until the troops were repositioned. Even so, "to rectify this alignment consumed some precious time." The Federals were also planning to renew the fighting. Hancock was ordered by Meade, that if the enemy left him "undisturbed," to rest the men and then, in conjunction with Burnside, make a forceful attack at 6:00 p.m. Although unsuccessful so far, Meade felt that such a move would "overthrow the enemy."[46]

Results/Impact

Lee had no intention of leaving Hancock's troops "undisturbed." The renewed Confederate assault on the Brock Road started at 4:15 p.m., four hours after Longstreet's flanking attack had stalled. Just as Confederate general Ewell had used Warren's delay in attacking his position on the Orange Turnpike to reinforce his own defenses, Hancock now used the four hours to reinforce his already strong position. Thus Hancock utilized the short time frame in which his troops were most vulnerable immediately after the flank attack to his advantage. His men were ordered to "hold these works at all hazards."

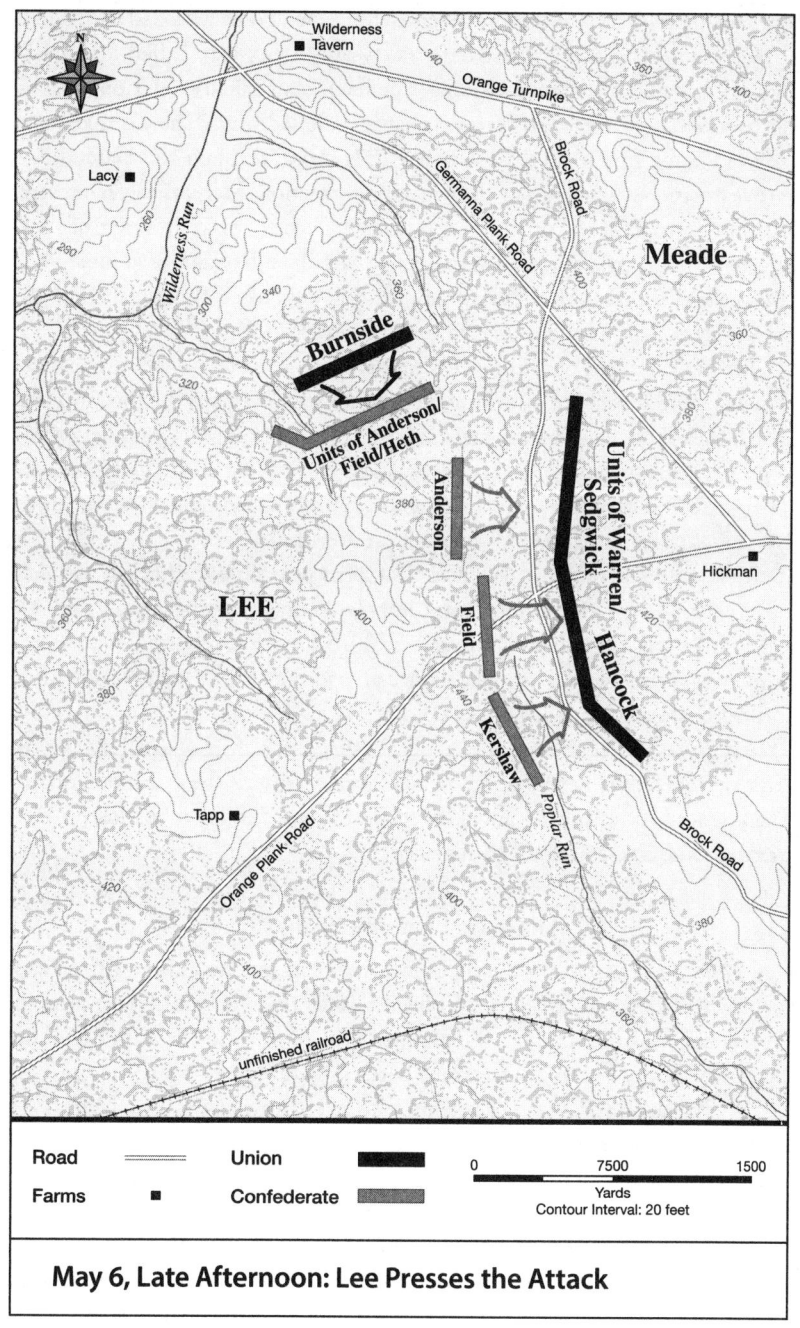

May 6, Late Afternoon: Lee Presses the Attack

Sgt. Patrick DeLacey of the 143rd Pennsylvania would be awarded the Medal of Honor for his effort at these defensive works. Lee's plan called for an all-out attack using the two divisions (Kershaw's and Field's) of Longstreet's Corps to "achieve the breakthrough that had eluded him." The Confederates achieved some initial success, but they only broke through the Federal defensive positions just south of the crucial intersection. Here, the ever-present fires had burned some of the Federal log defensive works in front of Col. Robert McAllister's position. An intense artillery barrage of double-shot canister and subsequent infantry support quickly halted this breakthrough. The Confederates had to retreat to their previous positions.[47]

The Federals had not been expecting an attack. "The enemy anticipated" Hancock, and instead of attacking, he "had to repel perhaps the most wicked assault thus far encountered." Hancock had reported that all was quiet on his front "until 4.15 p.m., when the enemy made a heavy assault." The Confederate "assault" on his front was "continuous . . . and exceedingly vigorous" until five o'clock, when the Federals finally beat it back.[48]

Meanwhile, three brigades of Hill's Corps were able to check Burnside's movement against the Confederate left flank north of the Orange Plank Road. Colonel Hartranft of Burnside's Third Division reported that severe Confederate fire had halted his advance at 4:30 p.m., and "he was soon after notified that no farther advance would be made."[49]

After the war, E. Porter Alexander, Longstreet's chief of artillery, wrote, "This attack ought never, never to have been made." The late hour of the attack would have limited any success. The Confederate troops lost in this renewed offensive could not be replaced.[50]

Alternate Decision Scenario

Had Lee chosen to go on the defensive, likely in the area around the Tapp Farm where Poague's artillery had been so effective, he would have been in a reasonable position to meet the planned 6:00 p.m. Union assault. Based on the Army of the Potomac's previous battles against Lee's army in Virginia, history did not favor Meade's intention to overthrow the enemy. With Lee positioned defensively, the combatants at the conclusion to the fighting in The Wilderness would likely have been in a different location. Grant would have continued seeking his overall objective, and most likely only the two armies' movements to Spotsylvania Court House would have been affected. The location of the military road that the Confederates would construct and use for their initial movement south would no longer have been viable.

When Meade became aware of Lee's forces in The Wilderness on May 5, he initially declared, "The rebels have left a division here to fool us while they concentrate toward the North Anna." From a defensive position near the Tapp Farm, Lee could have moved troops via the road into Parker's Store—the same route Longstreet had used to come to the fighting in The Wilderness. By traveling south via Parker's Store, linking with the Catharpin Road, bypassing Spotsylvania, and moving to North Anna, Lee would have been able to construct excellent defenses. He might also have been in a position to strike Grant's movements south. The costly fighting at Spotsylvania Court House would not have occurred.[51]

If, instead, Lee had still moved to Spotsylvania Court House, the opposing forces' revised position would probably have altered the troops' arrival times and the deployment of their respective fighting positions. Regardless, Grant's leadership style and tenacity was unlike any the Army of Northern Virginia had faced before. "Although Grant suffered numerous tactical setbacks in 1864 . . . daily operations did not dictate his strategic view." The Battle of Spotsylvania would still have occurred, but it might have been fought in an entirely different way.[52]

Ewell Decides to Delay a Critical Flank Attack

Situation

The north end of the battlefield (Confederate left flank, Union right flank) proved particularly enticing for a flanking movement. Believing his troops extended beyond the Confederate line, Union Maj. Gen. John Sedgwick (Sixth Corps commander) ordered Brig. Gen. Truman Seymour's brigade of Brig. Gen. James Ricketts's Third Division to attack on the evening of May 5. In fact, the Confederate entrenchments extended beyond the Union line, and the inexperienced brigade was repulsed. Earlier that day, Union assaults by Brig. Gen. James Wadsworth (Warren's left flank) south of the Orange Turnpike against positions manned by Brig. Gen. John Gordon's troops had failed. Union troops in this immediate area were subsequently withdrawn. With no troops in his immediate front, Gordon, who was on Ewell's far right flank, received orders on the night of May 5 to move and assume a position "on the extreme left of the Confederate line."[53]

The Sixth Corps' daytime attacks on May 5 had been against Brig. Gen. Harry Hays's brigade. When Gordon's brigade was moved, it rested on the left of Hays, north of where the Union had attacked.

Early on the morning of May 6, Gordon sent out scouting patrols revealing that his position outflanked Seymour's unprotected position, and that no viable supporting troops were situated behind the Union line. Gordon personally verified the accuracy of the reports, reported the situation to his division commander Maj. Gen. Jubal Early, and proposed an attack on the exposed Union flank. Early in turn reported the information to Ewell at about 9:00 a.m. But Early opposed the plan. He had reports from cavalry scouts that the Confederate left might be threatened and that the Union Ninth Corps under Maj. Gen. Ambrose Burnside was in the rear of the Union right. Burnside's troops had moved down the Germanna Plank Road the night before, but they were actually near the Wilderness Tavern by 9:00 a.m. The first of Burnside's troops to arrive (Brig. Gen. Robert Potter) was sent to support Warren's left in an area south of the Orange Turnpike Road.[54]

Lee often gave discretionary orders, relying on the officers closer to the fighting to make specific tactical decisions. Lee did not want to bring on a general engagement until Longstreet would be up on May 6, when all of Lee's army would thus be on the battlefield. The previous day, after the first attack on Ewell's line, Lee learned from Ewell's aide and new son-in-law Maj. Campbell Brown that Ewell had interpreted Lee's orders as a command to fall back to Mine Run "if pushed." Lee informed Brown that the order had been misinterpreted. Ewell should only fall back to Mine Run if he could not hold his position. Subsequently, Ewell told Brown that he could hold the ground "with ease," and this information was communicated to Lee. That evening at 7:00 p.m., Ewell received correspondence from headquarters re-

Lieut. Gen. Richard S. Ewell, CSA, Second Army Corps commander (USMA class of 1840). Library of Congress.

Maj. Gen. Jubal Early, CSA, division commander (USMA class of 1837). Library of Congress.

The Battle of The Wilderness

Brig. Gen. John B. Gordon, CSA, brigade commander. Library of Congress.

garding the fighting on Hill's front and Lee's desire for effort against the Federals in front of him "if an opportunity present[ed] itself." The dispatch ended with the order "Be ready to act as early as possible in the morning."[55]

Options

Was the supposedly exposed Union right flank the "opportunity" Lee was looking for? Ewell now faced the dilemma of deciding between Gordon's recommendation for an immediate attack or Early's advise not to attack. The thirty-two-year-old Gordon was a rising star in the Army of Northern Virginia's officer corps, and Lee considered him one of the best brigadiers. In a letter to Lee written the night before, Ewell had cited Gordon for "special praise" for his efforts on May 5. But Early outranked Gordon and had become one of Ewell's most trusted subordinates. Ewell relied heavily on Early's judgment.[56]

Option 1

If Gordon's assessment was correct, immediately making the attack he had advocated would likely inflict serious damage on the Union right flank. Any pressure brought to this side of the fighting might limit the manpower sent to support Hancock's movements against Longstreet's recently arrived corps, which was about to make its own damaging flank attack. An immediate attack would also be in line with Lee's wishes as he tried to gain the advantage. An unknown advantage to the Confederates was the fact that inexperienced, poorly led troops held the Federal position in front of Gordon.

The problem was the potential of a Federal counterattack. If Burnside's troops were as near as Early stated and able to respond to the attack, the tables could quickly be reversed. Not only might Ewell's Corps be threatened, but the rest of the army might be as well. Gordon, however, stated that he had personally made a reconnaissance of the area and verified the situation. The disparity between the armies' troop sizes remained a factor. No Confederate reserves were available to maximize any advantage gained from the attack, or to stop counterattacking Federals from rolling up the line.

Option 2

By relying solely on Early's recommendation based on unsubstantiated information and not authorizing the proposed flank attack, Ewell's defensve line would remain intact. There would then be no chance that the rest of the army might be threatened following a failed attempt. Early was Gordon's senior officer. He indicated that a real threat to the Confederate flank was possible, but he had done nothing to verify his intelligence information. In fact, Burnside's troops were located much farther south and attempting to threaten the gap between Lee's two corps. Ewell felt that an attack made so early in the morning and over open ground would expose the "smallness" of Gordon's forces. Ewell's rationalization was biased toward Early, since the time of day was irrelevant to whether soldiers were being observed.[57]

However, if the Confederates did not make the attack, they would miss a major opportunity. Previous battles attested to the lack of success head-on attacks achieved against defensive positions. Flanking attacks, however, yielded dramatic results such as those attained the year before at the Battle of Chancellorsville, just a few miles from Ewell's current location. The magnitude of the possible impact that such a flanking movement might have did not often present itself.[58]

Option 3

Given the conflicting opinions of the two seasoned officers, Ewell would be able to make a more informed decision if he could get more information. If Early was right, a possible disaster could be avoided, and if Gordon was right, the attack could proceed. The problem with this option was time. Opportunities like this one typically had a narrow window for execution, and delay could rapidly close that window.

Decision

Given the contradicting accounts of Lieut. Gen. Richard Ewell (corps commander), Maj. Gen. Jubal Early (division commander), and Brig. Gen. John Gordon (brigade commander), the pen of time has written a confusing history of the decision to delay Gordon's attack on the Union right flank until the evening of May 6, 1864. Ewell decided to wait and get more information, declaring that the situation "necessitated a personal examination." Unfortunately, "other duties" and "other unavoidable causes" postponed his review until almost sunset, when it was decided to launch a flank attack. In his official report, written almost one year later, Ewell neglected to state what "other duties" and "causes" had prevented him from more quickly assessing and acting on such an important situation. Warren, whose corps opposed Ewell, had been ordered to suspend operations and send support to his left at 9:30 a.m. Thus Ewell's front faced virtually no threat after this time.[59]

Accounts vary as to who actually initiated/ordered the attack:

- In his official report, Ewell stated, "After examination I ordered the attack."[60]
- Early wrote in his autobiography, "At my suggestion, General Ewell ordered the movement."[61]
- Gordon included this statement in his official report: "I received orders from Major-General Early."[62]
- Gordon's postwar *Reminiscences* differ from his official report. In the former, he described "[Lee's] prompt order . . . to move at once to the attack."[63]
- E. Porter Alexander related in his memoirs, "Lee listened in grim silence to his [Ewell's] reasons for non-action, and answered only with orders to Gordon to proceed immediately to make the attack."[64]

As to who actually gave the order, the most plausible answer, combining the information in Ewell's March 1865 report and Gordon's July 1864 report, would be that Ewell issued the order to Early, and Early then communicated it to Gordon. Ewell had initially been in favor of Gordon's plan, an opinion shared by Lee, but Early persuaded him to delay. Early had been completely wrong in his assessment of the Union position, and he maintained his view of the enemy's position well after the war, even when the facts proved otherwise. For his part, Ewell deferred to Early's judgment as he often did, waiting until late in the day when Early felt the possible Union threat was gone and concurred with the attack.[65]

The timing of Lee's presence on the Orange Turnpike front is uncertain. He probably first visited Ewell's line on the afternoon of May 7. His aide, Lieut. Col. Walter Taylor, stated that "Lee spent the time [May 7] in visiting all parts of his line of battle. . ." That day, a Confederate soldier in the Thirteenth Virginia of Pegram's brigade in Early's division wrote in his diary, "In afternoon loud cheering toward the Pike and presently Gen. Lee came in sight. We left breastworks and went and stood by the road until he passed. We pulled off our hats and yelled as loud as we could."[66]

In addition, a great deal of uncertainty surrounds Lee's involvement with the actual decision to order the attack. Writing almost 40 years after the fact, Gordon stated in his reminiscences that Lee had visited the area, and that he had ordered Gordon to attack once Gordon had apprised him of the situation and opportunity. Even if Lee had been there, it would have been unusual for him to so blatantly ignore the chain of command with Ewell and Early present. While corresponding with Lee in the winter of 1867–68, Gordon wrote, "I am positive that I conversed with you on the morning of the 7th. Do not remember having seen you [on] that flank prior to that time. Indeed I was not aware of your desire to make a movement on that flank until after the 6th. I am glad to know that such was your wish."[67] Lee responded as follows: "I am not certain whether I saw you before your attack on the 6th. I visited your flank, but you might have been then engaged in you[r] reconnaissance. I may have confounded our conversation subsequent to your attack with my visit to Genl Ewell before it took place—."[68]

As the chief of artillery for Longstreet's Corps on the Orange Plank Road, E. Porter Alexander had no reason to be on the Orange Turnpike with Ewell's troops. Therefore, Alexander could not have based his statements that Lee was present and that he "listened in grim silence" on his own personal presence. Alexander wrote his memoirs after Gordon's, and he likely used Gordon's work as the source for his depiction of events.

Finally, adding to the confusion is a conversation Lee had in 1868 with former Confederate officer William Allan, a teacher at Washington College, where Lee was then president. After reiterating Ewell's "want of decision," Lee went on to say that he had "urged Ewell to make the flank attack . . . several times before it was done."[69] This correspondence, written within only a few years of the actual event, is more likely to be accurate than something written many years later. Such accounts indicate that is highly unlikely that Lee gave a direct order to Gordon.

The Battle of The Wilderness

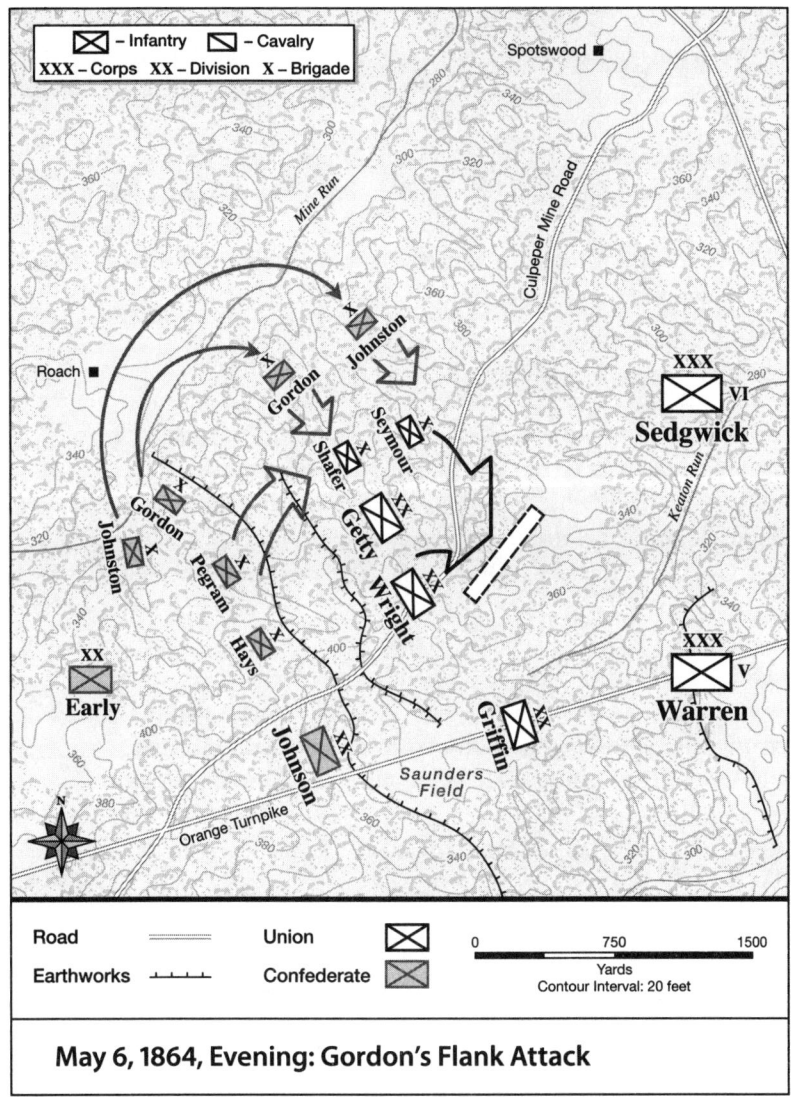

May 6, 1864, Evening: Gordon's Flank Attack

Results/Impact

Once the decision had finally been made and the orders issued, Gordon, supported by Brig. Gen. Robert Johnston's brigade, moved against the exposed and unprepared Union right flank. The attack completely surprised the Union troops. Brig. Gens. Alexander Shaler's and Truman Seymour's brigades were quickly routed, and both generals were captured as they tried to rally their panicking troops. Grant's aide Lieut. Col. Horace Porter described Union troops' alarm at this attack: "Aides came galloping in from the right, laboring under intense excitement, talking wildly, and giving the most exaggerated reports of the engagement."[70]

Confederates pressed the attack until the Federals were able to re-form along the Culpeper Mine Road and darkness stopped the advance. As so often happened in the dark, tangled woods of The Wilderness, Gordon's men had become disorganized, and they could not support the assault. As with other aspects of this attack, opinions as to its effectiveness and potential differ widely.[71]

Early, never an enthusiastic supporter of the plan, characterized the lateness of the attack "fortunate," for it prevented the Federals from realizing how inadequate the Confederate troops were. Otherwise, Union troops might have brought up reinforcements and reversed the limited success of the attack. Grant somewhat refuted this assessment in his memoirs, written after Early's. Grant stated that, even after Sedgwick's line had been rectified, thus stopping the Confederate advance, many officers had continued to come to his headquarters "with news of disaster" and the enemy's impending arrival.[72]

Gordon felt that making the attack earlier in the day "would have resulted in a decided disaster to the whole right wing of General Grant's army." In his opinion, "one hour more of daylight . . . would have insured the capture of a considerable portion of the Sixth Army Corps." Given that there were virtually no Confederate reserves in the whole of the Army of Northern Virginia, let alone in this sector, and given the limited number of brigades making the attack, this was an unrealistic assessment.[73] In his report to Halleck, Grant was more in concert with the first part of Gordon's assessment, stating in his report to Halleck the next day that if there had been more daylight, the "enemy could have injured us very much."[74]

One widely noted incident associated with this attack illustrates Grant's challenge in reshaping the character of the Army of the Potomac. It was only the second day of fighting with Grant in charge, and he and the army were far from in sync. As Gordon's flank attack was unfolding, a general officer spoke to Grant of the severity of the situation, adding that he knew Lee's methods and thought Lee would put his whole army behind the Federals and cut them off. At this point, Grant, "with a degree of animation" (unusual for him), rebuked the officer: "Oh, I am heartily tired of hearing about what Lee is going to do. Some of you always seem to think he is suddenly going to turn a double somersault, and land in our rear and both of our flanks at the same time. Go back to your command, and try to think what we are going to do ourselves, instead of what Lee is going to do."[75]

Alternate Decision and Scenario

An intriguing alternate scenario considers the timing and support of Gordon's flank attack. This was the only time during the campaign that Grant would have both of his flanks turned. The turning movements occurred at different times of the day and came to an end for very different reasons—the fall of Longstreet on the Union left flank and darkness on the right—thus limiting their success. Of the possibility that Gordon could have made his attack in concert with Longstreet's, one of Warren's aides stated, "Nothing, I think could have saved the Army of the Potomac." With darkness, a "disaster" had been averted. The limited action on Ewell's front meant that Maj. Gen. Robert Rhodes's division was in position to move against Warren's left flank as Gordon was pushing on Sedgwick on the Union right. But this would have required the decisive action by Ewell that was lacking. Ewell failed as the critical decision-maker because he could not choose between two subordinates' differing views. In a May 1868 private conversation with one of his department chairs, Lee remarked that Ewell "showed vacillation" at The Wilderness and therefore could not get the most out of his troops. With Jackson commanding, Lee felt he "would have crushed the enemy."[76]

Another alternate scenario involves the vast difference in corps-level endorsement for the two Confederate flanking attacks. Longstreet showed support for the offensive in his sector through his decisiveness in making the attack and his committing additional troops to maximize the attack's impact. Ewell did neither. By relying so heavily on Early's biased judgment, he missed an incredible opportunity.

CHAPTER 3

TRANSITION TO SPOTSYLVANIA COURT HOUSE

For the first time in the Civil War, Grant implemented the strategy of continuously engaging the enemy. In June 1862, Lee's and McClellan's troops had been engaged in fighting for a week during the Seven Days' Battles, which concluded McClellan's Peninsula Campaign. It was almost two months until the Battle of Second Manassas and then two weeks until Antietam. But as Brig. Gen. Andrew Humphreys, Meade's chief of staff, observed, the fighting would now be different: "From the 5th of May, 1864, to the 9th of April, 1865, they [the adversaries] were in constant close contact, with rare intervals of brief comparative repose." Once the fighting started, there would be no opportunity for either side to pull back and recover, and the struggle would continue until one side was ultimately defeated. In the week after the fighting in The Wilderness, Maj. Gen. William T. Sherman would initiate his part of Grant's overall coordinated strategy and move against Gen. Joe Johnston. After initial maneuvering, the Battle of Resaca occurred northwest of Atlanta. Sherman's campaign, like Grant's, would end a year later, only at Bennett Farm, near Durham, North Carolina, rather than Appomattox Court House in Virginia.[1]

Grant Decides to Move South

Situation

As darkness fell on the evening of May 6, 1864, the Battle of The Wilderness came to a bloody close. The battle was fought to a draw, and Grant stated that he could not claim a victory. Neither side had achieved what it had hoped. Grant had failed to destroy the Army of Northern Virginia, and Lee had failed to drive the Army of the Potomac back from the battlefield. Although Grant initially estimated his casualties at about 12,000, later data indicated the substantially higher figure of 18,400. His inference that Confederate losses would be heavier than his was also incorrect—the Rebels lost 7,000 fewer soldiers. Percentagewise, the losses for each side were about the same. However, Grant could replace his losses, while Lee would face the grim reality of attrition. The 18,400 Union casualties in the two days of fighting was second only to the 23,049 casualties in the three days' fighting at Gettysburg the previous summer. Union losses at The Wilderness were greater than Grant's 13,047 losses at Shiloh.

When Meade took charge of the Army of the Potomac just before the Battle of Gettysburg the previous year, he was the fifth Union commander in fewer than two years to lead troops against Lee and his Army of Northern Virginia. These five commanders' efforts to eliminate the threat Lee's army posed disappointed the Northern population and frustrated Lincoln. Federal forces had been defeated, pushed back, or unable to follow through on a battle. Even after Gettysburg, the Union troops could not capitalize on their victory, and ten months passed until the Battle of The Wilderness was fought. The history of the post engagement activities of Federal forces operating against Lee was not encouraging:[2]

Battle	Union Casualties[3]	Results
Seven Days' June 1862	15,000	McClellan retreats.
Second Manassas August 1862	13,824	The Union loses. Pope is relieved of command of the Army of Virginia, which disbands.
Antietam September 1862	12,401	The Union holds the field and does not pursue. McClellan is replaced.
Fredericksburg December 1862	13,353	The Union loses, and Burnside is replaced.

Chancellorsville May 1863	17,304	The Union loses. Lincoln is unwilling to meet Hooker's demands, and he is replaced.
Gettysburg July 1863	23,049	The Union holds the field but does not effectively pursue. Meade is retained as the Army of the Potomac's commander, and he retains this position under Grant.
Wilderness May 1864	18,400	The fighting ends in a stalemate.
Mine Run November 1864	N/A	The Federals maneuver, but there is no attack.

The overwhelming size the Union could bring to battle should have ensured success. As much as the tangled terrain of The Wilderness caused problems, the new lieutenant general's command structure presented its own set of challenges. The execution of the battle itself showed the consequences of Grant's divided command, and a lack of confidence in the army's leadership emerged.[4]

Burnside reported directly to Grant and his Ninth Corps and was considered a separate corps from Meade's Army of the Potomac. Burnside's inability to meet Grant's timetable squandered the opportunity to split Lee's army in two. Meade and his three corps commanders had not performed as expected. Warren's attacks at Saunders Field along the Orange Turnpike had not gained any ground. Along the Orange Plank Road, Hancock had not been able to deliver a coordinated attack, and he had missed the significance of the unfinished railroad on his left flank that allowed Longstreet to roll up his flank. In several instances in the Orange Plank sector, generals commanded troops that they were not familiar with adding to the confusion. Sedgwick's flank had been turned and his corps threatened. Sheridan's cavalry had failed to warn Grant of Lee's approach, thereby allowing Lee to attack and hold the much larger Union force's advance on ground more favorable to the Confederates.[5]

The Confederates held excellent defensive positions all along their lines. The vital Chewning Farm, which commanded the gap between the two Confederate corps, was now part of their defenses. Troops of Longstreet's Corps had constructed continuous earthworks south of the unfinished railroad bed and had established a secure Confederate right flank.[6]

Options

Although Grant had stated in correspondence with Halleck that the enemy had eventually been pushed back in all cases, he was not inclined to continue attacking here. On the evening of May 6, Grant told his aide Lieut. Col. Horace Porter that he had expected "excellent results from Hancock's movement" earlier in the day. However, the initial success could not be exploited because of the dense terrain. Grant felt that he had the capability to drive Lee back into his works, but he would not be able to "gain any decided advantage from fighting in this forest."[7]

In Grant's view, continuing to fight at this location was not an option. He needed to decide if he should pull back and refit, stay where he was and see what Lee would do, or once again try to maneuver around Lee's army.

Option 1

Grant could pull back across the Rapidan River or, more likely, to the nearby town of Fredericksburg. The town had the advantage of supply lines where he could refit his troops. Remarkably, Lee felt that Grant might move to Fredericksburg from this location in The Wilderness, and also from Spotsylvania Court House on the night of May 11. With a lull in the fighting, Grant could work out the command structure issues and put officers in place that would endeavor to implement his preferred style of fighting. After the Battles of Fredericksburg and Chancellorsville, the Army of the Potomac had retreated across a protective waterway. But Grant had "superstitions" about turning back or stopping "until the thing intended was accomplished." In his mind, this would not have been a viable option. The only effective way to ascertain which officers would emulate his trusted men in the West was to send them into action and evaluate their performance—something that could not be done by pulling back. In retreating after sustaining this level of casualties, what message would Grant be sending to his troops and the people in the North? Unlike Burnside at Fredericksburg and Hooker at Chancellorsville, Grant understood that another retreat would have a negative impact on Northern political and public opinion.[8]

Option 2

Grant could remain in place and see what Lee might try in the dense terrain. But the army would gain little advantage by remaining where it was, and it was not in Grant's nature to wait. Grant was gaining respect for Lee's ability and comparing him to another Confederate general, told Meade, "Joe Johnston would have retreated after two days' of such punishment." The Chewning

Farm, located in the center of the Confederate line where Burnside was supposed to attack, was now controlled by the Rebels. In addition, the farm ominously faced the Union's center. Having both flanks turned had shown that leaving the initiative with Lee was not a good idea.[9]

Option 3

The huge supply train the Union army required had limited its maneuverability in terms of both speed and available routes. But Grant's maneuvering of his army had resulted in victories at Vicksburg and Chattanooga. Ten miles south of The Wilderness lay the town of Spotsylvania Court House. Relative to the terrain in The Wilderness, the area around Spotsylvania was fairly open. The crossroads leading in and out of the town would provide several options for continuing the campaign. Resupplying could easily be accomplished through Fredericksburg, which lay along the Rappahannock River. Moving south would create no perception that the army was retreating. And it would put the Army of the Potomac between Lee and Richmond, forcing Lee to move.

Decision

On the evening of May 6, Grant decided to move south to Spotsylvania Court House, thereby validating Lincoln's choice to appoint him general-in-chief. During the fighting in The Wilderness, Lincoln was asked how Grant compared to previous commanders of the army. He cited Grant's "persistency of purpose" and likened him to a bulldog, saying, "Once let him get his 'teeth' in, and nothing can shake him off." Grant's decision allowed him to move to the left, thus forcing Lee to move outside his breastworks and fight in more open country . In a private conversation with *New York Tribune* correspondent Henry Wing, who would be traveling to Washington that night, Grant clearly articulated his choice: "Well, if you see the President, tell him for me that, whatever happens, there will be no turning back." When Wing personally delivered the message in a private meeting with Lincoln, it gave the president "a newly kindled hope," and he was so moved that he kissed the reporter on his forehead.[10]

Speaking with his staff the next morning after breakfast, Grant acknowledged that the Federals had not achieved a positive victory. Voicing his typically positive perspective, he also observed that the Confederates had only reached the Federal lines twice in their many attacks, and that they had not gained any advantage.[11]

The challenge of implementing this decision would be to covertly extricate Union troops from their positions in The Wilderness. To ensure that the

Grant's objective: Spotsylvania Court House in 1864, where the Brock and Fredericksburg Roads intersected and then continued south. Library of Congress.

Confederates could not react effectively, Grant envisioned a night march. On the morning of May 7, he ordered Meade to "make all preparations" for a night march to Spotsylvania Court House and stated that Burnside's independent Ninth Corps would also move south. Grant's orders demonstrated that he would now be more directly involved in orchestrating movements than he had previously been.

Once he had stated his overall objective of Spotsylvania Court House, Grant issued orders to Meade that were sensitive to the delicate command structure his personal presence with that officer created. Grant couched his commands in suggestions concerning details such as the order the corps should move in and the roads they should take, as well as the need to move the vehicles quietly before moving the troops. In a more direct manner, Burnside got his orders via Grant's chief of staff Brig. Gen. John Rawlins, who sent a copy of these orders to Burnside with a note: "Your orders are embraced therein, and you will act accordingly." During the march into The Wilderness, the troops were spread out over different routes and therefore unable to support one another. Conversely, the march deeper into Confederate territory to Spotsylvania Court House would have the corps moving along the roads closer together, allowing them the ability to provide mutual support if needed.[12]

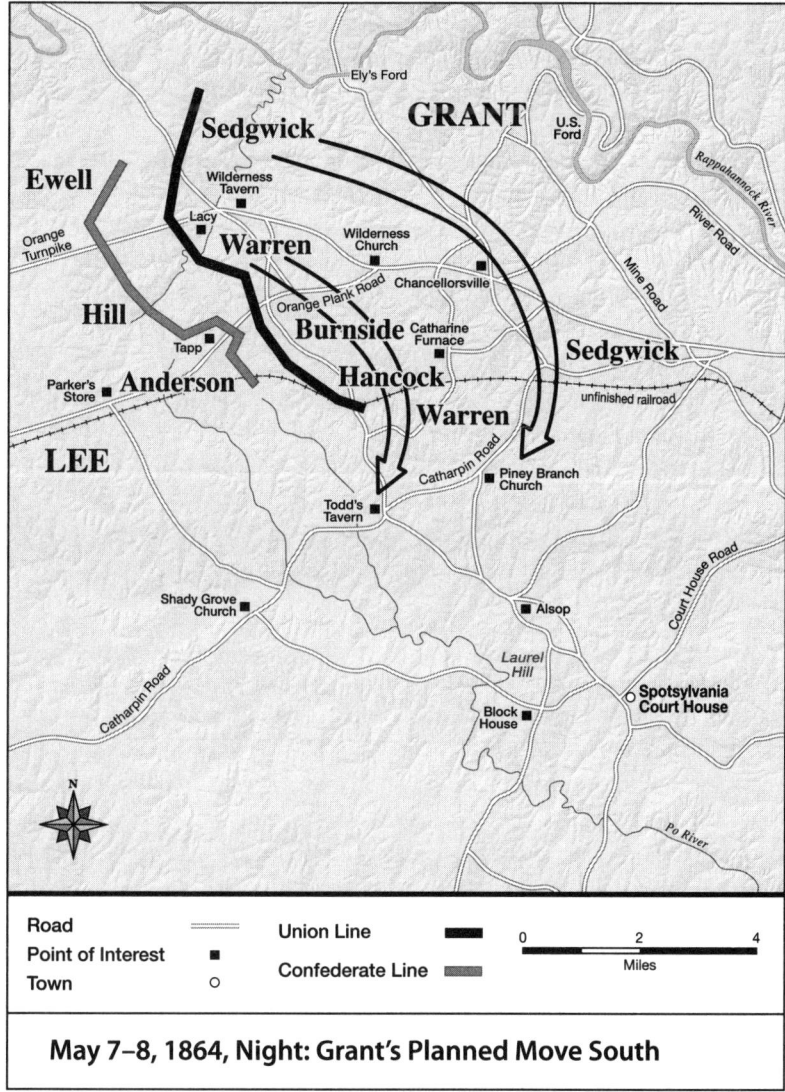

May 7–8, 1864, Night: Grant's Planned Move South

A May 7 reconnaissance along the extreme left of the Confederate line at The Wilderness provided a forewarning of Grant's resolve in executing his overall campaign strategy. Ewell's chief of artillery, Brig. Gen. Alexander Long, led the reconnaissance and found that the ford and Germanna Road, the route Warren's Corps had used to initiate the campaign just a few days earlier, was virtually abandoned. Grant had no intention of turning the Federals back.[13]

Results/Impact

At 3:00 p.m. that day, Meade issued the details of the movement, which reflected Grant's suggestions. The relevant items of the order are as follows:

> At 8.30 p.m. Warren's Fifth Corps would move by way of the Brock Road and Todd's Tavern.
> At 8.30 p.m. Sedgwick's Sixth Corps would move by the Pike and Plank Roads to Chancellorsville. Thence they would travel by way of the small hamlet called Alrich and then to Piney Branch Church to the intersection of the road from Piney Branch Church to Spotsylvania Court House.
> Hancock's Second Corps would move to Todd's Tavern by the Brock Road, following the Fifth Corps closely.
> Sheridan's Cavalry Corps was to have a sufficient force on the approaches against the Union right flank to advise the corps commanders of the appearance of the enemy.
> Grant's orders to Burnside had him following Sedgwick's troop movements.

Warren's Fifth Corps was expected to be in Spotsylvania Court House by daylight on May 8. However, the execution of this audacious plan was a debacle. The coordination required for four Union corps to march at night along the same roads as ambulances and slow-moving wagons was nonexistent. Sheridan failed to clear the roads south, and Meade grew furious with the performance of the cavalry. Once again the army lacked urgency in its movements. As the timing of the march degenerated, the corps became separated. The Army of the Potomac's inability to disengage from a battle and effectively move to a new location meant losing the race to occupy the strategic village of Spotsylvania Court House. Grant had not yet achieved the nimbleness with which he wanted and needed to wield this army. The two previous days of hard fighting and what future Supreme Court justice Oliver Wendell Holmes described as very a fatiguing march impacted Grant's army more than Lee's.[14]

The Wilderness demonstrated the Federal army's inability to perform coordinated movements to meet Grant's goals. The additional coordination challenges that a night march after heavy fighting would entail needed to be addressed. Grant's propensity for ignoring inadequacies in those he liked blinded him to Sheridan's failures the previous two days, and he seemed oblivious to the growing friction between Meade and Sheridan. If Grant was to remedy these issues for the impending move south, he needed to gather Meade, Meade's infantry corps commanders, Burnside, and Sheridan to hear him clearly articulate his plans and expectations.

Had such a meeting occurred, Sheridan, who had pushed south past Todd's Tavern along the Brock Road on May 7 and then pulled back for the night, would likely have continued to push south. Instead, he would have to fight the next morning to retake the road. The Federal withdrawal from this advanced position allowed the Confederates time to improve their position during the night and make it more difficult for Sheridan's troops the next day. Much of the traffic congestion that occurred could have been minimized with a common understanding and objective in mind.

Anderson Decides to Push On

Situation

Although both sides skirmished to ascertain one another's position, no major action occurred on May 7. Lee held a good defensive position, but the armies' disparate numbers would force him to react to Grant's movements rather than initiating actions, his preference. Grant felt that Lee would retreat, but Meade's aide Col. Theodore Lyman told Grant that Lee was not Pemberton. According to Lyman, Lee would retreat far enough south to get across Grant's lines, "then . . . retreat no more." Based on findings from reconnaissance, Lee informed Confederate secretary of war James Seddon that he felt the Federals were moving toward Fredericksburg. Meanwhile, the Army of Northern Virginia would move to Spotsylvania Court House, a position that would allow Lee to block a direct Federal move from Fredericksburg to Richmond. Both commanders had more to learn about their opponent.[15]

On the Confederate right flank at 9:30 a.m., cavalry commander Maj. Gen. "Jeb" Stuart reported to Lee on the Union activity around Todd's Tavern. Lee directed Stuart to reconnoiter the roads that would be needed should the enemy continue moving toward Spotsylvania Court House. Union troops could move directly toward Spotsylvania Court House along the Brock Road. In contrast, the Confederates would need to move back to Parker's Store, then south on a back road to Shady Grove Church Road, east to the Block House Bridge across the Po River and thence to Block House Road, and then turn north to the intersection of the Brock Road, which the Union would be using. Brig. Gen. E. Porter Alexander, Longstreet's former chief of artillery and Anderson's current one, estimated that the Union had twelve miles to travel versus the Confederates' fifteen miles.[16]

On the morning of May 7, Stuart notified Lee that the Federals were again operating around Todd's Tavern on the Brock Road, located roughly halfway between The Wilderness and Spotsylvania Court House. Considering the possibility of Grant's moving to Spotsylvania Court House and the longer route the Confederates would have to take, Lee ordered Brig. Gen. William N. Pendleton, chief of artillery for the Army of Northern Virginia, to construct a military road to facilitate a rapid movement south. This road would subsequently be referred to as the Pendleton Road, and it would connect with the Catharpin Road just above the Corbin's Bridge crossing the Po River. Lee's order would prove to be very astute.[17]

With the loss of Longstreet, his most trusted and experienced commander, Lee had to immediately make the important choice of who would command Longstreet's Corps. At sunrise on May 7, he summoned Lieut. Col. G. Moxie Sorrel, the corps' chief of staff. Lee stated that Sorrel's insight would be helpful because he had been with the corps since its beginnings as a brigade. Lee had three possible candidates in mind: Major Generals Early and Johnson, both division commanders in Ewell's Corps, and Brig. Gen. Richard H. Anderson, a division commander in Hill's Corps. After expressing gratitude for Lee's confidence in him, Sorrel provided his insight. He stated that Major General Early was probably the ablest commander, but that he would likely be unpopular with the corps due to his irritable disposition. Lee then mentioned Major General Johnson. Sorrel acknowledged Johnson's excellent reputation but argued that someone personally known to the corps would be preferable. The final candidate was Brigadier General Anderson, who had been a division commander in Longstreet's Corps at the Battle of Chancellorsville, and who had been moved to A. P. Hill's newly created corps after the death of Stonewall Jackson. Sorrel stated that Longstreet's Corps knew Anderson, and he felt they would be satisfied with him.

Although Lee's parting comments led Sorrel to anticipate that Major General Early would be put in command, Anderson was selected for the position. Ironically, Maj. Gen. Lafayette McLaws, who had been relieved of command following a dispute with Longstreet while in Tennessee, had been ordered to return to his command on May 7. As the senior commander in Longstreet's Corps, he would typically have been the one to replace Longstreet. But Lee apparently was unaware of Special Orders No. 107, and McLaws never returned to the Army of Northern Virginia.[18]

With the conviction that Grant would be moving to Spotsylvania Court House, sometime before 7:00 p.m. on May 7, Lee ordered Anderson to quietly move his troops out of line, rest them, and start for Spotsylvania at 3:00 a.m. along the newly created military road. Lee's start time of 3:00 a.m. suggests that he did not fully appreciate the urgency of this move to Spotsylvania Court House. To ensure the move went as smoothly as possible, Pendleton, who had supervised the road's construction, was ordered to provide a staff officer to guide Anderson along the new road. Pendleton himself went to Anderson, described the route, and left an officer as guide.[19]

Unknown to Anderson and the rest of the Army of Northern Virginia, a race to Spotsylvania Court House had begun. The Union's Major General Warren was implementing his orders and starting to move his troops toward Spotsylvania Court House along the Brock Road at 9:00 p.m. But Grant's troops would once again had difficulty in executing a speedy maneuver.[20] The movement called for the coordination of four corps of tired men traveling roads at night. These roads were also clogged with ambulances carrying the wounded, further impeding the troops' progress. At one point, Meade's cavalry escort caused about one and one half hours' delay for the infantry. Two miles beyond Todd's Tavern, Union cavalry caused a three-hour delay while trying to clear the Brock Road for Warren's Corps—critical hours that Grant could not afford to lose.[21]

Options

Anderson moved out between 10:00 p.m. and 11:00 p.m. to try and find a resting place for his men. But the smoke from the fires the fighting had ignited and the stench of the dead and burned bodies was incredible. Anderson had to decide if he should try to find an encampment as his orders had stated, or to try and press on and get out of the smoky area.[22]

Option 1

By quietly withdrawing the troops, Anderson had fulfilled the initial part of Lee's orders. Following a forced march the previous day, the men had participated in the intense fighting for the critical Orange Plank and Brock Roads intersection. Now, the exhausted soldiers would start the long, hard march to Spotsylvania Court House before dawn. The second part of the order was to let the men sleep, if only for a brief time. This option would demonstrate that the newly appointed corps commander could be counted on to follow orders. The problem was that the severe fighting had started many fires in the thick, dry woods in the area, making it difficult to find a suitable place to adequately rest the troops.

Option 2

If Anderson pushed his troops out of the fire-ravaged area, they might be able to find a more suitable place to rest. However, the men had started their movement to the rear at 10:00 p.m., and marching troops in the dark was always difficult. Finding a suitable encampment at night increased Anderson's difficulties. The Pendleton Road was rough and would not accommodate artillery, which would have to go another way.

Decision

The road that Anderson used to pull his men from the line was narrow and full of obstructions. When the guide that Pendleton had provided noted that the remainder of the road to Spotsylvania was similar, Anderson decided to press on through the terrible night to find a proper camping place. Coupled with his decision to start withdrawing the men before midnight rather than

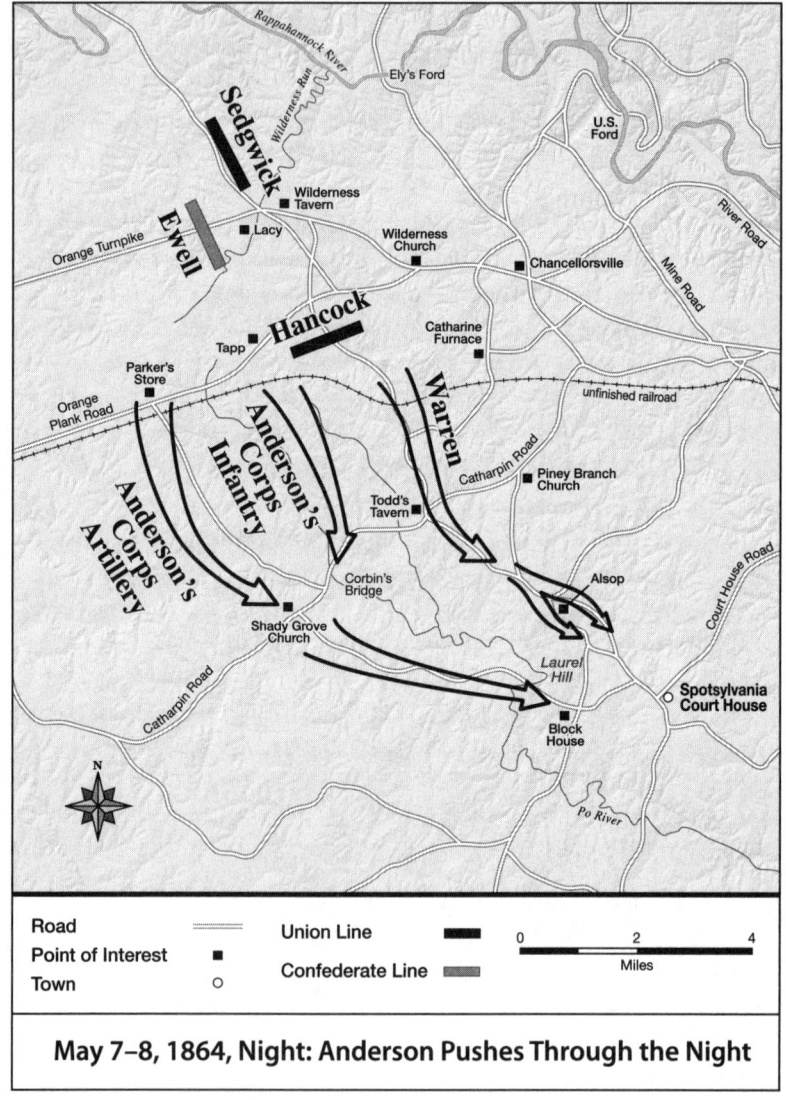

May 7–8, 1864, Night: Anderson Pushes Through the Night

at 3:00 a.m. as Lee had ordered, Anderson's choice to continue on was one of the most significant of the campaign.[23]

Results/Impact

Sorrel characterized Anderson as having a "rather inert, indolent manner of commanding troops," but the new corps commander displayed no such behavior in making and implementing his decision. Anderson marched his men toward Spotsylvania Court House, but once clear of the fires, he could not find a suitable place to bivouac the corps without separating the divisions. He then continued to march south until almost daylight, when he was able to rest his men and close up his troops. Unfortunately, their respite lasted only about an hour. A courier from Maj. Gen. Fitzhugh Lee arrived with a message that the Confederate cavalry was being pushed back at Spotsylvania Court House and needed immediate support. Although Fitzhugh Lee had sent one regiment of his command to meet the threat of the advancing Federals, it could not hope to stem Wilson's cavalry division threatening from this direction. Anderson's men could hear the cavalry fighting two miles away. Almost at the same time, Anderson got word that Federal infantry was pressing down the Brock Road. He reacted quickly and decisively, hastening Confederate artillery to Laurel Hill with infantry support and sending infantry to the town to thwart the Union cavalry.[24]

By the time Warren's Fifth Corps reached the Spindle Field near Spotsylvania Court House, Anderson's men were in a position to support Fitzhugh Lee's delaying cavalry action and prevent the Federals from gaining the vital crossroads. Warren's Fifth Corps was just about to reach the area known as Laurel Hill outside of Spotsylvania Court House. In an incredibly close race, both armies pressed on through the night to Spotsylvania Court House. Anderson's decision to continue on to an area near the Block House Bridge, aided by his earlier departure time and quick response to Fitzhugh Lee's call for help, enabled the Confederates to barely win the race. In fact, the race was so close that the US assistant secretary of war, who was traveling with Grant, thought that the Federals had arrived first. Anderson's performance on May 8 has been called the most significant contribution of his military career.[25]

Alternative Decision and Scenario

Had Anderson decided not to push on, but rather camped nearer The Wilderness Battlefield and then left at 3:00 a.m. as ordered, Warren's troops would have made it to the Brock and Block House Road intersection well before Anderson's men and would have easily won the race to Spotsylvania

Court House. Grant would then have had control of the critical intersection of the roads at Spotsylvania Court House where supplies could be directed, and access to a direct route that would place Grant between Lee and Richmond. Any defensive measures that Lee might have taken to respond to the Federals' advantage would not have included the battlefields landmark feature—the Confederate defensive line called the Mule Shoe.

The argument can be made that Anderson's decision to move on to Spotsylvania Court House resulted from the lack of adequate places to bivouac rather than military expertise, but he did decide to move, and the results proved him correct. Once he reached the Block House Bridge, he continued to make the most of the situation by quickly moving to support the Confederate cavalry. Anderson gave cavalry commander Maj. Gen. "Jeb" Stuart full authority to place the troops once he realized Stuart had a better knowledge of the ground.

Fitzhugh Lee Decides to Stay and Fight

Situation

On the evening of May 8, Grant's plan to disengage from The Wilderness and move to Spotsylvania Court House began. To accomplish this, Federal troops needed to clear Brock Road of any opposition. Todd's Tavern stood about halfway between the Union-controlled Brock Road and Orange Plank Road intersection and Spotsylvania Court House. The Catharpin Road intersected the Brock Road at this location, extending from Confederate-controlled territory to the west, and running eastward toward Piney Branch Church and then on to the settlement of Alrich, where Sheridan had his headquarters. Thus,

Todd's Tavern on the Brock Road, used by Federal infantry in their move to Spotsylvania Court House. Library of Congress.

the Catharpin Road could provide Lee a location to either block Warren's Fifth Corps as it moved down the Brock Road or threaten his flank.[26]

Sheridan had orders to maintain a sufficient force on the approaches from the west to alert the corps commanders of any enemy's appearance. The Union cavalry had to ensure that the Brock Road was clear, and that no surprise attack came from the right flank via the Catharpin Road. The Union cavalry had to ensure that the Brock Road was clear. About 7:30 a.m. on the morning of May 7, Sheridan deployed Brig. Gen. Wesley Merritt's First Division along the Furnace Road to the Brock Road north of Todd's Tavern.[27]

On May 7, Lee ordered Stuart to examine the roads that his infantry and artillery would need to take. Lee felt that Grant's intention was not to retreat, but rather to proceed toward Spotsylvania Court House. The implementation of Lee's orders placed Maj. Gen. Fitzhugh Lee's cavalry division near Todd's Tavern and Maj. Gen. Wade Hampton's cavalry division near Corbin's Bridge, located about two miles west of the Brock Road intersection where the Catharpin Road crossed the Po River.[28]

Union and Confederate cavalry had fought on May 6 north of Todd's Tavern, with Sheridan boasting that the Confederates had been driven from the field in great disorder. Ordered not to pursue the enemy, Federal troops remained on the field until just before dark. They were then commanded to

Maj. Gen. Fitzhugh Lee, CSA (USMA class of 1856), cavalry division commander. Library of Congress.

withdraw and encamp near the "Furnace", presumably Catharine Furnace near the intersection of present-day Furnace Road and Sickles Drive.[29]

On the morning of May 7, Merritt's division's First Brigade under Brig. Gen. George Custer and its Second Brigade under Col. Thomas Devin were ordered back to the area they had occupied the previous day. These units encountered Brig. Gen. Williams Wickham's brigade of Fitzhugh Lee's cavalry just north of Todd's Tavern. The Federals initially drove these advance units back, but their efforts stalled when they encountered dismounted Confederates behind barricades they had constructed during the night. Sheridan received orders at 10:00 a.m. authorizing him to detach any portion of his command for offensive operations. He dispatched his Second Division under Brig. Gen. David Gregg to move along the Catharpin Road to Todd's Tavern. Fitzhugh Lee had Federal cavalry both in his front and moving toward his right flank.[30]

Options

Having established contact with the enemy, Fitzhugh Lee had three options: disengaging and uniting with Hampton to his west near the Corbin's Bridge, pulling back toward Spotsylvania Court House and forming a defensive line, or actively contesting the Union movement down the Brock Road.

Option 1

By pulling back and moving west on the Catharpin Road, Fitzhugh Lee could unite with Hampton's division and threaten Union troops moving south on the Brock Road with a much stronger force. In addition, Confederate infantry would be closer to support any gained advantage. The disadvantage to this course of action was that Fitzhugh Lee would forfeit a position that would enable him to directly impede the Union movement along the Brock Road.

Option 2

With two Union cavalry divisions threatening him, Fitzhugh Lee could move down the Brock Road and establish a good defensive position closer to Spotsylvania Court House, the Confederate army's final objective. Hampton's cavalry division was in a strong defensive position on the Po River. Moreover, the unit had infantry support behind it, and it should not need additional assistance from Lee's division. Again, the drawback to this option would be giving up ground and allowing Union troops to move down the Brock Road uncontested.

Option 3

By conducting a delaying action, Fitzhugh Lee could gain time for the Confederates to move to Spotsylvania Court House. He was unaware of how much force was behind the advance of the Union cavalry in his front. Nor did he know how quickly Gregg's troops might advance down the Catharpin Road, get behind his current position, and cut him off.

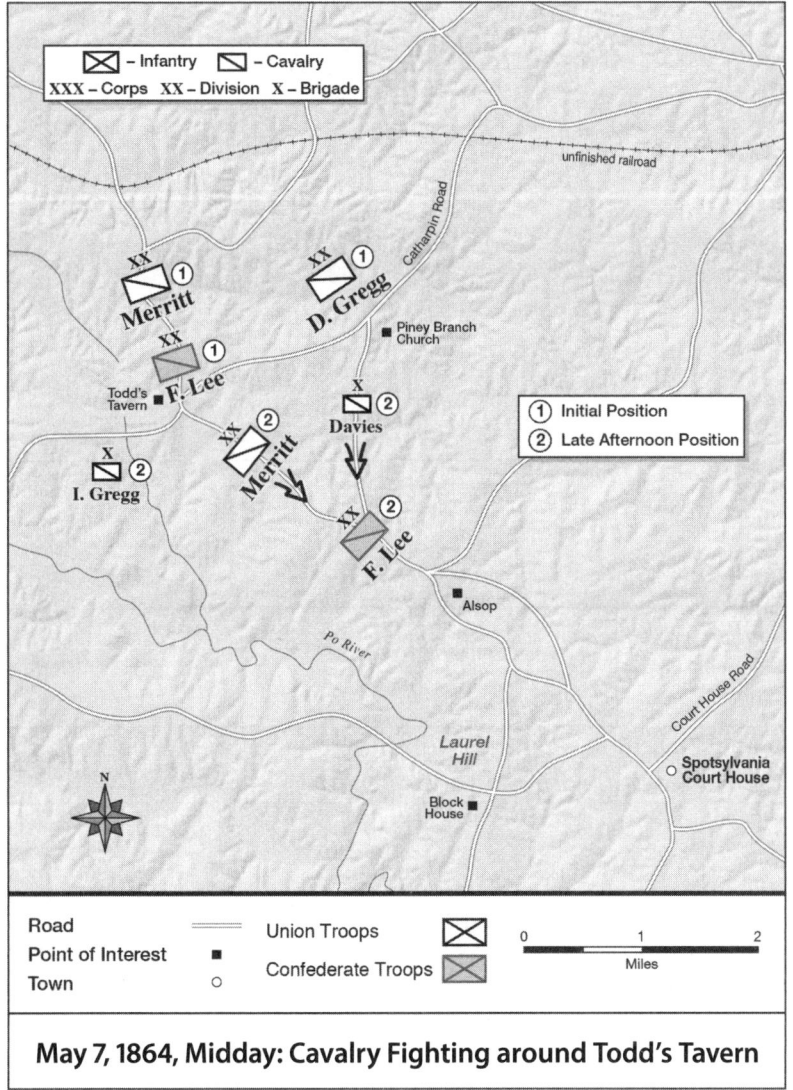

May 7, 1864, Midday: Cavalry Fighting around Todd's Tavern

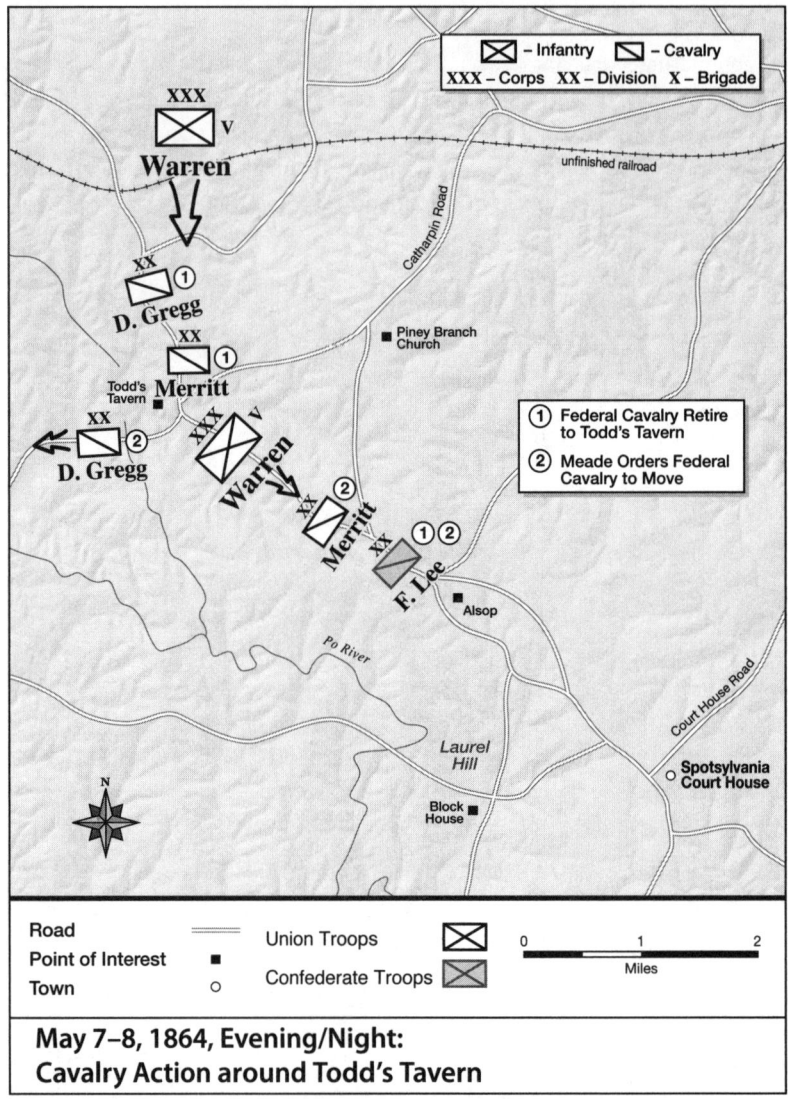

May 7–8, 1864, Evening/Night: Cavalry Action around Todd's Tavern

Decision

Unlike the evolving rocky relationship between Meade and his cavalry, commanded by the newly arrived Sheridan, Lee and his Cavalry Corps were very much in tune with each other. The Army of Northern Virginia's overall aggressiveness was part of the Confederate cavalry's culture. Fitzhugh Lee decided to fight a delaying action and impede the Union advance as much as possible.

Results/Impact

About 3:00 p.m. on May 7, Sheridan's advancing Union cavalry converged near Todd's Tavern. Todd's Tavern was important because the east–west Catharpin Road intersected the north–south Brock Road at this location. Any Confederate forces west of Todd's Tavern could use the Catharpin Road to threaten Federal troops moving south on the Brock Road. Fitzhugh Lee had already shifted his troops south of the tavern to a location near the Hart House. The wooded terrain in this area provided better defensive positions. Once again, the Confederate cavalry dismounted and erected barricades, this time in two subsequent positions a half mile apart. The first was manned by Wickham's brigade and the second by Lomax's. Fitzhugh Lee would continue to have one brigade fight in the first position until the enemy had put it in jeopardy, then move the brigade behind the second position that had its own set of defensive barricades. This approach made the most of the terrain and barricades on the road that the Federals had to clear. Fitzhugh Lee's tactic clogged the Federals' advance to fighting one brigade at a time, delaying the fulfillment of their mission and buying time for the Confederates.[31]

Once at Todd's Tavern, Sheridan sent David Gregg's Second Brigade under Col. Irvin Gregg to continue west along the Catharpin Road, where it became engaged with cavalry units of Hampton's division. The remaining troops pressed on down the all-important Brock Road and ran into the waiting Confederates. Intense fighting ensued until about 4:00 p.m., when Union reinforcements arrived and fires began to break out along the initial line of the Confederate barricades. Fitzhugh Lee was thus forced to retreat to the second set of barricades. Fighting between Merritt's and Fitzhugh Lee's troops continued until dark, and it has been described as one of the bloodiest cavalry engagements of the war.[32]

Sheridan decided not to press the attack and had Merritt's troops move a mile back up the Brock Road to Todd's Tavern, thereby averting a defeat for Fitzhugh Lee's troops. When Meade arrived at Todd's Tavern before 1:00 a.m., he found the Federal cavalry camped there and apparently lacking orders for the morning. Exasperated, Meade sidestepped the normal chain of command and issued orders directly to Merritt to immediately move out and open the Brock Road beyond Spotsylvania Court House. Merritt moved out and once again ran into stiff Confederate resistance behind barricades. The head of Warren's Fifth Corps caught up with the Federal cavalry about 3:30 a.m. By about 6:30 a.m. Merritt was still unable to break through, and Warren was ordered to use his infantry to clear the way.[33]

The Confederates used the night of May 7–8 to bolster their barricades across the Brock Road. While they succeeded in stopping the Federal cavalry the next morning, the Federal infantry was able to force the Confederates out of position. Withdrawing to near the Alsop Farm where the Brock Road split, Fitzhugh Lee once again established a defensive position. The Federals continued to encounter barricades before reaching the farm, sustaining considerable losses during the movement. The dense woods rendered artillery ineffective. But Maj. James Breather, commander of Fitzhugh Lee's horse artillery battalion, suggested that placing the guns on the line of battle would allow the Rebels to inflict great damage. Fitzhugh Lee agreed, and the guns were so positioned. They fired canister and short range shell into the Union troops and were very effective in delaying the overall Union advance. As the Confederates were forced back once more, they utilized the guns in this manner again. Finally, when Fitzhugh Lee's troops were once again driven back, they linked up with Stuart, who was quickly directing Anderson's onrushing infantry into defensive positions along the south side of the Spindle Field.[34]

At 6:45 a.m. Warren informed headquarters that he was at the front, where he said Merritt's cavalry had "made no advance." Warren's infantrymen then assumed the lead of the advance to force their way down the Brock Road. By 8:30 a.m., Warren's lead division under Brig. Gen. John Robinson gained the open ground around the Alsop Farm and soon pressed on toward the Spindle Field. Up to this time, Warren and his men thought they were only fighting cavalry. After the initial fighting at Laurel Hill, Warren sent Meade a report at 10:15 a.m. stating that some of the just-captured prisoners were from Longstreet's corps. Meade responded, "I hardly think Longstreet is yet at Spotsylvania." In fact, it was elements of Anderson's Corps that had been directed into the rapidly developing defensive positions of Laurel Hill. Fitzhugh Lee's stubborn resistance along the Brock Road had made it impossible for Meade to meet Grant's timetable. Against long odds, Fitzhugh Lee's decision and execution of delaying the Union movement to Spotsylvania Court House had worked.[35]

Troop congestion and road conditions contributed to the slow Federal progress down the Brock Road. One officer stated that in seven hours of marching, his troops had covered only eight to ten miles rather than the typical fifteen to eighteen miles. Although Sheridan had failed to clear the Brock Road for the Federal infantry, Meade's chief of staff Maj. Gen. A. A. Humphreys felt that Warren's infantry would not have fared much better in the task than Merritt's cavalry. Simply stated, Fitzhugh Lee's presence on the Brock Road prevented the Federals from capturing Spotsylvania Court House. On May 7 and 8, Fitzhugh Lee demonstrated the advances defensive fieldworks had made in delaying army corps through effective use of log barricades and dismounted cavalry.[36]

Sheridan's reports to Meade's headquarters on the evening of May 7 were full of praise for his cavalry's accomplishments and success in driving the enemy. This boasting contrasted sharply with Meade's expectations that the Brock Road would be free of enemy troops and the road to Spotsylvania open. The Fifth Corps' planned evening movement down the Brock Road required Sheridan to clear the way. Sheridan's orders were intended to ensure the infantry's smooth transition to Spotsylvania Court House. Sheridan had abandoned his initial hold on the road below Todd's Tavern and had no plans to head south until daylight. On the morning of May 8, Wilson's cavalry had driven Wickham's brigade of Fitzhugh Lee's division out of Spotsylvania Court House. Union cavalrymen held the town for two hours before Sheridan recalled them to a position near Alsop Farm, where they were of little value.[37]

Alternative Decision and Scenario

If Fitzhugh Lee had either moved to support Hampton near Corbin's Bridge on the Po River or shifted to a defensive position closer to Spotsylvania Court House, Grant would almost certainly have won the race to the critical village. By moving to support Hampton, the Confederates would have been in an extremely strong position. A river would have protected them to the front, and only a single bridge across the river would have served as an avenue for a Federal attack. Thus, they could have threatened Grant's flank in events similar to those that had occurred earlier on the Orange Turnpike and Orange Plank Roads at The Wilderness. If Grant had learned his lesson from stopping his advance in The Wilderness to confront Lee, and if he had retained his goal of Spotsylvania Court House, he could have easily stationed a strong but relatively small defensive force west of Todd's Tavern to protect troop movement along the Brock Road. The single bridge that would have been an impediment in a Federal advance would now be an advantage in protecting Grant's flank. Grant then could have proceeded south unchallenged.

If Fitzhugh Lee had moved farther south, Federal cavalry would have been able to travel down the Brock Road much more quickly and engage the Confederate cavalry practically at the Federal goal. Meanwhile, the unopposed Federal infantry could have rapidly followed the Confederates, outflanked them, forced them to abandon their defensive position (as they had in reality done on several instances on the Brock Road), and taken control of the vital road junction. It follows that the Battle of Spotsylvania Court House as we know it today would have been dramatically different and favorable for the Federals if it would have happened at all. Grant would have been in a much better position to dictate the Army of Northern Virginia's future movements.

Grant Decides to Let Sheridan Pursue Stuart

Situation

The afternoon of May 8 saw major personality differences between two members of the Army of the Potomac's command structure come to a head. As general-in-chief, Grant had been instrumental in placing two officers with little to no cavalry experience in key leadership roles of the Cavalry Corps. As commander of the Army of the Potomac, Meade had virtually no input into these decisions. Now, less than a week into the campaign, Meade felt that the cavalry had failed his army on the Orange Plank Road, on the Orange Turnpike, and, finally, on the Brock Road to Spotsylvania.

Unlike the outwardly somewhat amicable rapport between Meade and Grant, the Meade-Sheridan relationship started out strained and never improved. Sheridan reported to his new command with the Army of the Potomac on April 5, 1864. He found the horses in poor condition due to what he considered Meade's unnecessary assignment of the cavalry to perform picket duty for the infantry. Sheridan was able to eliminate this assignment and restore the animals' health for the upcoming campaign. In a report written over a year after Appomattox—a time lapse that must have biased his account—Sheridan stated that the cavalry should fight the enemy's cavalry, and the infantry the enemy's infantry. He felt that using cavalry to protect trains misused the cavalry's capability. Ironically, on May 6 Sheridan had played to Meade's concern for protecting the supply trains as a rationale for pulling back from his advanced position around Todd's Tavern. To Sheridan, cavalry's role of gathering intelligence and screening Union troop movements from the enemy did not seem to be a factor.[38]

These views clashed with Meade's, and disagreements were inevitable. Always anxious about protecting the train of over four thousand wagons, Meade considered cavalry an adjunct to the infantry. As well as protecting the supply trains, it would locate the enemy and guard the infantry's flanks during movements. In his orders, Meade often cited both the manner in which the trains would be deployed and the need to protect them. In his correspondence with Halleck, Grant complained that efforts to defend the extensive supply train negatively impacted his ability to hit Lee's army as hard as he would have liked.[39]

Careless Union cavalry operations on May 4 resulted in Confederate infantry unexpectedly emerging on the flank of Meade's southbound columns. This happened not once but twice on the Orange Turnpike and Orange Plank Roads on the opening day of the Battle of The Wilderness. As a result, the Union lost its opportunity to exploit its superior numbers with a planned quick movement through the tangled terrain of The Wilderness and into open ground farther south.

For the movement to Spotsylvania Court House on the night of May 7–8, Sheridan's cavalry was expected to clear the route for Meade's infantry and ensure that any enemy movement threating it would be quickly identified and communicated to the appropriate commanders. When Meade arrived at Todd's Tavern around midnight on May 7, he found Gregg and Merritt camped there without orders. The fuse to the powder-keg relationship between Meade and Sheridan was then lit.

Clearly, pulling the cavalry back to Todd's Tavern that night would not facilitate the infantry's objective of being in Spotsylvania Court House in the morning. It is not known where communications concerning the urgency of the infantry's movement south broke down. The already frayed interface between Meade and Sheridan could not have helped. In Meade's mind, Sheridan had failed him in The Wilderness, and had now failed him again.[40]

At 1:00 a.m. on May 8, Meade ordered Gregg and Merritt to immediately move out. At this same time, Sheridan was issuing commands to Gregg, Merritt, and Wilson from his headquarters at Alrich, some five miles away from Todd's Tavern. Sheridan's orders had his cavalry moving at 5:00 a.m.:

> Gregg's Second Division was to move west out the Catharpin Road, across Corbin's Bridge, and thence to Shady Grove Church, where the road intersected Block House Road extending east to Spotsylvania Court House.
> Merritt's First Division would follow Gregg's men and move down the Block House Road.
> Wilson's Third Division, from a different Alsop shown on the Federal maps near the Fredericksburg Road, would travel west to Spotsylvania Court House and then south for two and a half miles to Snell's Bridge on the Po River.

Sheridan's orders clearly reveal that he neither appreciated Grant's aggressive timetable for the infantry's movement to Spotsylvania Court House nor understood the actual military situation near Todd's Tavern. He made no mention of the Brock Road or of clearing it of the Confederate cavalry delaying the Federal movement. Meade would have expected Sheridan to be aware of all of this. Writing after Meade's death, one historian tried to justify Sheridan's actions, stating that if Sheridan's orders had been followed, all Confederate routes to Spotsylvania Court House would have been closed. This was clearly not the case. If Gregg and Merritt had moved out to Corbin Bridge on their way to Shady Grove Church at 5:00 a.m. on May 8, they would have met Hampton's division in a strong defensive position on the opposite side of the Po River. There, only a single bridge was available to cross the river and get at the enemy. If Gregg and Merritt had broken through the Confederate cavalry, they would have found Confederate infantry in control of the road to Shady Grove Church and already at Spotsylvania Court House.[41]

Shortly thereafter, Meade moved about two miles east of Todd's Tavern to Piney Branch Church and established his headquarters near Grant's. Sheridan, summoned by Meade to his headquarters, arrived between 11:00 a.m. and noon on May 8. The tall, sparse Meade arguing with the 5-foot-6, 115-pound Sheridan must have presented an odd picture. Despite their differences in physical appearance and their divergent views on the use of cavalry, the men shared a bad temper. The tension building between the two since Sheridan had arrived in April erupted into "a very acrimonious dispute" displaying Meade's towering passion and Sheridan's impetuous nature. The officers angrily accused each other of keeping the infantry from meeting Grant's timetable to move to Spotsylvania Court House. At one point, Sheridan declared that he could not command the cavalry under such conditions. If he could have his own way, he would concentrate all of the cavalry, move out in force against Stuart's command, and whip him. Immediately after this argument, Meade went to Grant's tent and related the substance of the meeting. When told about Sheridan's boast, Grant had to decide how to resolve the growing rift between these senior commanders and how to best utilize them.[42]

Options

Following Sheridan's public insubordination toward Meade, Grant could set the tone of appropriate military chain of command and support Meade as army commander. Or Grant could ignore the argument and continue to conduct the business of the army as he had done since taking over. Yet another option for Grant was clearly articulating his view of how the cavalry should be utilized, then requiring his subordinates to act accordingly. Finally, Grant could acquiesce to Sheridan's boast and let him have his way.

Option 1

The military chain of command is not a democracy, and it is needed to ensure the proper flow and execution of orders. By supporting Meade, Grant would be telling Sheridan and everyone else who heard the intense argument, how he felt about military protocol, and how Sheridan needed to express his differences in a more appropriate manner. But one of Grant's personal traits was supporting friends to a fault. Sheridan was one of his favorites, and the only field-grade combat general officer he had bought from the West. Grant could easily view the recent cavalry shortcomings in a different light than Meade.

Option 2

By simply listening to Meade vent and using his calming demeanor to defuse the situation, Grant could continue to focus his energy on fighting and defeating Lee's Army of Northern Virginia—a task harder than he might at first have envisioned. As for the downside, Grant would be ignoring an underlying problem in his command structure that could have major and unforeseen consequences in the hard fighting that lay ahead. The previous political intrigues plaguing the Army of the Potomac had impacted its performance, and this situation could be even more problematic. More importantly, ignoring the argument would be a tacit approval of Sheridan's apparent insubordination.

Meade and Sheridan argued so loudly that others in camp heard their contentiousness and insubordination, and this fact was relevant to both of these options. Sheridan's language throughout the meeting had been "highly spiced and conspicuously italicized with expletives." During the first day of fighting at The Wilderness, Brigadier General Griffin had been very vocal about the lack of support on his flank in his attack across Saunders Field. Brigadier General Rawlins, Grant's chief of staff, had urged Grant to court-martial Griffin, and only when Meade had stepped in and calmed things down was the situation defused. Sheridan's tirade with Meade was a much more serious exchange, and leaving it unrecognized would be a mistake. But Grant knew and liked Sheridan while Griffin was completely unknown to him, and it would be a factor in Grant's response.[43]

Option 3

Grant could take a more proactive attitude, get Meade and Sheridan together, and state his views of the cavalry's roles in the army and his expectations that they be fulfilled. This would be a major step in clearing up any miscommunications between the infantry and cavalry. Grant would also get the two branches of the army on the same page and subordinate two strong individuals' personal conflicts to the task of defeating Lee and his army. Also, this option would send a clear message that dissention was neither productive nor acceptable.

However, Grant would have to clearly state what he expected from the two branches of the army, something he had not done or felt the need to do before. The nearly impossible task of getting Meade and Sheridan to work together might take some time. And time was a commodity that Grant did not have.

Option 4

If Grant acquiesced to Sheridan's assertion, he would encourage and reward a more aggressive culture in the army. When Grant came east, he had observed that the Army of the Potomac had never fought to its capabilities, and he had expressed concern about its lack of spirit. This course of action would address both problems.[44]

The negative impacts stated above would still be applicable, and they would also undermine Meade's authority. More importantly, as poorly as the Federal cavalry had performed in The Wilderness, the army would be going into the next battle with virtually no cavalry support.

Decision

At the end of his meeting with Meade, Grant made the seemingly impetuous decision to let Sheridan try his idea. Grant told Meade, "Well, he generally knows what he is talking about. Let him start right out and do it." With these nonchalant words, the general-in-chief made a critical decision just before a major battle without carefully considering the implications. Grant had been singularly impressed by Sheridan's performance at Chattanooga, when his troops stormed Missionary Ridge, and he seemed to be hoping for a similar outcome here. Grant expressed the extent of his esteem for Sheridan over ten years later, calling him one of the great soldiers of the war and stating, "No better general ever lived than Sheridan."[45]

Maj. Gen. James E. B. Stuart, CSA (USMA class of 1854), cavalry corps commander. Library of Congress.

The effectiveness of Maj. Gen. Nathaniel Bedford Forrest was perhaps an influencing factor in this decision. Bedford's Confederate cavalry had caused much trouble for Grant and other Union leaders in the Western Theater. In January 1864, before his promotion to lieutenant general, Grant had sent Sherman additional cavalry with the intention to "go against" Forrest. Grant considered Forrest an able and effective leader of a well-disciplined cavalry, a leader whose successes led to further successes. A critical factor of these triumphs in Grant's mind was how the troops were officered. He considered Forrest the most effective commander for "the particular kind of warfare" he carried out, and he thought that the aggressive Sheridan could provide similar results for the Federals in the East if he went against "Jeb" Stuart's outnumbered Confederate cavalry.[46]

Results/Impact

At 1:00 p.m. that day, Sheridan received orders to immediately concentrate his troops and proceed against the enemy's cavalry. It can be inferred from these orders that Grant did not fully understand the situation, or that he did not plan on utilizing cavalry in the upcoming battle. Sheridan's orders further stated that when he ran out of provisions, he was to link up with Major General Butler on the James River at Haxall's Landing, about thirteen miles southeast of Richmond, to be resupplied. He was then to return to the Army of the Potomac.[47]

Grant also conveyed his misapprehension of the situation in his 3:00 p.m. orders to Major General Burnside to move to Chilesburg, well south of the actual Spotsylvania Battlefield, as soon as his trains could get started. Grant even contemplated the need for pontoons. Warren had informed headquarters at 12:30 p.m. that his advance to Spotsylvania Court House had been stopped and that troops of Longstreet's Corps were in his front. In Warren's ever-cautious style, he had added that he held a good enough position if he "was not attacked in some unprepared point on [his] flanks" by Lee's forces. He went on to state that he was out of ammunition and that he could not take Spotsylvania Court House with his current force.

When Sheridan received permission to go on his escapade, Warren's morning reports should have informed Federal headquarters that events were not going as planned. By nine o'clock that evening, Union military leadership better appreciated the situation. Burnside then received revised orders directing him to Gayle, about three miles northeast of Spotsylvania Court House on the Fredericksburg Road. Sheridan meanwhile ordered all three of his divisions to concentrate at his headquarters at Alrich's. He departed the next day, May 9, at 4:00 a.m. to carry out his orders and left Meade with only three cavalry regiments. In contrast, Stuart, knowing the value of his cavalry to provide intelligence to Lee's army, would employ only three cavalry brigades to oppose the Federals. Sheridan was so confident that he even considered the following possibility: if Stuart should interpose between him and Haxall's Landing, reminiscent of Stuart's famous ride around McClellan and the Army of the Potomac, Sheridan would ride around Lee's army and reemerge near Gordonsville, where Longstreet had been encamped before the Battle of The Wilderness.[48]

The absence of the Union cavalry would be immediately evident. The same morning that Sheridan left with basically all of Meade's cavalry, Hancock reported from the Union right flank near Todd's Tavern, "The enemy have left my right and front." On the Union's left flank, Brig. Gen. Orlando Willcox of Burnside's Ninth Corps, located east of Spotsylvania Court House along the Fredericksburg Road, reported, in his words, a large "force of cavalry, and I think infantry." Grant surmised that Lee was moving east and therefore planned to advance on his supposedly exposed left flank. But without cavalry, Hancock's infantry had to probe later in the day to find Lee's line. With Hampton's Confederate cavalry posted south of the Po River, the Rebels noted the Union movement. During the night, Lee moved Mahone's and Heth's divisions from his right flank to strengthen the area on his left flank. The next morning, Hancock was unable to move forward, and the planned attack did not occur. Nor was any cavalry present to ascertain the force in front of Willcox. As a result, the weakened Confederate right saw no action against it.[49]

Thus, in a battle that would cost him over eighteen thousand casualties, Grant would have no cavalry to determine the deployment of the Army of Northern Virginia's troops. Sheridan's adventure did nothing to determine the fighting's outcome. Much as Lee had been deprived of battlefield knowledge at Gettysburg without his cavalry, so too was Grant disadvantaged at Spotsylvania Court House. In the absence of Sheridan's cavalry, Grant would be forced to use infantry or to guess at the enemy's positions.[50]

In his personal memoirs written over twenty years later, Grant did not even mention the argument between Meade and Sheridan. He recalled verbally directing Sheridan, thereby bypassing Meade, who was also present, to cut loose from the Army of the Potomac and pass around the left of Lee's army. In fact, Sheridan went to Lee's right in pursuit of Stuart. Always Sheridan's supporter, Grant stated that the effort was "brilliant" and "accomplished more than was expected."[51]

Alternative Decision and Scenario

Had Grant taken a moment to consider the impact of his decision and then conveyed his expectations for Sheridan's cavalry to both Meade and Sheridan at the same time, he might have avoided some of the problems resulting from his actual choice. If Grant, at a minimum, had wanted the cavalry to provide intelligence on Confederate troop dispositions, at least a brigade of cavalry commanded by one of the more experienced officers, such as Brig. Gen. Wesley Merritt, could have stayed with the army. Sheridan's force outnumbered the Confederate cavalry that he would face by three divisions to three brigades. Also, the soldiers carried Spencer repeating rifles, thereby greatly increasing their firepower over the Confederate cavalry. Thus, Sheridan could have easily left Grant with some cavalry. As previously noted, the departure of Sheridan's cavalry had immediate consequences. Had Grant had at least some competent cavalry force to inform him of the Confederate troop movements' origins, Burnside, stationed below Gayle and close to Spotsylvania Court House, might have attacked the weakened Confederate position. Instead, Grant could only make guesses, often wrong, as to how to best utilize his forces.[52]

CHAPTER 4

THE BATTLE OF SPOTSYLVANIA COURT HOUSE

Having failed in his attempts to maneuver past Lee to Spotsylvania Court House, Grant would now endeavor to attack and attrite the Army of Northern Virginia wherever and whenever he could. After days of intense fighting in The Wilderness, the weary armies made a difficult night march south. "Never did a night's march seem harder," one Federal soldier reported. After marching all night, the opponents' lead units would march right back into combat. Some of the war's most horrific fighting would occur here, and the Mule Shoe and Bloody Angle would be added to the lexicons of the Civil War.[1]

Warren Decides to Redeem His Reputation

Situation

On the evening of May 7, Warren's Fifth Corps was once again the lead unit as it departed The Wilderness and commenced the long, slow march to Spotsylvania Court House. Warren was moving south with a mixed reputation that the previous two days' fighting had tarnished. The thirty-four-year-old West Point graduate had participated in several of the Army of the Potomac's major battles with diverse results. A year earlier at the Battle of Chancellorsville, Warren had suggested disobeying orders related to leaving a good defensive position east of the Chancellor house. At Gettysburg, where he had no line responsibility, Warren recognized the importance of holding Little Round Top and ordered troops there. In a council of war after Gettysburg, Meade, who favored continuing to pursue Lee, received support from Warren, his chief engineer. Unfortunately for Meade's reputation, four of his corps commanders opposed his plan and carried the decision. Meade rewarded Warren with temporary command of the Second Corps while Hancock recovered from wounds received at Gettysburg. When Hancock resumed command of the Second Corps, Warren was assigned command of the Fifth Corps that he would lead in the Overland Campaign.

In October 1863, Warren bested A. P. Hill at Bristol Station and contributed to Hill's mixed results as a corps commander. At Mine Run in November 1863, Warren delayed carrying out orders to attack a Confederate position after judging it too strong. He wanted to inform Meade of his concerns before proceeding. Meade agreed with his assessment, and the non-Battle of Mine Run was the result.[2]

When General Ewell's Confederates were discovered on the Orange Turnpike on the morning of May 5, Warren was ordered to attack immediately. Even Grant wanted the attack to begin before identifying the strength of the enemy position. But Warren delayed, trying to get all of his troops into position and wanting to wait until Sedgwick was in position on his right. Warren's First Division commander Charles Griffin and his First Brigade commander Romeyn Ayres were at the front, and they both agreed that they needed to wait for Sedgwick. The attack did not begin until about 1:00 p.m. Federals launched additional assaults during the day but were unsuccessful in forcing the Confederates from their position.[3]

Warren's action revealed a very different view of fighting tactics than Grant wanted. Seeing the heavy losses in these frontal movements, Warren felt that attacking earthworks required preparation. He did not want to rush into any combat. Grant believed that he had to make rapid assaults and apply constant pressure in order to beat Lee. Setbacks such as Saunders Field were acceptable in Grant's mind as long as the army kept to the overall strategy. This was an entirely different type of fighting than the Army of the Potomac had waged in the past. Some felt that it was the beginning of the reckless, sometimes brutal battle strategy of hurrying troops into action one division, brigade, or regiment at a time. This method would characterize the fighting from the crossing of the Rapidan to the Battle of Cold Harbor in June. Warren would have to demonstrate adaptability if he wanted to stay with the Army of the Potomac.[4]

Thus, Warren's Corps marched south while fighting traffic jams with Sedgwick's Corps on the Orange Turnpike, Union cavalry impeding his progress, questionable roads, and finally Confederate cavalry. Grant's only first-hand exposure to Warren's abilities had raised questions. When the Federal cavalry got out of the infantry's way on the Brock Road south of Todd's Tavern, they were able to push Fitzhugh Lee's Confederate cavalry back. Finally arriving on the north side of Spindle Field about 8:30 a.m., Warren faced an opportunity to possibly regain some of his personal status with the army's commanders.[5]

Options

As the lead elements of Warren's Fifth Corps approached the Spindle Field south of the Alsop Farm, they thought they were only facing Fitzhugh Lee's cavalry. The exhausting marching conditions and the enemy's effective delaying actions found the Union troops arriving in the early morning at an area north of the critical junction of the Block House Road from the south and the Brock Road from the north leading to Spotsylvania Court House. Given the high command's concerns about his slowness and the negative impact his perceived lack of aggressiveness was having on his reputation, Warren would not have considered assuming a defensive position. Instead, he needed to quickly decide how to get at the fleeing Confederates as they approached the south side of the Spindle Field, a ridge the soldiers would refer to as Laurel Hill. Warren could feed his troops in as they arrived across the open field. He could also concentrate them to achieve sufficient force to capture the vital junction of the Block House and Brock Roads. But time was a critical element. Warren knew that the Confederates needed to be driven from Laurel Hill if he was to reach Spotsylvania Court House.[6]

Option 1

As Warren's lead units moved south, they had to deploy constantly to engage Fitzhugh Lee's cavalry. The fatigued Federals also experianced a large amount of straggling, so a jumbled formation evolved. It would take time for the regiments to re-form their brigades. But by pushing the troops in as they arrived, no time would be lost. Warren could deploy the arriving brigades across a wide field, demonstrating his ability to be aggressive in combat. In The Wilderness a few days before, Grant had ordered Warren to pitch into Lee's army "without giving time for disposition," and this option would certainly be consistent with the underlying intent of those orders. An attack along the whole front could move across an open field covered with short grass, and nothing would obstruct the troops' view of the objective.[7]

The open field, though, might also expose the men to the type of fire they had withstood at Saunders Field only a few days earlier. A major drawback would be that regiments would be sent in piecemeal, thus negating the Federal advantage in troop strength. Unknown to Warren at this time, two lead regiments of Brig. Gen. Joseph Kershaw's Confederate brigade were just arrived and would basically be opposing the six regiments of Col. Peter Lyle's First Brigade and one regiment of Col. Andrew Denison's Third Brigade one at a time in succession as they approached down the Brock Road. The remaining regiments of Denison's brigade, arriving just a little bit later, would be to Lyle's right and have slightly farther to travel to reach the Confederate line. The last of the leading brigades coming from Alsop's was Brig. Gen. Joseph Bartlett's Third Brigade, still farther to the right. These later arrivals traversing longer distances would allow the Confederates more time to create and man their defenses. Additionally, the wider front would make it more difficult for Federal regiments to support each other if a breakthrough occur, or if a part of the advance stalled and required assistance.

Option 2

Focusing his attention and troops on the intersection, Warren could put the maximum amount of force where it was needed the most. Warren had just told his Second Division commander Brigadier General Robinson that Spotsylvania Court House was his objective. The junction of the Block House and Brock Roads was essential to that goal. The junction was closer, and Warren's men would be moving mostly along a road rather than open fields. They could therefore advance more quickly. Finally, the three brigades would be close enough to support each other, and they would have enough force to push the limited number of Confederates out of position, take possession of the intersection, and control the Brock Road leading to Spotsylvania Court House.[8]

Concentrating the assault solely along the Brock Road might not appear as aggressive as Warren might have desired. In the unlikely event that Confederate troops that were just arriving and hastily constructing fieldworks should try their own assault on his troops, Warren's flank could be threatened.

Decision

As his lead units reached the base of Laurel Hill after 8:30 a.m., Warren rode up. The Confederates were visible, and a decision needed to be made right away. Warren concluded that the best plan was to immediately attack with the troops on hand down the Brock Road, and subsequent troops would continue to attack all along Laurel Hill as they arrived. This is exactly what Warren did not want to do a few day earlier at Saunders Field in The Wilderness. General Robinson asked for time to bring his other brigades up. But Warren was adamant; concerned about the strengthening defenses, he

Modern-day view of Warren's objective at Laurel Hill. Courtesy of the author.

ordered an immediate charge on the enemy's line. To spur the attack on, he declared, "We must drive them from there, or they will get some artillery in position." Soldiers stated in numerous accounts how tired they were from the continuous fighting and marching over the last four days, and now their performance would reflect their fatigue. The Confederates, equally tired, held the advantage in that they were taking a defensive position. Grant's overall intent was offense—to advance. Warren would conform. His initial attack would have three phases: Lyle's brigade, then Denison's brigade of Brig. Gen. John Robinson's division, and lastly units of Bartlett's brigade under Brig. Gen. John Griffin's division, all attacking different parts of the Confederate defenses at slightly different times.[9]

Results/Impact

At about 8:30 a.m., Col. Peter Lyle was sent down the east side of the Brock Road with six regiments, all heading straight for the intersection, one behind the other, followed closely by one regiment of Denison's brigade. About fifteen minutes later, Denison's remaining three brigades farther to the west went down the right side of the Brock Road through the open field. Finally, at about 9:00 a.m. leading regiments of Bartlett's Third Brigade advanced into the far west side of Spindle Field.[10]

Johnson's Confederate horse artillery battery that had been part of Fitzhugh Lee's delaying action down the Brock Road was now positioned next to the critical intersection. Kershaw's lead Second and Third South Carolina Regiments, the rapidly advancing portion of Lee's infantry, supported Johnson. More importantly, two regular artillery batteries were fast

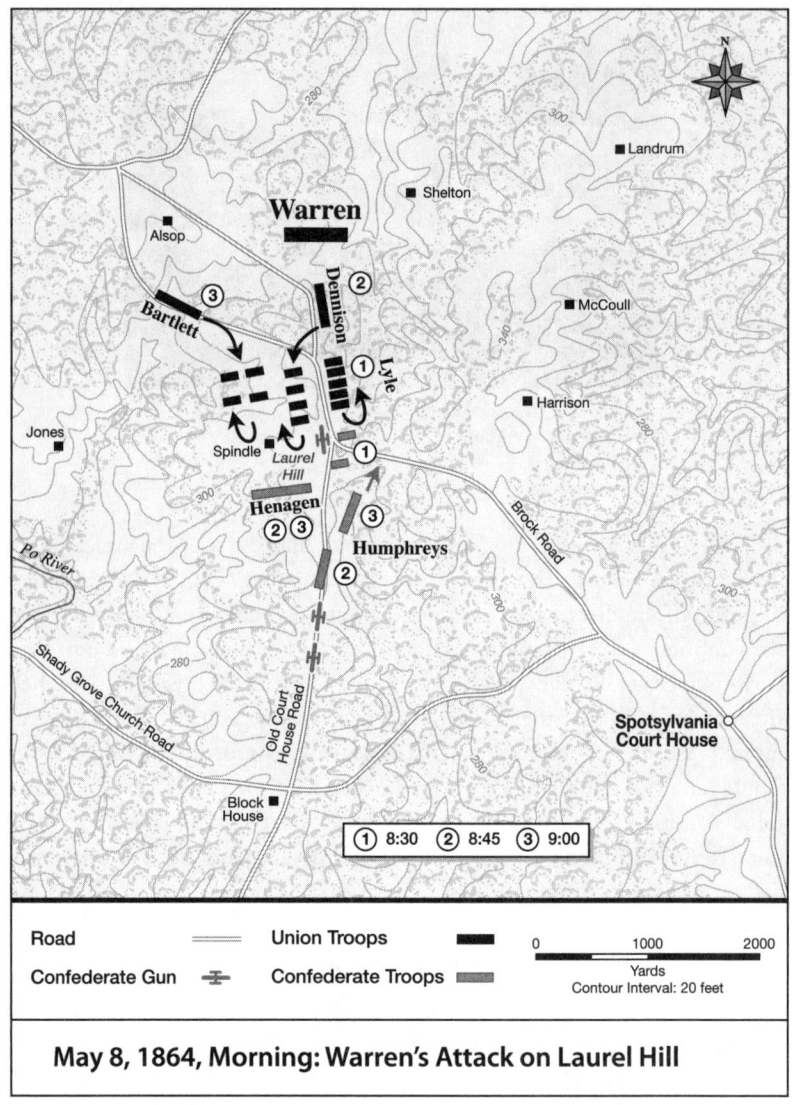

May 8, 1864, Morning: Warren's Attack on Laurel Hill

approaching the intersection and would significantly increase the Confederate firepower.

The Confederates were well known for their ability to quickly construct defensive earthworks. Within one hour, they could construct a fortification to protect a kneeling infantryman from bullets. Stuart, commanding both his cavalry and the limited infantry then on hand, ordered the men to hold fire until the enemy was well within range. When Lyle's brigade had advanced to within one hundred yards of the Confederates, they rose and delivered a withering fire.[11]

Col. John Henagan's brigade of Kershaw's division was rapidly deploying to the left. At about 8:45 a.m. Denison's brigade was the next to advance on the Confederates. Once again, Warren's troops would be just minutes late. General Warren himself encouraged them on, declaring, "Never mind cannon! Never mind bullets! Press on and clear this road. It's the only way you'll get your rations." Men went in on the right and behind Lyle's advance, just where Henagan's troops had moved to.

Sharpshooters fired as the Federals broke cover, and when the enemy soldiers came closer, the entire Confederate line opened on them. The head of the advance erred in trying to return fire. Struggling to reload slowed the progress of the attack, and the soldiers in the rear, impatiently pressing forward, caused the ranks to become intermingled and less effective. Brigadier General Robinson was wounded and had to be taken to the rear. His absence was keenly felt—it left the men without their commander at a critical time and affected the assault.[12]

Next, four regiments of Bartlett's brigade attacked the Confederate defenses to the left of where Denison's brigade, now commanded by Col. Charles Phelps, had struck. Even as the intense fighting was under way on his left, Bartlett sent word to the commander of the Forty-Fourth New York that there was no force in his "front but cavalry and to march right up the road by fours." More Confederates had moved into the area, and "the idea that the position was held only by cavalry was soon exploded." Adding to the Federal problem was the arrival of Brig. Gen. Benjamin Humphreys's Mississippian brigade, which had just moved up and to the right of the junction, enabling soldiers to flank Lyle's brigade and eventually force it back. Phelps's troops would also be forced to retreat. The Seventh Maryland Regiment of Denison's (Phelps's) brigade had made the greatest penetration into the Confederate lines, and it is now identified by a marker on the field.[13]

The various Federal units had retreated to the north side of Spindle Field by 9:45 a.m. Warren tried to rally the retreating Marylanders by waving a battered flag—poignantly, the flag of the Thirteenth Massachusetts of Lyle's decimated brigade. At 10:15 a.m., Warren informed Meade that his first attempts had failed, Robinson's troops had fought with hesitancy, and part of Griffin's command had also fallen back in confusion. In contrast to Warren's assessment, one officer in Griffin's command stated, "It was the common impression that had the affair been properly managed at the start, the position could have been carried and held." Instead, a great part of the attack was repulsed in detail. Another participant reported that soldiers had never exhibited a more resolute and determined spirt.[14]

As this initial set of attacks was occurring, Lieut. Col. Frank Huger's Confederate artillery battalion was rapidly advancing north along the Block House Road. With the Federal withdrawal and the subsequent brief break in the fighting, Huger was able to position his artillery around and to the east of the intersection, making its defenses almost insurmountable.

Warren's decision not to direct a sufficient force at the junction and the subsequent failure to quickly secure the junction would have a twofold effect on the future of the Battle of Spotsylvania Court House. First, the Confederates would control Laurel Hill throughout the rest of the battle. Lee would be able to move his troops farther to the east and take up a defensive position that would be called the Mule Shoe. Two subsequent attacks were made against Laurel Hill on May 8, and both failed.

Second, Warren's unsuccessful effort to restore his reputation affected his state of mind in subsequent decisions. Grant and Meade no longer tolerated his propensity for questioning and delaying orders. Later that day, Meade had Sedgwick moved to Warren's left. When Meade asked Warren to cooperate with Sedgwick, who outranked him, Warren exploded in a loud, profane manner and stated that he would not cooperate with Sedgwick or anybody else. On the afternoon of May 10, with preparations being made for the Upton assault, Warren told Meade of an immediate opportunity for attacking. Warren convinced Meade, and Meade uncharacteristically agreed and ordered the assault. Thus, Warren's attack proceeded earlier than originally planned, providing no support for Upton's offensive. When it was then repulsed, Warren's attack disrupted the timing of Upton's assault. Thousands of soldiers paid the price for Warren's ambition.[15]

In only a few weeks, Grant's opinion of and confidence in Warren had changed dramatically. In Grant's mind at the start of the campaign, Warren would have been in line to replace Meade had Meade been killed. On May 10 and again on May 12, Warren was ordered to attack Laurel Hill to prevent Lee from pulling troops from that location to reinforce Confederate defensive positions where Grant was attacking. However, Warren questioned the orders and was slow to engage. On May 12, Grant gave an order for Warren to be relieved from command if he did not act promptly. Grant had started to lose confidence in Warren after his failure on May 8, and that same day, Meade told Warren that he had lost his nerve. Lieut. Col. Theodore Lyman of Meade's staff felt that Warren was not up to corps command.[16]

Warren precluded the capture of the Brock and Block Road intersection when he chose to demonstrate aggressiveness by sending in arriving forces piecemeal against a relatively wide front rather than concentrating sufficient force and focusing on the junction. This untimely failure allowed the Confed-

Confederate defenses at Laurel Hill with abatis. Library of Congress.

erates to turn Laurel Hill into a well-defended position that was never taken. In so doing, they justified Lee's belief that the area was secure. This security would allow Lee flexibility in deploying his outnumbered forces to meet the evolving Federal threats. When Warren was ordered to attack on May 12 to support the Federal assault on the Mule Shoe, his soldiers covered the same ground that they had on May 8 in their failed effort to take Laurel Hill. The infantry and artillery held their fire until Warren's troops were within one hundred yards. The men then fired into the attackers, driving them back "with heavy loss to them, and but little to ourselves."[17]

Alternative Decision and Scenario

If Warren had "managed the affair" by focusing his attack on the Brock and Block House Roads intersection, the added weight and support of the additional brigades might have made it possible to take the position. An officer in Lyle's brigade said that after his troops had covered "badly gullied" terrain, they got to within thirty feet of the makeshift Confederate works before taking cover in the slope of the hill. Another advancing regiment was too far to the right to assist them. According to the officer, "Had there been any support for the brigade which got up to the rebel works, the enemy's line would have been broken."

The race to Spotsylvania Court House was repeated on a smaller scale near the important Brock and Block House Roads junction. Had Warren sent his additional troops to this area rather than farther west, they might have prevented Humphreys's Mississippians from gaining such a strong flanking position. The narrative of Spotsylvania Court House would then have been different. With the artillery that had supported the initial assaults close at hand, Warren's men could have moved into the position the Confederate army's artillery would occupy. This Union force would have been located such that it could greatly impede Lee's troops heading to the eastern part of the battlefield. Grant would have controlled the Brock Road to Spotsylvania Court House. Lee would have had to cross the road to deploy his incoming troops to the east. In the 6:30 p.m. assault on May 8, Warren's Third Division under Crawford would have met far less resistance, if any, in its movement to the east toward what would eventually become the west side of the Mule Shoe. Thus, there likely would have been no Mule Shoe or Bloody Angle, and history would have recorded a very different story about the Battle of Spotsylvania Court House.[18]

Grant Decides How to Use His Corps

Situation

Fourteen years after the Battles of The Wilderness and Spotsylvania Court House, Grant provided insight into questions that he might have contemplated after May 8, 1864. He stated that he was new to the Army of the Potomac and did not yet have a feeling as to what he could do with the generals or men. Grant had spent his entire Civil War career in the West and would have "had it in hand." Shortly after he was confirmed as a lieutenant general, he returned to the West to meet with some of his longtime and especially trusted officers—Gens. James McPherson, John Rawlins, John Logan, Grenville Dodge, Phil Sheridan, and William Sherman.[19]

McPherson had been Grant's chief engineer at Fort Donelson and Shiloh and had commanded a corps at Vicksburg. Rawlins had been on Grant's staff from the beginning, and John Logan had led McPherson's Third Division at Vicksburg. At Chattanooga, Dodge, an excellent railroad man who would have preferred to stay in command of field troops, quickly assented to Grant's request to revitalize the railroad to help supply Union troops. Phil Sheridan's aggressive style had won Grant's heart at Chattanooga, and Sherman and Grant had established a lifelong friendship and trust at Shiloh.[20]

Grant's trust and confidence in the members of this group had grown over their time serving together. When Grant had come east, he had no ready replacement for Sherman as his confidant. With Sheridan currently off pursuing

Stuart, only Grant's longtime aide Brig. Gen. John Rawlins remained, and he had never commanded troops in the field. The Overland Campaign would not afford Grant the luxury of time to develop a corresponding inner group of officers. After the fighting in The Wilderness, Grant would have been mentally appraising his corps commanders and evaluating which of them could implement his vision and style of fighting. The eastern army behaved in a more formal manner than Grant had been accustomed to in the West. This change of atmosphere likely inhibited the development of his inner circle. Grant had been brought east because he was a fighter, and he needed to infuse his style of combat into the Union army. When the soldiers were moving south to Spotsylvania Court House, Grant had not yet learned what he referred to as the special qualifications of his different corps commanders.[21]

The press had described the Army of the Potomac as being directed by Grant, commanded by Meade, and led by the Second Corps' Hancock, the Sixth Corps' Sedgwick, and the Fifth Corps' Warren. Meade said of this portrayal, "[It] about hits the nail on the head." The last component of Grant's forces in the Overland Campaign was Maj. Gen. Ambrose Burnside, commander of the Ninth Corps and equivalent to the other corps commanders. Because he was Meade's senior by date of rank, Burnside reported directly to Grant. He had previously held Meade's position as commander of the Army of the Potomac for three months in late 1862–63.[22]

At The Wilderness, Hancock's left flank had been turned, partly because he had held troops back while expecting intelligence from Sheridan on Longstreet's location. However, Hancock had done well repairing and holding his line. Official reports often mention his prompt response to orders. In discussing the May 7 movement to Spotsylvania Court House, Grant stated that Hancock's capacity to use his whole force when necessary would have enabled him to crush Anderson before he could be reinforced. Grant's decision to have Warren rather than Hancock lead the movement to Spotsylvania Court House was a tactical one. Hancock's entrenched troops were well positioned to defend against a Confederate assault, and the men passing behind Hancock could be used as a reserve.[23]

Sedgwick had been slow in the fighting in The Wilderness, and on May 9 he was killed by a Confederate sharpshooter. Although Brig. Gen. James Ricketts was next in line to command, Sedgwick had wanted the position to go to Brig. Gen. Horatio Wright, his First Division commander. Orders were subsequently issued to that effect. Thus, Wright was a complete unknown as to how he would perform at the corps command level.[24]

Warren's May 5 delay at Saunders Field in The Wilderness while trying to get everything just right before attacking had concerned army headquarters.

His habits of seemingly lecturing his superiors on military matters and questioning orders were not the type of actions Grant wanted. Warren displayed this type of behavior again on May 12, when he was ordered to support Hancock's attack on the salient. Warren was so slow in making his dispositions that his orders had to be repeated frequently and with emphasis.[25]

As Burnside was not colocated with Grant, their communications were sometimes awkward. Burnside's performance as head of the Army of the Potomac at Fredericksburg in late 1862 had called his ability to command troops, even at the corps level, into question. His actions had also led to some political intrigue with other officers in the Army of the Potomac. In the West, Burnside had successfully defended Knoxville and regained some of his reputation. But Grant's first assignment for Burnside and his Ninth Corps in the Overland Campaign was picket duty to guard the rail line supplying Meade and his army. The congenial commander seemed to operate at his own leisurely pace. At The Wilderness, Grant's aide Rawlins had tried to encourage Burnside to move with one push and drive the enemy from Hancock's front. Hancock had been expecting Burnside for three hours and attacking all the while. Grant ordered Meade's aide-de-camp Lyman to inform Hancock of the progress of the fighting in other areas. When Lyman told him of Burnside's lateness, Hancock vehemently exclaimed, "I knew it! Just as I had expected."[26]

Grant now had to quickly calibrate the abilities of these four corps commanders and decide how they would implement his strategy and style of fighting—aggressively pursuing the enemy and capitalizing on any opportunities that might arise. His decision would significantly impact how he would wage the upcoming battle.[27]

Options

Grant could view the four corps as just that, four corps of a single army, and coordinate them as such. He had utilized the four corps in The Wilderness in basically the this way. Secondly, Grant could use Meade's and Burnside's troops as two separate forces. The larger Army of the Potomac and its three corps could be Grant's main force against Lee's army. Burnside's Corps could act as an independent command, much like Butler's troops at Bermuda Hundred on the James River, to strike farther south and threaten Lee's supply lines. Finally, Grant could use part of this force to hold Lee in position while the remainder attacked Lee at the time and place he chose.

Option 1

By treating all four corps as a single army and issuing orders as such, Union forces could more effectively utilize their numerical advantage—in theory.

Corps had been awkwardly coordinated in The Wilderness. The terrain had contributed to the troops' disarray, but it was not the sole factor. Units had gotten mixed up, and commanders were at times leading units outside their chain of command. Grant had referred to himself as a "stranger" to the Army of the Potomac when he had gone east. His interface with Burnside and his Ninth Corps while they both were in Tennessee was such that the same description would be applicable now. With Burnside senior to Meade in rank, typical military protocol would put Burnside in command of the single army. But it had been fewer than two months since he had decided to leave Meade in command of the single largest army in the Union—the Army of the Potomac. Grant's most practicable solution would be circumventing protocol by putting Burnside and his Ninth Corps under the Army of the Potomac and having him report to Meade while waging his first campaign in the East.

Option 2

If Grant utilized Meade's and Burnside's commands as two separate forces, he could have Meade focus directly on Lee's front, and send Burnside farther south to threaten and attack Lee's rear and supply lines. This may have been Grant's initial thinking on the afternoon of May 8 when he ordered Burnside to Chilesburg, located well south of Spotsylvania Court House. The two major drawbacks of this option were that the force directly attacking Lee would be reduced, and Burnside had shown no signs of being a capable and aggressive independent commander. At this time, Grant was not aware of the failures of Sigel in the Shenandoah Valley or Butler on the Virginia Peninsula.[28]

Option 3

By using part of his forces to pin Lee in place, thus limiting his maneuverability, and reserving the remainder to move as he chose, Grant could offset his lack of cavalry support to locate the Confederate positions and possibly determine their strengths and weaknesses. This option would allow him to more effectively employ his aggressive fighting style by maneuvering one set of forces into the most advantageous position before launching a coordinated overall assault. Finally, by using this same set of forces in this way, Grant would be dealing with a consistent unit's fighting capability and leadership.

The downside to this plan was that it would limit the ability to promptly move soldiers to the desired location of attack. The terrain made maneuvering both hazardous and difficult. Targets of opportunity could emerge on any part of the battlefield, and troops would likely have a limited time frame in which to exploit them. Also, the men of the corps would be doing more marching and fighting than their counterparts.[29]

The Battle of Spotsylvania Court House

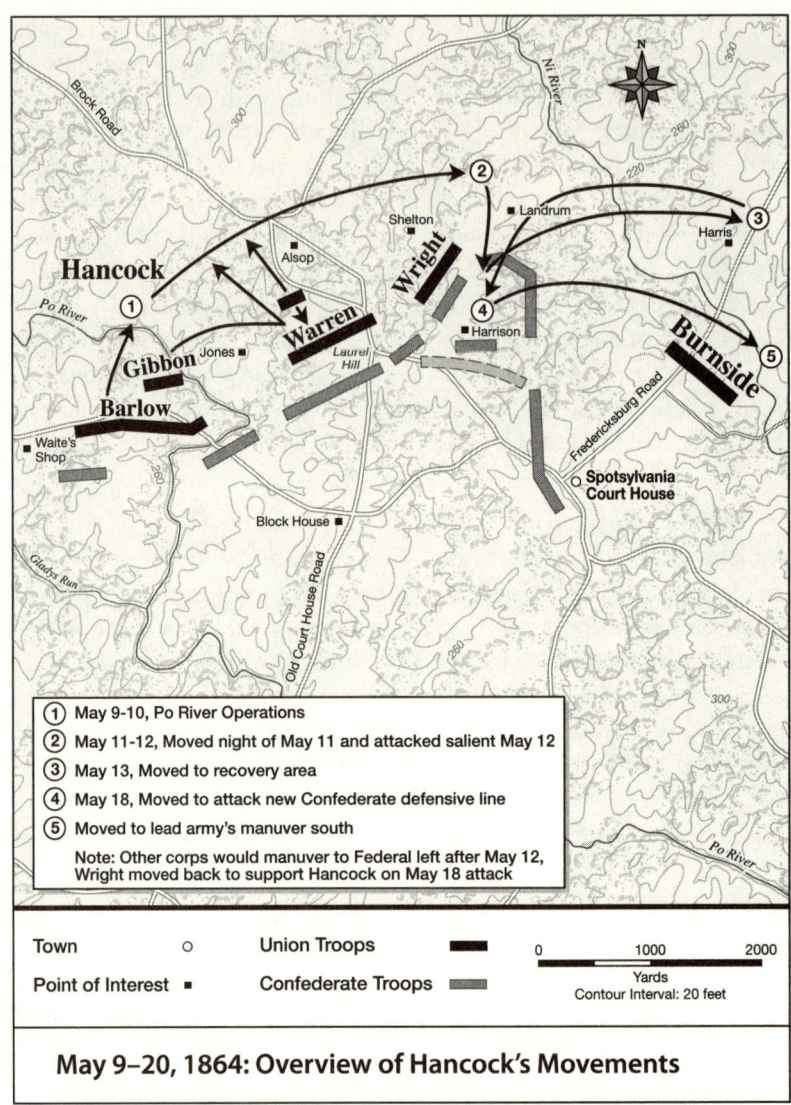

① May 9-10, Po River Operations
② May 11-12, Moved night of May 11 and attacked salient May 12
③ May 13, Moved to recovery area
④ May 18, Moved to attack new Confederate defensive line
⑤ Moved to lead army's manuver south
Note: Other corps would manuver to Federal left after May 12, Wright moved back to support Hancock on May 18 attack

May 9–20, 1864: Overview of Hancock's Movements

Decision

Typically, critical decisions are explicitly made and executed by specific orders and/or immediate actions. For example, when Grant decided to move to Spotsylvania Court House, he issued commands for the journey south and provided detailed instructions as to how it should be made. His troop dispositions over the next ten days reveal that he usually relied on Warren, Sedgwick/Wright, and Burnside to pin the enemy in place, and he dispatched Hancock to be his lead attack force.[30]

Results/Impacts

Of the four corps, Hancock's would move the most. While the Union enjoyed interior lines at Gettysburg, the Confederates held the interior lines at Spotsylvania Court House. Hancock was being used at different parts of the battlefield, so he would have to travel a greater distance than Lee's troops would to counter him.

On May 8, Warren's Fifth Corps moved to attack the Confederates as they arrived at Laurel Hill. The corps' various assaults that day proved unsuccessful, and the Confederates continued to improve their defenses in the meantime. Warren would remain in front of Laurel Hill until he was moved to the Union left on May 14. At various times, most notably on May 10 and 12, Warren received orders to attack Confederate defenses and keep Lee from pulling troops from Laurel Hill to support the areas targeted by Grant's attacks. At times, units of Warren's Corps would be taken from him and moved to support other operations.[31]

When Sedgwick's Sixth Corps arrived at Spotsylvania on May 8, it moved to Warren's left and assumed a position facing roughly east toward the developing Confederate defensive line forming the west side of what would be called the Mule Shoe. After Sedgwick was killed, Wright took over the Sixth Corps. Under Wright, the corps would move about a mile to the northeast of its initial position on May 12 to support Hancock's assault on the Mule Shoe. On May 14 it moved farther east as part of the overall Federal shift in that direction. Later, portions of Wright's Corps would return to assist Hancock's main assault on May 18.

Burnside had arrived on the east side of Spotsylvania Court House along the Fredericksburg Road south of the Ni River on May 9. His operations would stay basically in this area and until they moved on May 19.

In contrast, Hancock's Second Corps was continuously being redeployed to areas where Grant considered the Confederates most vulnerable. On May 8, Hancock's Corps was near Todd's Tavern to guard against any Confederate movement toward the army's flank. The next morning, Hancock informed headquarters that he believed the enemy had left his front—Grant's right flank. Grant had also received a report that morning from Brig. Gen. Orlando Willcox, Burnside's Third Division commander, asserting that the enemy showed considerable force of cavalry and infantry. At the time of the report, Willcox would have been near the Fredericksburg Road where it crosses the Ni River—Grant's far left flank. Grant erroneously thought that Lee was moving toward Fredericksburg to cut off the Union supply lines. As a result, he wanted to exploit the supposedly weakened Confederate left. Hancock's would provide Grant's lead troops in his plan to attack Lee's perceived vulnerability.

Hancock was ordered to move to the right of Warren. Less Mott's Third Division, which was located near Alsop, Hancock was about one mile from the Block House near the Po River that evening. He had stopped because of darkness and the challenge of crossing the river, and he had received orders to move at daylight. Note that in this area, the Po River makes a sharp bend south such that the Federals would have had to cross the Po twice—once on the evening of May 9, and again on the morning of May 10—if they were to proceed with their advance. Clashes with units of Wade Hampton's cavalry warned Lee of Hancock's movement. The Confederate general took advantage of his interior lines. That evening, he moved Mahone's division, now under Early, who had replaced the ill A. P. Hill, from a location near Spotsylvania Court House to the Confederate left flank. In the morning, Hancock would face an entirely different situation than he had the previous evening.[32]

Just after dawn, Hancock's troops were involved in fighting west of the bend in the Po River, but they accomplished little. With no realistic chance of success, Hancock's men were moved back to the north side of the Po River in the afternoon. Hancock himself was sent back to oversee the withdrawal of his First Division under Brig. Gen. Francis Barlow. When Hancock returned to his main body of troops, he found them supporting the attack Warren had recommended to Meade. This attack also targeted Laurel Hill west of Spindle Field near the J. Perry House. Brig. Gen. John Gibbon, one of Hancock's seasoned division commanders, reviewed the ground to be covered with Warren. He informed Warren and Meade of his concern that no line of battle could move through such obstacles to produce any effect. Surprisingly, Meade deferred to Warren, but once the attack was made, it failed. Another part of Hancock's troops, Mott's Fourth Division, had been moved to the left of Wright in an unsuccessful effort to support Upton's attack in the late afternoon of May 10.

Following Col. Emory Upton's attack on May 10 and Grant's subsequent decision to use the same tactic with a full corps on May 12 (both events discussed later as critical decisions), Grant chose Hancock. To conceal this movement from the enemy, Hancock's troops did not leave their position on the Union right flank toward the Brown House, where the attack was supposed to start, until after dark. The rain and darkness made the journey difficult, and the tired troops arrived just hours before the planned assault. With the timing of the attack, the officers charged with leading the assault would have no foreknowledge of the ground to be covered. In fact, the Federal high command was not completely aware of the exact location of the Confederate defenses.[33]

The May 12 assault on the Mule Shoe lasted through the day and into the night. As the leading assault force, Hancock's Second Corps would sustain twice the casualties of either of the other two corps in the Army of the Potomac. On May 13 and 14, Hancock's Second Corps had withdrawn to near the Landrum House, where it occupied defensive positions. Hancock's troops were still recovering from the brutal assault on the Mule Shoe that had just ended the day before; consequently, they had limited ability to maneuver at this time. On May 15, Hancock's Corps once again shifted to the east, this time to the Fredericksburg Road north of the Ni River. With the weather clearing after rain on and off for the last few days, Grant decided to assault Lee's new defensive position south of the original Mule Shoe's defensive line. As before, Grant moved Hancock's Corps to the west to lead the planned dawn assault with support from Wright's Corps, which also was moved. The new Confederate defensive line ran roughly along the base of the Mule Shoe and had been well fortified. The attack was a total failure. Twenty-nine pieces of Confederate artillery delivered what was described as a murderous fire of spherical case and canister on the Union troops, who were forced to retreat.[34]

Grant's final gambit at Spotsylvania Court House was to try and force Lee out of his strong defensive positions. To this end, he would send a single corps as bait. Once the Confederates had moved, Grant would strike. Hancock had spearheaded most of his operations, and he felt no need to change this routine. On the night of May 19, it was planned for Hancock's Corps to move south to Milford Station, over twenty miles away. If Hancock encountered any enemy, he was to "attack vigorously." Such were the high expectations for Hancock's Corps. Grant felt it was a good gamble, believing Lee would not be able to attack Hancock with more than one of his own corps. In addition, Grant was certain Hancock could defeat such a force. Hancock's movement was delayed by a day when Lee sent Ewell on a reconnaissance mission to Grant's right flank, resulting in the Battle of Harris Farm. Grant's efforts to trap a part of Lee's army were unsuccessful, and the armies would continue moving south.[35]

Being Grant's mobile strike force had taken a heavy toll on Hancock's Corps. Charles Dana's report to Secretary of War Edwin Stanton on May 16 stated aggregate casualties for the Second Corps as 11,553. This total was higher than that for any other of the three corps, and it was twice that of Burnside's Ninth Corps. Hancock's losses at the ill-fated assault on May 18 would only add to this grim total.[36]

Upton Leads a New Type of Attack

Situation

By May 9, 1864, Lee had successfully moved his three corps from The Wilderness and reunited them in defensive positions from the area west of Laurel

Hill near the intersection of Brock and Old Court House Roads to territory south of Spotsylvania Court House. Two of his corps commanders were new—Anderson acting for the wounded Longstreet, and Early for the ill Hill. Even so, Grant was forcing the Confederate army into a more defensive posture that somewhat alleviated the challenge these commanders faced.[37]

Grant's overall strategy of a coordinated push on all Confederate fronts was developing. He had received news from Washington that Sherman was successfully moving against Joseph Johnston in Georgia. At the same time, Benjamin Butler had landed at City Point to threaten Petersburg, and Franz Sigle was preparing to move up the Shenandoah Valley. These Federal movements would ensure that Lee would not be receiving reinforcements from these threatened areas.[38]

By noon on May 9, the Union army was aware that Confederates held a continuous defensive line around Spotsylvania Court House. Hancock, Warren, and Wright occupied positions on the west and north sides of the defenses, and Burnside was about two miles from Spotsylvania Court House on the Fredericksburg Road. Under the mistaken impression that Lee was strengthening his right with the intention of getting between Grant and Fredericksburg, Grant planned to attack Lee's supposedly weakened left flank.[39]

By the time Hancock was ready to move on the morning of May 10, Lee had reinforced his position near the Po River to the point that the advance was called off. Grant was informed by Wright at 9:30 a.m. that enemy troops were moving to his right. By 10:30 a.m. Grant had decided to make a major overall assault on Lee's defenses. It would require a level of coordination that the Federal troops had not successfully executed thus far in the campaign. The original plan was for the assault to begin at 4:00 p.m. But the start time changed to 5:00 p.m., and some parts of the intended coordinated assault would not be launched until after 6:00 p.m., when darkness was approaching. The previous day, Hancock's failed flanking maneuver along the Po River had demonstrated that the Confederate defenses at Spotsylvania Court House did not lend themselves to this type of movement. Therefore, a frontal assault would have to be made.[40]

Meade informed Wright of the assault against the whole Confederate line planned for May 10. He wanted Wright to form a column of twelve to fifteen picked regiments, reconnoiter the enemy's line, and select a position to attack. Wright, who had assumed charge of the Sixth Corps upon Sedgwick's death the previous day, selected Brig. Gen. David Russell, who had just replaced Wright as commander of the First Division, to take overall charge of this movement. Lieut. Ranald Mackenzie of the US Corps of Engineers had been doing reconnaissance in the area where the Sixth Corps was operating. He

Col. Emory Upton (USMA class of 1861), commander of the Second Brigade, First Division, Sixth Corps. *Library of Congress.*

had found a path leading from the Sheldon (sometimes referred to as Scott) House behind Federal lines to the Harrison House behind Confederate lines. With Russell, Mackenzie found a spot where the Confederate lines could be approached unseen within two hundred yards, and this location was selected as the point of attack. With a place to launch the Sixth Corps' portion of the offensive determined, Russell took the suggestion of Wright's assistant adjutant general Colonel McMahon and chose Second Brigade commander Col. Emory Upton to lead the assault. Upton was an arrogant but up-and-coming young officer in the Sixth Corps. He was ambitious to gain a general's star, and he could be counted on to be aggressive.[41]

Options

Upton had to decide how he would execute the attack. As an 1861 West Point graduate, he would have been taught the traditional tactic of assault along a wide front, firing at the objective position, reloading when possible, and rushing the enemy to engage them. But Upton had had success with a different approach the year before at Rappahannock Station. Although on a much smaller scale than planned for at Spotsylvania Court House, using a nontraditional method of forming on a more compact front and making a rapid bayonet charge had worked.

Option 1

By attacking in the traditional manner, the troops would execute the assault as they had been trained to—it was simply the way things had been done. Striking across a wide front would increase the chance of finding a vulnerable spot in the enemy's defenses, and the rest of the assault force could be funneled to that area to exploit the opportunity.

Improvements in field fortifications during the Civil War had rendered this approach almost obsolete, but traditions die hard, and the wide-front assault was still a common tactic. It required a much larger force, such as Grant had at his disposal, to have any chance of success. It also relied on the ability to literally overwhelm the defensive position with numbers. An attacking force needed time to fire, move forward while reloading, and then typically fire again. The excitement of the advancing troops would do nothing to improve their accuracy of fire. At the same time, the defenders would be protected and have more time to deliver their own counterfire.

Option 2

With an assault in column, a concentrated force could be focused at a single point to maximize the pressure and break the defensive position. Trailing columns could then exploit the breakthrough. By not firing, Upton and his men could cover the ground to the enemy position much quicker. In November of the previous year at Rappahannock Station, Upton had successfully led his Fifth Maine and 121st New York Regiments as part of a Federal assault against strong Confederate defensive works. Upton was among those Sedgwick had commended for their conduct, and he called attention to the fact that the enemy entrenchments were defended by a force numerically equal to the attacking force.[42]

The biggest challenge to this option would be focusing the attack on the right place. If troops struck a defensive position that could deliver a converging fire at the point of attack, they would be ravaged. This option would make Upton's soldiers particularly vulnerable to artillery fire, as they would be tightly bunched in column formation. The forward units would be subject to the enemy's full defensive fire and would take the heaviest casualties.

Decision

Upton, considered extremely serious about soldiering and gaining a general's star, felt that he could reverse the Federals' past assault failures with a quick thrust at the enemy. Mackenzie showed Upton the selected point of attack and the ground his men would have to cover. The formidable Confed-

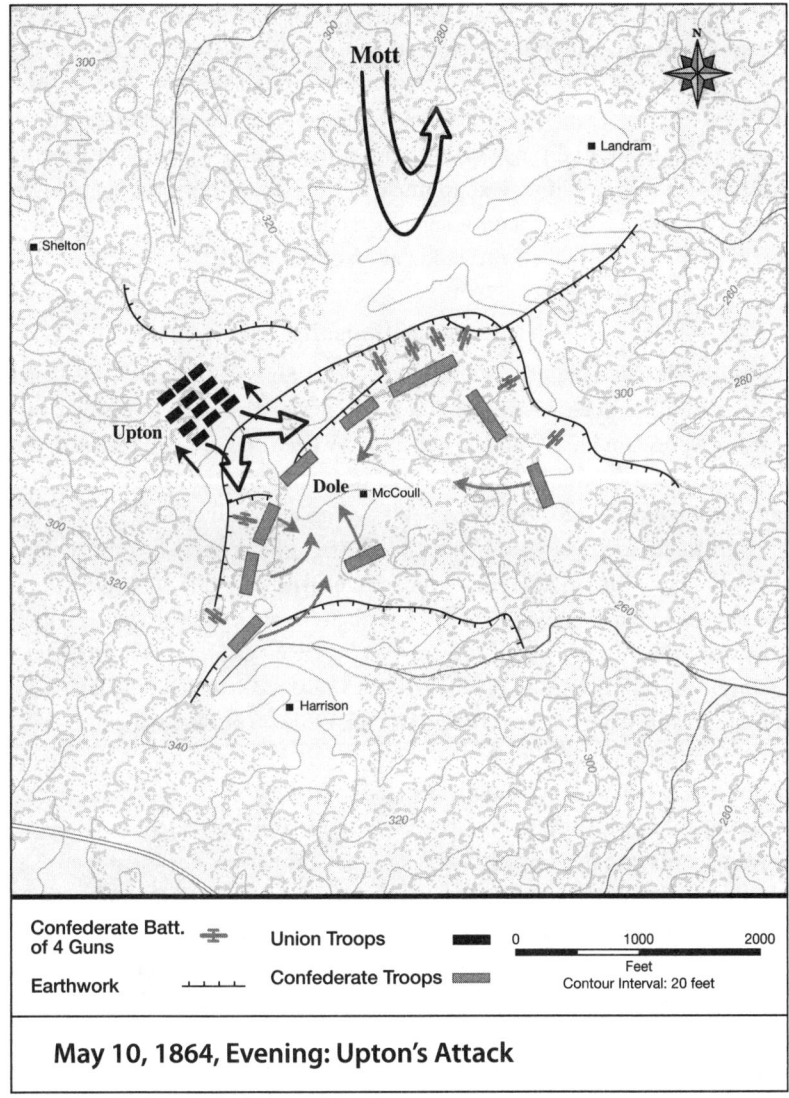

May 10, 1864, Evening: Upton's Attack

erate defensive works included abatis (in the Civil War era, typically brush and pointed logs facing out from entrenchments) and log-topped entrenchments. Traverses (earthworks perpendicular to the main trench line) further strengthened these works at various locations.

Next, Upton carefully reviewed the position with his commander, Russell, and finally with all twelve regiment commanders that would participate in the offensive. His attacking formation would be three regiments across and four regiments deep. The officers and men then reviewed maps of the defensive works and instructions for proceeding once the line was broken. Two of the regiments from the first line were to turn right and charge the batteries, and the other regiment was to charge front and left. The second line was to focus on the front, and the third to support the second. The final regiments making up the fourth line were to move to the edge of the woods and be in reserve. Finally, Upton's men drove back the Confederate picket lines to ensure that the enemy had no early indication of the attack.[43]

Results/Impact

As at the Wilderness, the coordinated simultaneous assault Grant had planned didn't happen. Warren had convinced Meade that an immediate opportunity presented itself, and at 3:30 p.m., Warren was ordered to attack the Confederates at Laurel Hill. The results were simply more Union casualties and disrupted timing for the intended coordinated assault.[44]

Burnside was encouraged to attack Lee's right flank at 5:00 p.m. An engineer was sent to him to help select the point of attack for this offensive, which was to be made in conjunction with Upton's proposed assault. Lee had pulled troops from Burnside's area, not from Wright's front as Grant thought. Once again, Burnside was late. Still, he was able to advance to within a quarter mile of Spotsylvania Court House, but no farther. Only six of Burnside's men were killed in the day's operation.[45]

In the center of the Confederate defenses, Brig. Gen. Greshom Mott, temporarily under Wright, was positioned north of the salient and was to coordinate his 5:00 p.m. attack with Upton's. Grant erroneously assumed that Mott's assault would strike the Confederates on the flank of Upton's attacking force. At 2:00 p.m., Mott received an order from Meade that only added to his dilemma. Should Burnside be attacked before 5:00 p.m., Mott was to aid Burnside by attacking vigorously from his current position. Meade made no mention of supporting Upton. At the same time, Wright, Mott's immediate commander, ordered him to attack promptly and vigorously at 5:00 p.m. to support Upton. That morning, Mott had been ordered to make a connection with Burnside and Wright, so his troops were now spread out over a two-mile front. Mott could only get 1,200 to 1,500 men in position to meet the planned 5:00 p.m. assault time. As previously ordered, however, Wright told him to advance with the soldiers he had.[46]

Upton was positioned to launch his attack on Dole's Salient, the part of the Confederate line manned by Brig. Gen. George Doles's Georgia brigade, at 5:00 p.m. The attack was moved to 6:00 p.m. because of Warren's earlier failure. Mott never received orders to delay, and he attacked at the earlier

Approximate point of Upton's attack against the Confederate position along the ridgeline called Dole's Salient. Courtesy of the author.

hour. When Mott's troops got into the open fields, the Confederates opened fire with artillery, enfilading his advance and forcing the men to fall back. Mott's supposed supporting movement had started and failed before Upton even moved out.[47]

Upton's assault was the one part of the grand overall plan that worked. Federal artillery targeted the intended assault point. The earlier artillery fire and Union squads forcing Dole's picket line back warned Ewell that an attack was expected. But before the Confederates could react, the Federal assault had begun. Upton attacked at 6:10 p.m., and the capability and discipline of the selected troops paid off.[48]

Yelling and not halting to fire, Upton's men rushed the works and quickly covered the short distance to Dole's Salient. The lead units sustained heavy casualties, but their overwhelming numbers made a half-mile break in the Confederate lines. Many Rebels were captured and sent to the rear. Initially, Confederate artillery fired canister into the Federal troops to stem the tide. Thinking the large number of Confederates moving toward the Union line was a counterattack, the artillerists stopped their firing. They quickly realized that these troops were in fact a large number of their captured comrades.[49]

Once the breakthrough occurred, the 121st New York and Ninety-Sixth Pennsylvania of the first line turned to the right to widen the breach as planned. They captured Confederate artillery but could not use them to fire on the rapidly re-forming Confederates. The quick-thinking Southern artillerists had taken the gun rammers with them when they had initially retreated. Federal efforts to disable the guns were unsuccessful, and when the

Rebel counterattack occurred, the guns were once again used to good effect on Upton's men.

The Fifth Maine of the first line turned left and pushed north, closely followed by the Fifth Wisconsin of the second line. At the same time, Upton's second and third lines poured into the break. Responding quickly, Confederate brigades of Ramseur and Gordon's divisions from the south and Johnson's division from the north and east were able to contain Upton's men to the section initially captured. Upton went back to bring his fourth line up, but he found that the troops had already been ordered forward. The expected divisional support on Upton's left to exploit the breakthrough did not arrive. As darkness fell, Upton withdrew his men due to the lack of support and the Federal failures on the other parts of the overall assault.[50]

Upton's focused assault had demonstrated the tactic's effectiveness in penetrating the enemy's defensive line. More importantly, it had revealed that timing and support had been lacking for the attack to achieve its full potential. To effectively exploit any success from an offensive like Upton's, the Union high command would have to do the following: Make the attack earlier to allow enough daylight to exploit a breakthrough, and provide reinforcements from a position closer to the point of attack.[51] In the case of Upton's assault, over twenty brigades in Wright's First and Second Divisions were close at hand but unutilized.

Union forces under Grant's leadership at Spotsylvania Court House continued to falter in executing a well-coordinated large assault. Attention to details and communication would be needed to realize the potential of Upton's tactic.

Lee Decides to Withdraw the Artillery from the Salient

Situation

Part of the Confederate defensive entrenchments at Spotsylvania Court House formed a salient called the Mule Shoe. Humphreys, accounting for the "artillery throughout" and the use of rifled muskets, estimated that the strength of this defense "was more than quadrupled" if adequately manned. This defensive line resulted from circumstances rather than a specific plan. After Anderson's division heroically marched through the night, its troops were in a position to block Warren's lead units of Grant's movements south. Lee's other two corps under Ewell and Hill/Early followed and moved into position. As the men of Ewell's Corps arrived, they were placed to Anderson's right, extending the Confederate defensive line northeast. Rodes's division arrived first, followed by Johnson's, and finally Gordon's, now commanding the division previously headed by Early. Johnson's division arrived at Spotsylvania in the late afternoon of May 8, 1864. Ordered to form on the right of Rodes's division and extend the line, soldiers at the head of Johnson's division reached their point of deployment about dark. They did not know the enemy's position until they spotted campfires. Seeing that the enemy campfires were slightly below him, Johnson moved his division along a ridgeline.

The topography in this area included a long swale running roughly northeast to southwest. This swale formed the ridgeline that the Confederates moved along as they developed the western edge of the Mule Shoe. Future Federal attackers would enjoy limited protection in parts of the swale, but more importantly, the terrain would tend to funnel troops along its primary axis to a point near the base of the western side of the Mule Shoe. This area of the swale where horrific fighting would occur would be named "the Bloody Angle." The last brigade to arrive was Brig. Gen. George Steuart's, whose original line ran across an offshoot of the main swale and on toward the Landrum House. The next day, this line was pulled back to the right along the swale spur ridgeline, thus forming the eastern side of the salient. Due to their shape, the defenses at the salient would be called "the Mule Shoe."[52]

From a military perspective, a salient is difficult to defend. The position is subject to deadly converging fire, and if a breakthrough occurs, defenders can be attacked on both their front and rear. A salient also makes it difficult for defenders to deliver interlocking fields of fire—their fire tends to diverge rather than converge, making it less effective. Steuart's position in the apex of the salient made him particularly susceptible to enfilading fire. When Lee saw the works on May 9, he said, "This is a wrenched line. I do not see how it can be held!" The saving grace was that the salient's higher ground provided many excellent positions for artillery that overlooked the Federals. If they pushed the Confederate line back, Union forces could move artillery to the higher ground and occupy a much more favorable position than they would if the Confederate line remained in place. Confederate engineers headed by Maj. Gen. Martin Smith assured Lee that artillery would make the salient defensible. Two batteries of four guns each were placed at the head of the salient, and other guns were located on the sides.[53]

Lee had received information indicating that Grant might once again be planning to maneuver his forces. The problem was the contradictory nature of this information. On the afternoon of May 11, Lee's cavalry reported a large force of Federal infantry moving toward Shady Grove and located west of Todd's Tavern. These troops were only a few regiments of Miles's brigade of Barlow's First Division of Hancock's Corps, and they were on a reconnaissance mission that returned that evening. Unfortunately for Lee, the report exaggerated the size of the Federal forces and incorrectly stated that there was "evidently a movement up the [Po] river."

Movement was also reported in the opposite direction on the Confederate eastern flank along the Fredericksburg Road. Burnside had ordered his entire corps to move to the north side of the Ni River that day. After 3:30 and 4:30 p.m., Lee received reports of this movement indicating that Federal forces might be making a general withdrawal across the Ni River. These accounts did not specifically mention movement toward the Telegraph Road. The Telegraph Road ran through Fredericksburg and then south. If Grant was attempting a flanking maneuver in that direction, troop movement along this road would be expected. Indications were that Grant might be trying to move back toward Fredericksburg. That same day, Johnson and members of his staff scouted outside their line past the Landrum House and saw no signs of the enemy until late afternoon. Ironically, Burnside's movement north of the Ni was a mistake he should not have made. When Grant's senior aide-de-camp Col. Cyrus Comstock traveled from Grant's headquarters to Burnside's, the mistake was discovered. The troops were ordered back to their previous positions. Confederates did not observe this vital aspect of Burnside's movement.[54]

Grant's superior numbers forced Lee to anticipate what Grant might actually attempt, and to be in a position in which Rebels could quickly intervene. To avoid alerting the Federals, Confederates would have to make preparations at night and avoid anything that might impede rapid movement. Lee erroneously thought Federals were moving to Fredericksburg. Even men of Hancock's Corps thought they were moving to Fredericksburg when they were in fact moving to make a major assault. Grant had successfully deceived Lee by having his troops move only so far as to get out of the Confederates' view and then march back undetected.

Lee was likely thinking of the close call in the race to Spotsylvania Court House just five days earlier. Anderson's actions allowed the Confederates to win the race to Laurel Hill and control the vital intersection. Confederate artillery had played a critical role in stopping and then holding all of Warren's efforts to break the defenses at Laurel Hill. Lee did not want another close call—one that he might not win—in a race to thwart Grant's next move.[55]

Options

Grant's large army and persistent aggressiveness continued dictating Lee's actions. On May 11, Lee had three possible options regarding the salient. He could decide to go with his initial assessment that the salient was a bad position and move Ewell's Corps to a better defensive location that would facil-

Apex of the Mule Shoe, where Confederate artillery was located the night of May 11, 1864. Note the open field of fire available for the artillery and the limited sight close to the line. Courtesy of the author.

itate any movements needed to counter Grant's. Another option was pulling the artillery back and positioning the guns so that they and the rest of the troops could quickly move to meet any threat from Grant. Finally, Lee could leave his troops in place and seek better insight into Grant's intentions.

Option 1:

By pulling the troops back out of the salient, Lee could form a stronger, more defensible line. The science of field fortifications had improved dramatically during the war. Engineers were adeptly laying out strong defensive lines, and soldiers were efficiently constructing them. A stronger defensive position would also help offset Grant's numerical advantage. Fewer soldiers would be required to man the improved defensive position, thus freeing more of them to counteract Grant's movements. These troops would be more consolidated so that the maximum force could be deployed.

The disadvantage of this option would be the difficulty of pulling men from the salient without triggering a Federal attack. An arduous withdrawal at night would be necessary. The original deployment that formed the salient had followed a relatively open ridgeline. Now, the Federals' closeness would not allow the Confederates to retrace their initial deployment to a new position. The salient's interior contained trees and gullies that would have to be negotiated, adding to the difficulty of quietly withdrawing all the troops.[56]

Confederate defenses showing traverses. Library of Congress.

Option 2

Moving artillery through the wooded areas in the salient was a difficult process. The area's pathways and trails were limited, and rain had made the ground soft and muddy. Moving the artillery out of the salient now would keep it from interfering with any later withdrawal of the infantry needed to counter Grant.[57] However, the artillery was the major factor in making the salient defensible. Removing it would make protecting the salient much more difficult.

Option 3

Unlike Grant, Lee had some cavalry available to provide intelligence of the enemy's actions. By delaying any movement of his troops currently in the salient, Lee could clarify the conflicting reports of Grant's possible intentions. Better insight into Grant's movements would be critical to helping Lee best counteract him.

The trouble with this option was that it would take time to get a clearer picture of Grant's intentions. Grant's aggressive movements would require a quick response, and he might initiate his plan before information could be gathered and communicated to Lee. If Lee guessed wrong or his artillery was not in a position to quickly act, the results could be devastating.

Decision

Based on the intelligence that he had at the time, Lee erroneously felt that Grant must have been moving toward Fredericksburg. He decided to have Ewell withdraw his troops from the salient that night. But Ewell convinced Lee that the exhausted infantrymen would be better off moving in the morning, so only the artillery was withdrew. Lee would later tell Ewell that he had been misled about the enemy in his front, and that removing the artillery had

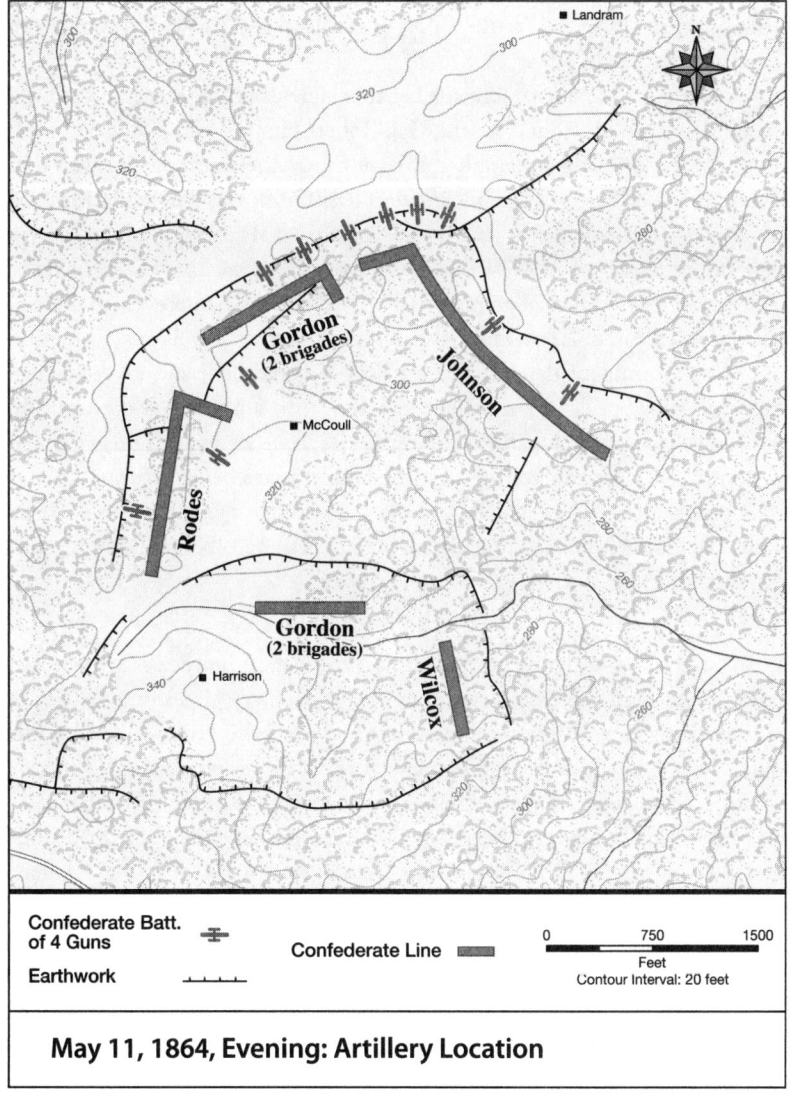

May 11, 1864, Evening: Artillery Location

been a fatal mistake. Concern that the artillery might retard troops' movement from the salient compounded the problem, so two batteries of artillery located at the tip of the salient were part of the initial withdrawal. Orders were issued and were carried out to varying degrees around dusk to remove the guns from the salient as cautiously as possible to avoid detection by the Federals.[58]

It was later learned that the movement on the Confederate left flank was unimportant. Mahone's infantry that had been moved there as a precaution was thus recalled to Hill's Corps (now commanded by Maj. Gen. Jubal Early) on the Confederate right flank. Unfortunately for the Confederates, this information was not relayed to the chiefs of artillery, and the withdrawal of the artillery remained in effect.[59]

On the night of May 11, Johnson became increasingly concerned about the enemy activity in his front. He sent Maj. Robert Hunter, his adjutant-general, to apprise Ewell of the situation and have the artillery returned. Ewell told Hunter that Lee had positive information that Grant was moving to turn the Confederate right. Reliable scouts had confirmed these reports, and it was therefore necessary to move the artillery. Lee's right was the weakest part of his force, and only Burnside's previously timid actions had prevented a Confederate misfortune. After midnight, Johnson went to Ewell and convinced him of the imminent danger. The artillery was recalled, and as the troops started to move back to the salient just after 3:30 a.m., they were unaware that Hancock would launch his attack in less than an hour. Johnson issued his command a circular stating that the artillery was being returned, that a daylight assault was anticipated, and that appropriate preparations should be made. All four of Johnson's brigade commanders signed the communication.[60]

Results/Impact

As the artillery was being moved back into place, Hancock's entire corps struck the salient. His troops captured twenty artillery pieces in only twenty minutes of fighting. Only two guns that had not been removed fired a few rounds before they and the remaining returning pieces were seized. Federals also apprehended Johnson and most of his command. The few guns that were not captured, along with reinforcements that were brought up, helped limit the extent of the breakthrough. Guns were posted at the rear of the salient just south of the McCoull House. Their fire, coupled with the infantry's efforts, prevented the captured guns from being carried off.[61]

The removal of the artillery left the salient virtually indefensible against the massive Federal assault the next morning. Grant had an opportunity to deliver a crippling blow to the Army of Northern Virginia that could have significantly altered the course of the campaign.

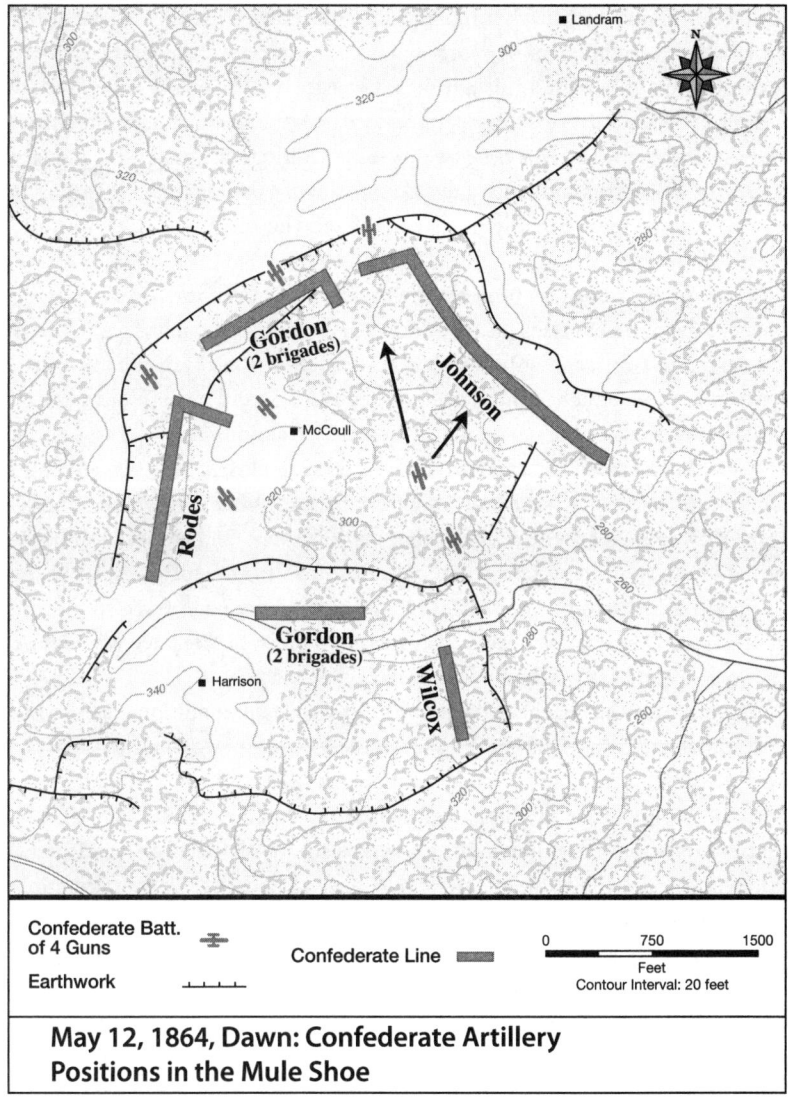

May 12, 1864, Dawn: Confederate Artillery Positions in the Mule Shoe

Alternative Decision and Scenario

Had the two artillery battalions been left in place at the apex of the salient, Hancock's Corps would have received a much stronger reception, and they might not have broken the Confederate line. The salient's artillery demonstrated its effectiveness on May 10 and again on May 18. On May 10, two depleted brigades of Mott's command had moved against the salient in the unrealistic expectation of supporting Upton's attack. Confederate artillery in the area quickly halted this advance. On May 18, Hancock, supported by Wright, moved across this same ground to attack the new Confederate line at the base of the salient. This time twenty-nine Confederate guns were in place to halt the Federal assault with a murderous fire of canister and case shot. The Federal commanders directly involved realized the futility of the effort and aborted the attack.[62]

Hancock's troops had been tightly massed to concentrate overwhelming power at the point of attack and break the enemy lines. Upton had successfully demonstrated this tactic only two days earlier. Once the advance had started, the men became bunched up such that they were a solid mass. Concentrated artillery fire would have devastated this tight formation. Not only would the front line of the troops be exposed, those following closely behind would also be vulnerable to whatever missed their comrades. When Brig. Gen. Francis Barlow, commander of the lead divisions on the salient, had proposed the tight formation he would use, he elicited objections to the anticipated effect of the artillery. Barlow had responded that he would have enough force to capture all of the artillery. His plan worked against two guns, but its price would have been much higher if Lee had left the artillery in place.[63]

A great deal has been made about the foggy, wet conditions during the morning assault. The weather did play a role in concealing the initial part of the Federal advance, and it did delay Hancock's assault until 4:35 a.m. However, the Confederate artillerists still had two hundred yards of visibility. That range would have allowed the gunners to greatly decrease the Federal momentum, and the Confederates could have reloaded the damp ammunition they initially tried to fire and added more effective infantry fire to the artillery. Hancock's staff member Maj. William Driver wrote that the left of the attacking column was "somewhat" sheltered from the infantry fire. But he added, "Had the artillery of the enemy been in position to open its fire when the column struck . . . or during the advance up the slope immediately in front" of the Confederate line, it would have had "great effect and consequent loss" as well as a very serious demoralization impact.[64]

Data from modern-day reenactments of canister fire and from modern Geographic Information Systems (GIS) and Computer-Aided Design and Drafting (CADD) software can illustrate the limited protection, or "Safe Zone," the terrain provided. Using the topography profile of a straight line from the Confederate position at the tip of the salient to the Landrum House, where Hancock's troops moved past, the following diagram shows a range of about 175 feet in which troops would not be exposed to canister fire. The side

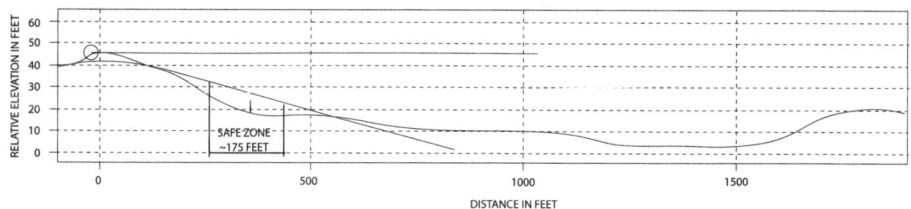

spread of canister from adjacent cannons would have reduced this area. Note that the scale is one vertical foot for every five horizontal feet for purposes of clarity.

Sgt. Cyrus Watson of the Forty-Fifth North Carolina Regiment of Daniel's brigade provided further evidence of the possible effect of the artillery in the apex of the salient. During the dawn attack on May 12, while pressing down the west side of the salient, Hancock's troops ran into this brigade. The North Carolina unit had taken the position held by Dole's brigade on May 10. Three batteries of artillery were still in place here to support the infantry and opened fire with canister. The "combined fire of infantry and artillery was more than human flesh could stand and it was impossible for them to reach our line," Watson stated.[65]

Grant Decides to Commit a Full Corps

Situation

After six days of continuous fighting, Grant had not effectively exploited his substantial numerical advantage. He characterized the results as decidedly in the Union's favor, but not decisively so. Casualties had been heavy on both sides. Grant estimated Union losses at twenty thousand, and Brig. Gen. Andrew Humphries, Meade's chief of staff, declared that Grant's army had lost over four thousand men on May 10 alone. On May 11, Grant received dispatches from Butler and Sheridan. Butler reported that he had strong entrenchments at Bermuda Hundred and had done damage to the Confederate forces in the area. Sheridan stated that he had destroyed ten miles of the Virginia Central Railroad, a vital supply line for Lee, and various military supplies. The news of the railroad's destruction pleased Grant the most. Lee would now have more difficulty bringing troops from Richmond to support him at Spotsylvania Court House. Although Lee's army was unable to initiate any strong movements directly threatening Grant's army, the Confederates maintained its position. Grant was more concerned about Lee dispatching troops against Butler or Sheridan than he was about attacks on his own army. With Sheridan detached from the army, Grant could not detect any major Confederate movement to fall back toward Richmond. Grant thought that the enemy was shaky, viable only because of its officers and position.[66]

On the Union right, Warren's Fifth Corps' repeated attacks had achieved nothing but increased casualties. On the left, Burnside had never fully engaged his troops, and little had been accomplished. The only bright spot had been Upton's temporary success at penetrating Dole's Salient the day before. Grant attributed Upton's lack of success to Mott's failure to advance. Upton felt that reinforcements from the left would have allowed him to move south toward the Brock Road. In his opinion, the extra support would also have given the Sixth Corps an excellent opportunity to exploit the situation.

Mott has been unfairly criticized for not reinforcing Upton's attack—an utterly unrealistic expectation. The original plan envisioned Upton attacking while Mott made a coordinated flanking assault to reinforce Upton's initial success. At the same time, the Federals would attack on all other fronts. Mott had between 1,200 and 1,500 men for the intended effort. They had to cover almost a mile of open ground exposed to artillery fire to reach a vaguely defined enemy position, and they were ordered to attack at 5:00 p.m. Upton had approximately 5,000 men from specially selected regiments. These troops had to cover fewer than 200 yards, and they attacked just after 6:00 p.m., not at 5:00 p.m. Even if the enemy had offered no resistance, and even if Mott could have coordinated the start of his movement with Upton's, the distance Mott had to travel would have prevented a timely convergence at Dole's Salient.[67]

Grant felt that Lee had not detached any of his troops to defend Richmond or support Confederate troops facing Butler. Thus there might be an opportunity to get Lee's army. Grant's stated objective on this front was Lee's army, and it was still strongly before him. Grant needed to determine his next move. Whatever he decided, Federals would have to contend with the rain that had been falling all day and turning the ground into mud.[68]

Options

That morning, Grant had sent correspondence to Halleck requesting fresh supplies and provisions in preparation for the fighting. He had also proposed fighting it out on this line if it took all summer. Unlike previous Union commanders in Virginia, Grant had no thought of pulling back and refitting the army. For him, the options were simply maneuvering or attacking.

Option 1

With Lee strongly entrenched, Grant could once again move around one of his flanks. Lee's defenses were now about the same as they had initially evolved. His stalemate after two days' fighting in The Wilderness was not unlike Grant's situation at Spotsylvania Court House after four days' fighting.

Grant had excellent supply lines that were not very vulnerable to Confederate disruption. He would maintain these vital routes by moving to his left flank. By maneuvering quickly, Grant could choose the next place to fight and force Lee to abandon his current defensive works. In addition, it would look good politically for the Union army to still be advancing, and with no additional casualties.

However, the Union army's movement from The Wilderness to Spotsylvania Court House had demonstrated the troops' inability to quietly and quickly withdraw and maneuver around Lee. Lee was still quite capable of exploiting any opportunities Grant might inadvertently provide in such a maneuver.

Option 2

Union troops' reconnaissance on their right flank showed no threatening Confederate force near the Po River, and Burnside's Ninth Corps protected the left flank. Upton's attack the previous day had broken through the Confederates' defensive line, if only on a limited scale. Grant felt that the officer's tactics on a portion of the salient had showed promise. In his mind, Upton had fallen short because Mott and Burnside had not vigorously supported the operation. If a larger force could be brought to bear on the vulnerable apex of the salient, Grant might be able to achieve the decisive success he wanted. However, as Upton's attack had shown, this type of assault was costly in terms of casualties. Also, all of its related maneuvering into position would have to be hidden from the enemy.[69]

Decision

By early morning of May 11, Grant had decided to attack Lee at the salient, and by 3:00 p.m. he had matured his plan of attack. Grant would launch a larger-scale version of Upton's May 10 assault, employing the same tactic from two days earlier that had temporarily pierced the Confederate defenses. But where Upton had attacked with about five thousand men, Grant would send Hancock's entire twenty thousand–man Second Corps supported by the other corps. To meet this onslaught, Johnson, who manned the apex of the salient, would have about one thousand men and almost no artillery.[70]

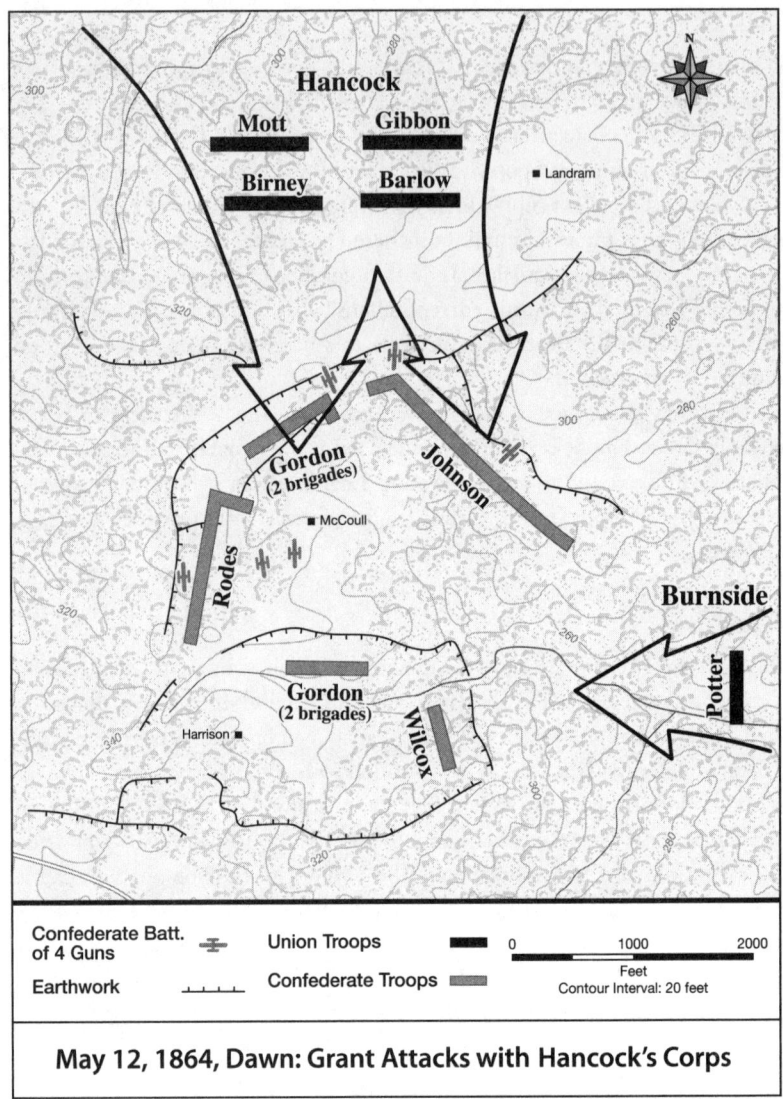

May 12, 1864, Dawn: Grant Attacks with Hancock's Corps

Results/Impact

Grant wanted to ensure that unlike Upton, Hancock would be well supported. Corps commanders were ordered to determine what forces would be sufficient to man the Federal works and how many would be available for offensive movements. Burnside was to vigorously attack at the same time as Hancock. Grant was once again going to use Hancock's Second Corps as the main attacking unit against the apex of the salient, and once again the corps would

View of the salient and the "Bloody Angle" to the right from Hancock's point of attack. Courtesy of the author.

have to be moved. To avoid detection by the enemy, it would move from the Union's far right flank, travel behind Warren's and Wright's troops, and assume a position in the center of the Union lines near the Brown House.[71]

Hancock's troops moved out in the rain and darkness for one of the most miserable of marches. Some of the last units leaving their position for the assault would march through the night and arrive just in time to move out without ever halting. Fog delayed Hancock from starting until 4:35 a.m. Once their journey was under way, the troops quickly moved through the incomplete abatis and into the enemy lines. Accounts of the Confederates' surprise conflict with one another, but they consistently state that damp powder caused the initial Rebel response to misfire, and that the few remaining artillery pieces were only able to get off a few shots before they were overrun. Barlow's First Division hit the east side of the Mule Shoe, and Birney's Third Division hit the apex and west side quickly afterward. Generals Johnson and Steuart were captured along with four thousand prisoners in the opening assault.[72]

After the initial breakthrough, more troops poured into the salient, Gibbon's Second Division behind Barlow and Mott's Fourth Division behind Birney. But the momentum soon stalled. Upton had issued carefully thought-out instructions to the troops two days earlier, telling them to move left and right once they pierced the enemy line. Grant, however, provided no guidance on how to best exploit early success. Grant's hasty decision to attack the salient and his timing of the strike to create a massive breakthrough did not allow the preparation needed for a devastating blow to Lee's army. Unlike Upton's assault, which Grant wanted to duplicate on a larger scale, the

View down Johnson's defensive works on the east side of the salient, where Barlow's troops attacked. Courtesy of the author.

general-in-chief made no extensive reconnaissance of the enemy position. Grant's commanders did not receive detailed instructions about what to do once the line was breached. Furthermore, the commanders had virtually no time for their troops to prepare—some had marched most of the night before going straight into battle. The only similarity between Upton's and Grant's attacks was that the soldiers would not fire and reload as they were advancing—they would only advance quickly.

With all of Hancock's men crammed into such a small space, there was no room in the salient for the Federal commanders present to regain operational control of the troops, re-form their lines, and move forward. Upton, who would once again lead an assault on the Mule Shoe as part of Wright's Corps' second wave, described how his twelve regiments became mixed up after the breakthrough during the May 10 attack: "There was not a single unit under my control. The same confusion occurred in the Second Corps on the 12th." Couriers were used to communicate between the front and headquarters, causing delays. With no designated commander at the front, immediate knowledge of the situational awareness was lost. In one instance when additional troops were sent to Barlow's position, he was forced to ride back to headquarters. He delivered this message: "For God's sake, Hancock, do not send any more troops in here."[73]

Gordon, now commanding Early's division, had been placed in reserve near the McCoull House and had constructed fieldworks that would help

Looking from Gordon's defensive line toward the ruins of the McCoull house. Courtesy of the author.

stem the Federal advance. Lee had previously ordered Gordon to support any part of the line about the salient that was attacked. Hearing the firing, he quickly advanced his troops to stop the Federals.[74]

Repeated Confederate counterassaults by Gordon and others pushed the Federals back to the outer works by about 8:00 a.m. Then, hand-to-hand fighting evolved to where the Federals were on the outside of the original defenses and the Confederates on the inside. Both sides continued to feed men into the fighting, which would last for the next twenty hours. The quarters were so close that that men in the rear would pass loaded guns to their comrades in front to fire. Troops in the front lines would then pass the fired guns back to be reloaded while firing the next round. The fighting was so intense that "one of the trees, fourteen to sixteen inches in diameter, was whittled down by the bullets about four feet above the ground." After Hancock's initial success, the Union troops were bottled up, and they advanced no farther from their position. After 7:00 a.m., Barlow's troops did little additional fighting.[75]

Burnside's movement was less than vigorous, and his troops never penetrated the Confederate works on the east side of the salient. Only Brig. Gen. Robert Potter's Second Division was engaged, and Maj. Gen. Cadmus Wilcox's division was able to hold the line. Grant stated that Burnside accomplished little on the left, only keeping Lee from moving reinforcements to the center. Humphreys reported, "The killed and wounded of the Ninth Corps between the 8th and 12th was very small."[76]

Wright's Second Division, under Brig. Gen. Thomas Neill, attacked near Dole's Salient about 8:00 a.m. By this time, Confederate reinforcements had moved up, and the defenses had held. As previously mentioned, the ravines that defined the topography here tended to funnel troops to the south end of the swale. This location of intense fighting would be called "the Bloody Angle." The next day, after the Confederates had withdrawn from this section, one participant surveying the area said the "sight was terrible and sickening; much worse than at Bloody Lane [Antietam]." On the Union right flank, Warren was ordered to attack at 8:00 a.m. But in typical Warren style, he objected until preemptive orders to attack arrived at 9:15 a.m. Grant was wrong in thinking that the enemy in front of Warren wasn't very strong, and the attack on the well-fortified Laurel Hill once again failed. In the salient itself, Lee was forced to sacrifice men for time as a new defensive line was hastily constructed across the base of the threatened position. At 3:00 a.m. the next morning, the Confederates pulled back to this new defensive line under cover of darkness.[77]

With the Confederates abandoning the Mule Shoe that night, this defining phase of the battle ended. Hancock reported that he had captured three thousand men, including two generals, and taken thirty to forty guns. Other estimates were as high as four thousand prisoners. Humphreys approximated the Confederate killed and wounded at between four and five thousand. He estimated Federal losses at about seven thousand. Updated totals place Lee's casualties at eight thousand for the day and Grant's at nine thousand. Grant would have replacements coming the next day, while Lee could not replace his losses so easily.[78]

Barlow, commander of one of Hancock's lead divisions to strike the Mule Shoe, attributed his initial success more to luck than anything else. With no information about the Confederate defenses or what they would have to cross, Barlow's men struck the angle by accident. The morning fog concealed their initial advance to an unusual degree, and it was only just before they moved that the open ground permitted the solid formation that made the breakthrough. Barlow did not cite the fact that most of the Confederate artillery had been removed and was returned just in time to be captured.[79]

The Federal high command thought the decision to attack the Mule Shoe had ended successfully. The next day, Grant wrote to his wife, "The enemy were really whipped yesterday but their situation is desperate beyond anything heretofore known." He included the caveat that it the battle had been a bloody one the likes of which the world had not seen. Meade described it as a decided victory over the enemy but with frightful losses. Northern newspapers concluded that the end was about to be reached, and that Grant would not give way until his army became a total wreck.[80]

Lee's May 12 letter to Confederate secretary of war James Seddon had a distinctly different viewpoint. The general lamented the loss of troops but stated that other than breastworks occupied by the enemy when Johnson's division was captured, his army had maintained its ground. The following day, Lee sent a letter to Jefferson Davis requesting that Robert Hoke's troops be sent from the Richmond defenses to reinforce him. Yet Lee provided no indication that he planned to move from his current position.[81]

The combined seventeen thousand casualties for the day's fighting left the adversaries substantially in the same position. Lee had pulled back to a better defensive line fewer than three-quarters of a mile from the apex of the Mule Shoe. Anderson's position on Lee's left was unchanged. Only a portion of Early's brigades on the south end of the east side of the salient had been repositioned, linking the new Confederate defensive line with Early's existing defenses toward Spotsylvania Court House. It was largely luck—with Grant hurling a massive force on the heels of Lee's artillery withdrawal—that accounted for the Mule Shoe's capture. Unable to either continue past the Confederates' secondary defenses or hold the ground they had initially captured, the Federals failed to deliver a decisive blow.

Second Corps commanders involved in the assault on the Mule Shoe consistently observed that they lacked information about the enemy position. Given that Grant made his decision in the morning, he had time to learn more details of the Confederate position and develop a plan to effectively implement the attack. Even so, he seems to have given no thought as to how to exploit the initial success, possibly because he had no insight concerning the defenses. Just as Grant's lack of knowledge about the actual defensive strength in front of Warren caused him to order the hopeless attack on May 12, troops continued to be fed into the breakthrough without actual knowledge of the situation. If Barlow had been designated as the on-site operations control officer at the salient, the actual situation might have been better known and orders issued accordingly. Command and control of the troops would more likely have been maintained and the flood of troops into the limited space avoided. Re-forming and continuing the movement down the edges of the salient and focusing on capturing the intricate cross earthworks could have been possible.[82]

E. Porter Alexander, Longstreet's chief of artillery, stated that the military lesson to be learned from the Federal assault on the Mule Shoe was that the size of a force to be effectively used in this type of operation was limited. Once that limit was exceeded, paralysis would set in. With the unfulfilled potential of Grant's decision to launch a massive attack on the Mule Shoe, fighting and limited maneuvering would continue for another week at Spotsylvania Court House.[83]

Alternative Decision and Scenario

As the recent fighting had shown, Grant faced a more capable foe in Lee and his army. Frontal assaults had resulted in heavy casualties with little results, but Grant could maintain the initiative by maneuver. If Grant had decided in mid-May what he eventually decided at the end of the month—to maneuver to his left and away from Spotsylvania Court House—he would have forgone the casualties and limited gains at the Mule Shoe and moved to a more advantageous location. Maneuver would play a larger role in both Grant's continued stay here and his unrelenting campaign. Following the fighting at the Mule Shoe, Grant would move troops to his left, extending the battle lines farther south of the town. With no success there, he moved troops back and again unsuccessfully tested Lee's new defenses at the base of the Mule Shoe. The glaring exception to Grant's increasing use of maneuvering was Cold Harbor. In June, he used another frontal attack against Lee that once again resulted in heavy casualties but without reaching the Confederate defensive works.

As inept as Burnside was, his corps was positioned where it could threaten Lee's right flank and hold it in place while another corps maneuvered south. Hancock's Corps, Grant's most likely choice, would not have been used up in the attack on the salient. It could have moved south, possibly to the North Anna River, and thereby threatened Richmond. Lee would once again have been forced to react to Grant.

Grant Decides to Move South from Spotsylvania Court House

Situation

The typical perception of the Battle of the Spotsylvania Court House was that it ended with Grant's previous critical decision to attack the Mule Shoe with a full corps on May 12. But many important decisions made over the next week laid the foundation for this final critical decision. "The ninth day of battle," as Grant described it to his wife, had taken a heavy toll on the troops on both sides. Decisions involving troop movement needed to account for the continued rainy weather, a major factor affecting the ability to relocate the weary men.[84]

Grant's first order of business on the morning of May 13 was to determine where Lee's army was. He and his troops thought the Confederates were retreating. Lacking cavalry support for reconnaissance, Grant wanted Meade to push "at least three good divisions" forward to determine the enemy's status. At 5:30 a.m., Wright informed Meade that the enemy had abandoned the position and that his troops were in possession of "the rebel works." Meade ordered his men to push forward to "feel for the enemy." That morning, Meade's headquarters also received word that the Confederates still occupied Laurel Hill. About one o'clock, Wright informed Meade that Lee had abandoned his original position and fallen back to a line that extended the line that had faced Warren's initial position. Lee's gamble to confront the Federal advance in the Mule Shoe to buy time to form a defensive line across the base of the salient had worked. He remained in force to confront Grant's next move. Grant's plan to deal a fatal blow to Lee by a massive attack on the salient had pushed the Confederates back "three-quarters of a mile in rear of the salient in a strongly-entrenched line occupied by infantry and artillery." Federal estimates of Lee's total losses in killed, wounded, and imprisoned men on May 12 were between nine and ten thousand troops, but Lee was strongly entrenched.[85]

The reconnaissance forces found horrific desolation. Assistant Secretary of War Charles Dana surveyed the scene with Grant's chief of staff Brig. Gen. John Rawlins and described the area as thickly covered with dead and wounded. The ground was also the consistency of pudding from all the trampling during the fighting. Grant's aide-de-camp Lieut. Col. Horace Porter described the scene as appalling; bodies were piled on each other "in some places four layers deep, exhibiting every ghastly phase of mutilation."[86]

Frontal assaults on Lee's left flank at Laurel Hill and his center at the salient had not worked. Therefore, Grant would try Lee's right flank. Hancock's was Grant's favorite offensive corps, but its soldiers were recovering from the brutal fighting the day before. As Grant had no faith in Burnside, he would be forced to carry out his plan with the uncooperative Warren and the inexperienced Wright. Orders were issued for the two corps to withdraw from their positions fronting Laurel Hill and the west region of the Mule Shoe after dark. The units would then move around behind Burnside on the Federal left flank, take up positions on his left, and make a surprise attack on the thin Confederate line south of Spotsylvania Court House at 4:00 a.m. Once again, Grant's expectations were unrealistic. The rain and bad roads so delayed the movement and exhausted the men that the attack was called off.[87]

Instead of a major assault on the Confederate line, the fighting in this area was for Myers Hill, located less than two miles east of the town and just south of the Ni River between the two lines. The back and forth fighting finally ended with the Union occupying the hill. Initially, Lee did not realize the significance of the fighting in this area. Then he received word from Anderson that Union troops had disappeared from his front. Only at this point did he realize that Grant had shifted to this flank, and he moved to bring troops from Laurel Hill to reinforce this sector while cautioning Ewell to be diligent should the enemy change his lines on Ewell's front. On May 14, Lee moved Field's division of Anderson's Corps to the right. The rest of Anderson's Corps was ordered to move to the Confederate right flank on May 15. With the open ground and artillery support in this location on the right, the Confederate line in this area was now too strong to be attacked.[88]

The rainy weather continued for several more days and effectively brought operations to a standstill. Grant told Halleck in Washington that offensive operations could begin after twenty-four hours of dry weather. By May 17, the weather had improved, and Grant was anxious to resume action. Wright at first suggested that his corps attack the supposedly weakened Confederate center. Concerned that the assault on May 12 had lacked sufficient support, Grant wanted as much force as possible and would include the Second Corps. Hancock received orders at 7:15 p.m. that evening. He was once again to move under cover of darkness from his position at Harris Farm north of the Ni River and near the Fredericksburg Road back to his original position near the Landrum House, where he had made his original assault on the salient. Afterward, Hancock was to attack at 4:00 a.m. Wright's Sixth Corps would move from its position at Myers Hill to Hancock's right flank to support the attack. Burnside, still in the same relative position he had occupied when his corps arrived at the battlefield, was also ordered to be ready to assist the planned attack from his position on the Union left flank.[89]

Once again, Grant could not execute his plan of maneuvering of troops at night before immediately going into battle. Some of Wright's soldiers were late getting into position, and Hancock delayed his attack until 4:35 a.m. This major assault on the new Confederate line would involve four divisions. Six days before, Confederate artillery had been withdrawn from their defensive position. Now, twenty-nine guns were focused on the advancing Federals. The Confederates were astonished that Ewell's strong position was being attacked, and they looked forward to settling the score from the salient. After an hour of fighting, Hancock informed headquarters that he could not penetrate beyond the abatis. It would take almost three more hours of futile struggle before the attack was called off. One member of Barlow's lead brigade described the Federal assault as a noble, absolutely hopeless attempt. Lee described the attack as easily repulsed and yielding few Confederate casualties.[90]

That night, Hancock's Corps was withdrawn from the area and ordered to Anderson's Mill near the Ni River and the Massaponax Church Road in the rear of the Federal left flank. Grant's latest costly attempt to smash the Rebels had failed completely. The Union general now had to decide his next move in this deadly chess match with Lee.

Options

On May 11, Grant had written Halleck that he intended to fight it out on this line if it took all summer. This statement indicated Grant's intentions, but not his plan for their implementation. If his words were taken literally, Grant would stay at Spotsylvania Court House and fight Lee for as long as required. Otherwise, Grant meant that he would maintain relentless pressure on Lee. As in The Wilderness, maneuvering was a viable option, but withdrawing was not.[91]

Option 1

By remaining at Spotsylvania Court House and continuing to attack Lee, Grant would be able to maintain the all-important initiative and prevent Lee from executing any major offensive operations. The Federals were now more familiar with the terrain, and their lack of good maps of the area would be less of a hindrance. Six thousand reinforcements had arrived to offset some of the losses, though they were raw and not as capable as those they were replacing. The additional stress on the troops that the large-scale maneuvers had inflicted would be eliminated. Grant felt that the men were in good spirits. Unknown to the Federals, this option's likelihood was validated by the Confederates, who were expecting to be attacked the next day, May 19, just south of the Spotsylvania Court House.[92]

The downside was that Lee still held a strong defensive position. Hancock's and Wright's May 18 assault on Lee's revised lines at the base of the salient had been a total failure. That day, Grant had also received word of Sigel's defeat in the Shenandoah Valley, and Butler had been pushed back from Drewry's Bluff. If Grant stayed at Spotsylvania Court House, Lee might receive reinforcements from the valley and/or send troops to help fight Butler.[93]

Option 2

One of the factors in Grant's move to Spotsylvania Court House was the availability of transportation options. The Telegraph Road in the rear of the Union lines ran south. Farther east, the Richmond, Fredericksburg and Potomac Railroad ran to the Confederate capital. Both offered excellent opportunities for Federal troops to relocate quickly. With Grant's previous movement, Lee had been forced to follow to counter the threat to Richmond. Only by the narrowest of margins had Lee been able to get there first. Another move south would once again force Lee to leave his earthworks.

But the Confederates also had access to roads leading south that would allow them to respond to Grant's movement. Lee's army had demonstrated its ability to both move quickly and exploit any openings a Federal move south might present. Other than being farther south, little would be gained by shifting from one Confederate defensive position to another. Finally, taking the troops south would lengthen Grant's supply line while shortening Lee's.

Decision

Thinking that Lee had been severely damaged by the assault on the salient on May 12, Grant continued to operate in the vicinity of Spotsylvania Court House. Though he had thought about moving south for several days, Grant felt compelled to try one last assault. But after the failed attempt on May 18 against Lee's new defenses at the base of the salient, he immediately ordered Hancock, commander of his best corps, to travel south toward Richmond under cover of darkness on May 19. Wright was dispatched to his former position on the Federal left. Knowing they could not wrest the Confederate entrenchments from Lee's army, the Army of the Potomac's enlisted men felt that this was the right decision.[94]

Results/Impact

Spurred by Grant's decision, both armies maneuvered to establish where the next battle would occur. Grant's initial step in implementing his decision to move south was repositioning Hancock's and Wright's Corps to the Massaponax Church Road section near the north–south Telegraph Road. Lee sensed that Grant was shifting his northern flank. He knew he needed more information about the Federals' situation, Grant's intentions, and the opportunities those two things might afford. For example, Grant's vital supply lines along Fredericksburg Road were now exposed. Lee initially intended for Ewell to demonstrate along his front to ascertain the enemy's strength. Ultimately, though, he took Ewell's suggestion to move to the enemy's right flank. Kershaw's brigade of Anderson's Corps moved up to occupy the position Ewell's troops had vacated, and Ewell moved out with six thousand troops.[95]

With no cavalry available to screen his flanks, Grant was unaware of Ewell's movement. At 1:30 p.m. on May 19, Hancock received orders to move to the south at 2:00 a.m. the next morning and vigorously attack any enemy he found. About 5:00 p.m. Grant and his staff were surprised to hear firing from their rear. Warren, closest to the fighting, was ordered to meet the attack. Hancock received orders to immediately move back to the Harris Farm on the Union far right and help counter Ewell's surprise move. Ewell attacked vigorously, but as Federal reinforcements arrived his movement stalled, and he soon found himself fighting for his survival. In such an exposed position, he retreated after dark. Ewell reported nine hundred casualties, but the poorly

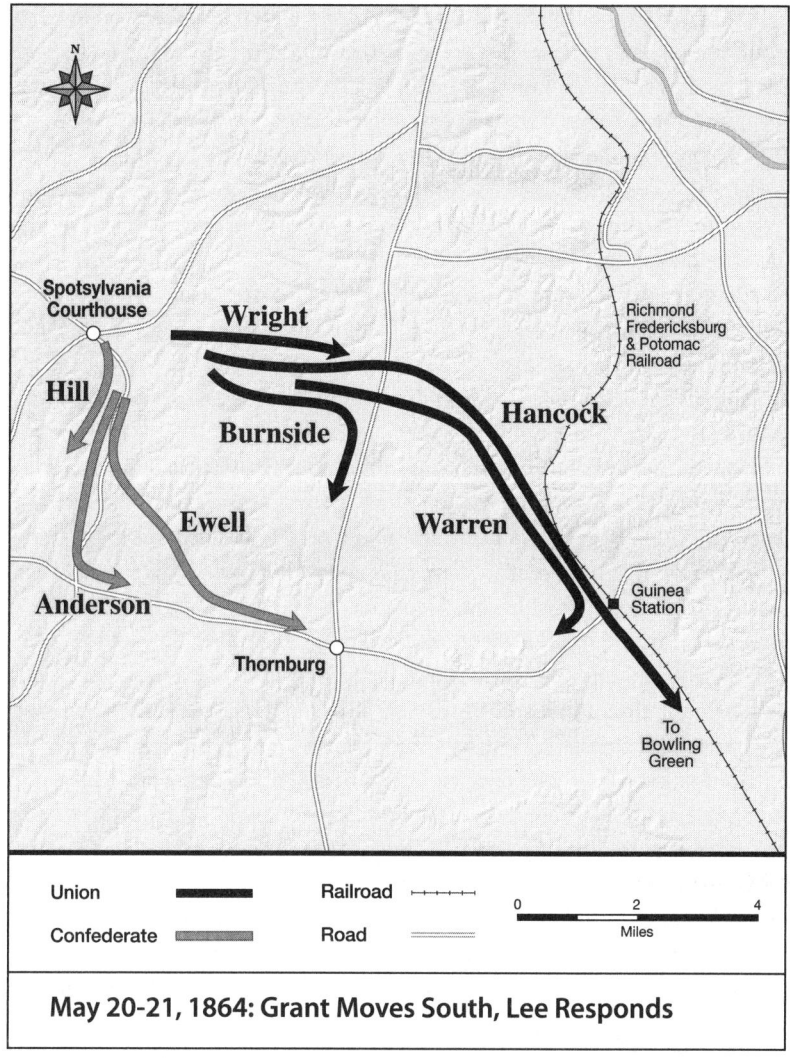

May 20-21, 1864: Grant Moves South, Lee Responds

managed affair also cost him some of Lee's confidence in his judgment. Federal losses were initially estimated at well over one thousand men, but this figure was later revised to between five and six hundred.[96]

That night, Hancock's troops in the area of Harris Farm were ordered to remain in position until the next day. The next morning, Meade informed Grant that Ewell's troops had completely withdrawn and returned to their entrenchments, and Grant would resume implementing his plan to move south. Other than the casualties, the only real result of the fighting at Harris Farm was that Hancock's movement south was delayed by one day.[97]

On the afternoon of May 20, Hancock was ordered to once again head south. He planned to leave at 11:00 p.m. that night to conceal his movement from the Confederates. Grant hoped that Hancock's maneuver would tempt Lee out of his defenses. Grant had recalled Sheridan to the army on May 17, and Hancock would now have cavalry support to clear his route. Warren's Corps was to be withdrawn in connection with Hancock's movement, and Wright and Burnside were ordered to take up revised positions east and southeast of Spotsylvania Court House.[98]

Grant was ever aware of the importance of logistics, and his movement south would require a new supply line. The current supply point at Belle Plain was to be abandoned and a new one set up farther south at Port Royal along the Rappahannock River. This location would be much closer to upcoming operations.[99]

Lee was not sure of Grant's overall intentions on May 20, but his movements forced him into a delicate juggling act with his troops. Confederate general P. G. T. Beauregard had submitted a plan to Jefferson Davis to unite Lee's army with his own army. The combined forces would face and defeat Butler, then move against Grant. Gen. Braxton Bragg, now Davis's military advisor, had voiced several objections to the plan, including the fact that Federal cavalry could get between Lee and his sixty-miles-distant destination, giving Grant time to crush him. Bragg also noted the possibility of losing essential supply lines.

Beauregard's plan was rejected, and Lee intended to keep Grant as far as possible from Richmond. With the enemy no longer on his front, Ewell was ordered to move south at daybreak on May 21 and take up a position on the Confederate right flank. Ewell would then be situated to block Grant if he tried to move down the Telegraph Road. Lee's contingency plan would be to fall back to the North or South Anna Rivers. By 8:40 a.m. on May 21, Lee was aware that Grant was changing his base, and he would regulate his movements by Grant's. With Ewell moving to block the Telegraph Road at Mud Tavern, Grant's gambit to use Hancock's Corps as bait to draw Lee out and attack him with Warren's Corps down this same road ended.[100]

The race to the next battle had begun.

CONCLUSION

Lincoln had selected Grant because he believed Grant had the resolve to fight and thereby win the war, something that had seemingly been lacking in the East. After over two weeks of horrific fighting, this resolve was as strong as ever. It had shaped Grant's decisions thus far in the campaign and would continue to do so—he would maneuver south and assault Lee's army whenever and wherever possible. With Grant ordering Hancock's movement south on May 20, fighting at Spotsylvania Court House had effectively ended, but Grant's relentless campaign had not. Starting with the fighting in The Wilderness on May 5, the two armies would stay in constant close contact until the surrender at Appomattox on April 9, 1865. Inactivity between the two armies during this time was rare.[1]

The fighting at The Wilderness and Spotsylvania Court House must be viewed in the overall context of Grant's Overland Campaign of 1864. A microcosm of the campaign was the fighting at the Mule Shoe on May 12; the South was fighting with desperation, the North with determination. By the time Grant's army crossed the James River on June 12 and the Siege of Petersburg and Richmond began, his losses approximated fifty-five thousand. In just over one month of command, Grant lost roughly the equivalent of Lee's entire army at the beginning of the campaign. Once the Federals crossed the Rapidan, all subsequent critical decisions would contribute to these casualties.[2]

Two key factors impacted these high casualty numbers: First, Grant remained determined to continually hammer at his main objective—Lee's army.

While in the West, Grant had not encountered forces with the capability of these Confederates. Grant's often impulsive style of generalship further impacted the casualties. At times, he based his spur-of-the-moment decisions on inadequate information about Lee's position, as at the Po River on May 9–10. A recurring theme in Grant's decision-making would be the essence of his May 5 orders for Meade to attack Lee if an opportunity presented itself, and to "do so without giving time for disposition." Grant impulsively ordered attacks against well entrenched Confederate positions, and with little to no knowledge of Lee's troop deployments. He correctly assessed the tip of the Mule Shoe as its weakest point, but provided no time for reconnaissance or preparation. It was lucky for the Federals that the Confederates had withdrawn their artillery at a most opportune time. Otherwise, Union casualties would likely have been higher. Only the substantial resources of the North were able to sustain this style of fighting.[3]

Grant was aware of the political aspect of the war and the impact of Northern public opinion. Unlike his predecessors in the East, he could now hear the 1864 presidential election clock ticking. Lincoln had selected Grant because he was a relentless fighter. Grant's use of the phrase "if it takes all summer" in his May 11 report to Halleck made headlines in the New York papers. It also energized Northern public opinion that the war would be won.[4]

The fighting in Virginia in May 1864 would validate Lincoln's assessment of Grant as a general who would fight. Now the war's two best-known generals would be facing each other. Grant was no longer fighting against "Joe" Johnston, who he thought would have retreated after the two days of fighting at The Wilderness. In Grant, Lee had also found an adversary who would not back down. Fighting mostly on the defense, Lee had incurred fewer casualties than Grant, but Lee's losses were basically irreplaceable.[5]

A common link between critical decisions is a combination of the situations that made them necessary and the magnitude of the decisions' results. Other less obvious contributors to the decisions' final impact are the ways in which they are implemented, the nature of the expectations, and the overall elusive element of luck. The dynamics of these factors would have a significant influence on the decisions' results. In Grant's first month of commanding the Federal forces confronting Lee, his expectation of smooth coordination between the four infantry corps was unrealistic. The personal, political, and combat heritage of these two Union armies had instilled in them a more cautious and respectful attitude toward their foe. Grant's aggressive instincts were contrary to the previous leadership styles. Warren felt that Grant's demand for a hasty attack at Saunders Field had needlessly sacrificed his troops. One officer in Warren's Corps stated that Grant's approach was "the beginning of a reckless . . . way of fighting battles by hurrying into action." Although he handled his duties professionally, Meade's "position was in some measure a false one" in which he functioned more as a staff officer than an army commander. And given Burnside's seniority and political ties, there seemed to be little that could be done to change his style.[6]

Much of the major marching of the campaign was done at night to hide the movements from the enemy which complicated any movements and were physically demanding on the troops. The campaign started with Warren's midnight movement across the Rapidan. Grant in particular seemed inclined toward night movements concluding with the troops immediately moving to attack. Warren left after dark to make his march to Spotsylvania on May 7 and went directly into fighting. Hancock moved after dark on May 11 for the assault on the salient at 4:35 a.m. the next morning. Grant attempted to move against Lee's new defensive line at the base of the Mule Shoe, and he had Wright's and Hancock's Corps leaving at night and initiating a morning assault.[7]

Decisions were made at various levels, ranging from President Lincoln, to army commanders, corps-level commanders, and to Colonel Upton's brigade command decision. Based on the outcomes, did the eighteen critical decisions have the intended results? Nine would have the desired impact, three would produce mixed results, and six missed the hoped-for outcome. These decisions will be organized into these three categories and discussed below.

Decisions That Achieved Intended Results

Lincoln Decides to Nominate Grant

In nominating Grant for the rank of lieutenant general, Lincoln's long search for a general "who could grasp the whole field and get out of the Army what he knew was in it" had finally paid off. Lincoln and Grant quickly developed a strong trust in and appreciation of each other. Lincoln had seen Grant's potential early on, and when the situation was looking bleak for Grant's Vicksburg Campaign in 1863, Lincoln stated that what was wanted and needed was a general who would fight. Grant had fought, and the president would stand by him. After the intense fighting in The Wilderness, Lincoln declared that if any other general had been leading the army, "it would have now been on this side of the Rapidan." Grant's tenacity, Lincoln correctly stated, would lead the Federals to victory.[8]

Grant understood and respected Lincoln's thoughts and policies for pursuing the war. He did not lecture Lincoln on military matters as some of his predecessors had attempted to do, but he would share his overall thoughts, thus gaining Lincoln's wholehearted trust. In return, Lincoln gave Grant unprecedented autonomy and diligently worked to meet his requests.[9]

Grant's overall strategy of having the various parts of the Federal army work together against the Rebels revealed his grasp of the whole picture of the war. This approach also proved to be the correct one. Sherman's capture of Atlanta in August 1864 would turn Lincoln's faltering presidential campaign into a second term in office.

Grant Decides to Locate with the Army of the Potomac

Grant's physical presence with Meade and the Army of the Potomac was the best option for command. It had the dual benefits of keeping Grant out of Washington, and providing him the means of change the character of the Army of the Potomac into a relentless fighting force. Being stationed with the army allowed Grant to quickly respond to the ever-changing military situation and provide his troops visible leadership and encouragement. Unfortunately, Grant's presence limited recognition of Meade's military contributions. The unflattering newspaper treatment of Meade only compounded this situation.

Lee Takes the Initiative and Attacks

Lee was aggressive by nature and had been described as an audacious commander. In March, Lee had confided to Longstreet, who was still in Tennessee, his concern about Grant's operating with the Army of the Potomac and the concentration of Federal troops "in this region." Of this eventuality Lee said, "We shall be obliged to conform to their plans" and move to "wherever they are going to attack us." In all previous fighting in Virginia, the Federals had retreated when Lee's Army of Northern Virginia had taken the offensive. The basic difference this time was Grant and his reaction. In the clash of the two commanders' aggressive tendencies, Lee's move obliged Grant to respond in the manner most advantageous to the Confederacy. Battle was inevitable, and Lee forced the fighting as far from Richmond as possible.[10]

Grant Decides to Move South from The Wilderness

Grant's moving the Union army south from The Wilderness was the first tangible demonstration that the character of the war had changed. This shift in location validated Lincoln's assessment that Grant was willing to continue fighting and do what was needed to win the war. Grant's movement also sent a strong message to the officers and men of both sides that henceforth there would be no letup to the Federal army's pressure on Lee and his Army of Northern Virginia.

Anderson Decides to Push On Through the Night

In the race to Spotsylvania Court House, minutes would literally make a difference. Luck can dramatically change the outcome of a decision, but the decision still has to be made. Anderson's determination to leave at 10:00 p.m. rather than at Lee's start time of 3:00 a.m. and then to keep moving was critical to disrupting Grant's move to Spotsylvania Court House. Anderson's choice had put his troops in a position to create a strong defensive line at Laurel Hill. His quick and decisive response to Stuart's request for help ensured the best outcome for the Confederacy.

Fitzhugh Lee Decides to Fight a Delaying Action

Coupled with Anderson's decision, Fitzhugh Lee's decision to fight a delaying action along the Brock Road resulted in one of the most dramatic outcomes of the campaign. Fitzhugh Lee's actions gained the Army of Northern Virginia time in its movement to Laurel Hill, and this time was vital in developing the defensive position that Grant's maneuver had forced the soldiers into.

While Sheridan's troops were pulled back and briefly resting at Todd's Tavern on the night of May 7–8, Fitzhugh Lee's troops used this time improving their makeshift defensive position. The limited opportunity for artillery in the fighting was effectively capitalized on by the Confederate Horse Artillery to add firepower to the outnumbered Confederates. Gen. Robert E. Lee realistically could not have hoped for a better outcome.

Upton Implements a New Attack Strategy

Upton's attack strategy clearly demonstrated that this tactic could penetrate an entrenched enemy in the right situation. The twenty-four-year-old Upton's intense ambition to become a general no doubt contributed to his enthusiasm for the opportunity and his meticulous preparation for the offensive. He reviewed the ground to be covered with regimental commanders and explained exactly what was to be done in the assault. The fact that the twelve regiments participating in the attack had been specifically picked for the assignment—they were top units—further contributed to the decision's qualified success. Grant incorporated the concept of a quick, concentrated direct thrust at the enemy into his future plans. But he ignored two critical elements of the initial success of the attack: the thoroughness of Upton's preparations and the uniqueness of the location Lieutenant Mackenzie had found.

Grant Decides to Move South from Spotsylvania Court House

For the North to win the war, Lee and his Army of Northern Virginia would have to be defeated. Grant made the right decision to move south from Spotsylvania Court House; it was consistent with his approach to the war and his resolve to see it through. The basic situation of not achieving the desired results in the fighting at this location by maneuver or direct assault merely meant changing location, not approach. Grant could change his supply lines, and the necessary material to wage war would continue uninterrupted and undiminished. Although his supply lines were being lengthened as Lee's were being reduced, Federal control of the extensive waterways in this part of Virginia offered multiple avenues to easily transport the needed supplies. With the return of Sheridan's Cavalry of approximately 10,000 troopers to the army, the ability to break up "the railroads on which General Lee would more or less depend . . ." existed.[11]

Decisions That Produced Mixed Results

Grant Decides on the Leadership of the Army of the Potomac

The two tightly coupled components of this decision, Meade and Sheridan, yielded mixed results for Grant. Retaining Meade would provide consistency and validate him in the troops' eyes. For Meade's soldiers, the value of the commander who had led them to victory at Gettysburg and made the hard decision not to needlessly sacrifice them at Mine Run was confirmed. The men had pulled back from the Mine Run line with confidence in their leader's decision. They had then gone into winter camp in good spirits, appreciating that Meade had taken care of them before the current campaign. Meade had already put in motion an organizational change from five to three corps, and he understood the impact it would have on the affected units' morale and pride. Maj. Gen. John Pope's personality and the results of his leadership had fueled the troops' lingering resentment for a western commander. Meade had been with the Army of the Potomac since its inception, and the men considered him one of their own. The benefit of his insight into the unique personalities of critical commanders became evident on the first day of fighting in The Wilderness, when he intervened with Grant on Griffin's behalf.[12]

As with many of Grant's decisions, his selection of Sheridan to head the Army of the Potomac's cavalry and, by implication, Wilson to be a division commander, seems almost impulsive. In responding to Halleck's suggestion of Sheridan, Grant appears to have set only two criteria—Sheridan's enthusiasm and the fact that both Grant and Halleck knew Sheridan. Halleck had no direct field leadership of any of the men in the affected Army of the Potomac. Rather, he only lacked confidence in Meade. Sheridan did bring positive results to upgrading the condition of the cavalry. But his inexperience and his ambition and friction with Meade would provide minimal positive results in the campaign's upcoming opening operations.

An intriguingly different scenario would likely have evolved if Grant had put the energetic infantry commander Sheridan in command of the Ninth Corps rather than the Cavalry Corps. Grant had not worked closely enough with Burnside in Tennessee to appreciate his limitations. With Burnside's seniority and political clout, it would have been difficult to appoint someone else in his place. But Sheridan probably would have exploited the opportunities that Burnside missed.

Grant Determines How He Will Attack Lee

Grant gets a positive mark for his decision to move on Lee's right flank, but a negative mark for that decision's implementation. By choosing this line of attack, Grant would have excellent supply lines to provide the necessary material for his massive army. Possible lines for the Confederates to threaten Washington would be blocked, and Grant would be in a position to dictate the flow of the upcoming fighting. Humphreys had developed an excellent operational plan that would accomplish Grant's goals if followed.

For his plan to succeed, Humphreys envisioned a quick negotiation of The Wilderness and a move to more open country that would facilitate the Federals' use of their superior number of troops. But the progress of the large supply train necessitated stopping. When Ewell's troops suddenly appeared along the Orange Turnpike, Grant ordered his men to pitch into them without regard to disposition, stopping his advance. Grant had neither intended nor wanted to fight in The Wilderness, where the terrain favored the Confederates. Ironically, by stopping and fighting here, Grant was doing exactly what Lee wanted.[13]

Grant's response to Lee demonstrated that his overall objective was making the Army of the Potomac more aggressive, regardless of the cost in casualties. The fighting ended in a stalemate. Grant moved on without accomplishing any advantage, only 40 percent more casualties than he had incurred at the two-day Battle of Shiloh.[14]

Grant Decides How He Will Use His Corps

Hancock's was Grant's best combat corps, and using it as his main strike force made sense. Hancock was prompt in obeying orders and aggressively worked to carry them out. He had good division commanders and good esprit de corps. But Grant was unable to provide the associated tactical support needed to fully realize the Second Corps' potential in this role. Without an effective cavalry presence to scout and provide good intelligence on enemy positions, Grant acted on his hunches. He would move this corps from one side of the battlefield to the other and then immediately send it into battle. In the most notable instance of this tactic, Hancock's Corps carried out a movement on the evening of May 11 and attacked the salient at 4:35 a.m. the next morning. This maneuver of moving at night and attacking at dawn was repeated during the May 18 assault on the revised defenses at the base of the Mule Shoe. But the Second Corps paid a heavy price for Grant's critical decision. After starting the campaign with twenty-seven thousand troops, the Second Corps reported only sixteen thousand troops at its May 13 muster—a 40 percent reduction in manpower.[15]

Grant Decides to Commit a Full Corps

The convergence of several critical decisions resulted in the defining moment of the Battle of Spotsylvania Court House—the May 12 dawn attack on the Mule Shoe. Grant had made a good decision based on the results of Upton's attack two days earlier. However, significant criteria that had allowed Upton's success were not met in Grant's attack. The attacking Federal troops were closely packed together, and they were extremely fortunate that most of the Confederate artillery had been withdrawn from the Mule Shoe during the night.

Grant's choice of Hancock's Corps to lead the assault and his timetable for the attack forced Hancock's troops to move into position on a rainy night. When they arrived at the Brown House to initiate the dawn assault, they were tired and wet. Accurate knowledge of the ground to be covered and the enemy position was virtually nonexistent, and the men had no time to determine exactly how the attack was to be executed.

Decisions That Did Not Produce Positive Results

Lee's Decision to Leave Troops in Place on the Night of May 5–6

Grant and Lee were still in the early stages of replacing perceptions of one another's reputations with assessments of their individual abilities and military styles based on actual head-to-head combat. Their past aggressive tendencies should have been a good indicator. The decision to leave the Confederates in place on the night of May 5–6 was predicated on Longstreet's predawn arrival. As the timing of Longstreet's arrival became more uncertain, the original decision should have been quickly changed and the men withdrawn.

The Confederates were in no position to withstand Hancock's assault, and the destruction of Hill's Corps was avoided by the narrowest of margins.

Lee Decides to Press the Attack

Lee and the Army of Northern Virginia found themselves in a problematic situation—they faced a larger force and had limited opportunities to strike a decisive blow to the enemy. Longstreet's flanking attack had been very successful but not totally decisive. Lee quickly chose to press the attack, but it took valuable time to regroup and align the tangled regiments and their commanders into a cohesive force strong enough for any chance of success. The Federals effectively used this time to improve their own defensive position. Now the factors that had led Lee to fight in The Wilderness played against him. Lee's decision to press the attack would cost him in the battle of attrition but leave him with nothing to show for his attempt.

Ewell Decides to Wait

By delaying, Ewell wasted the opportunity presented to him. Stonewall Jackson, Ewell's former corps commander, had translated Lee's sometimes discretionary orders to him into specific ones for Ewell with little latitude. Ewell was unprepared to fill this translation void, and the decisive leadership that Lee needed from him was fading. Ewell was mistaken in failing to verify Early's and Gordon's conflicting reports. Had he done so, he would have been prepared to make an informed decision rather one based on his emotional reliance on Early's opinions. Ewell's swings in leadership ability and emotional steadiness would eventually lead Lee to place him on medical leave. Lee would have to summon all of his tact to keep Ewell from returning to a command position in the Army of Northern Virginia.

Grant Agrees to Sheridan Pursuing Stuart

This apparently impulsive decision by Grant was perhaps his biggest mistake in the early days of his Overland Campaign. Eager to bring a more aggressive attitude to the army and personally fond of Sheridan, he overlooked his need for cavalry to gather information on the movements of Lee and his army. Grant never clearly articulated what he wanted from his cavalry. In addition, he had only a vague idea of what Sheridan might accomplish on his adventure. The Federals' notoriously inaccurate maps of the area where the infantry would be operating only compounded Grant's problem. However, effective cavalry could help alleviate this situation. Despite Sheridan's glowing reports to army headquarters, the cavalry's actual accomplishments under his command had been limited. In his zeal to chase after Stuart, Sheridan took all three of his cavalry divisions, while Stuart left Hampton's cavalry back with the army to continue to work with Lee.[16]

In the upcoming fighting at Spotsylvania Court House, Grant would have to use infantry to try and assess the strength of Lee's flanks. In the absence of good intel on the enemy, Grant resorted to making uneducated guesses about where Lee might be shifting troops from. The Union general was unable to detect the actual weaknesses in Lee's defensive line. Grant was more than generous with his friend when he said that Sheridan had "accomplished more than expected" in the sixteen days that he was absent from the Army of the Potomac. The most publicized result of Sheridan's movement was the death of "Jeb" Stuart at Yellow Tavern southeast of the main fighting at Spotsylvania Court House. But the Army of Northern Virginia maintained the quality of its cavalry leadership when Maj. Gen. Wade Hampton replaced Stuart.[17]

Warren Decides to Redeem His Reputation

In Warren's initial attempt to restore his reputation, his decision to attack the rapidly developing Confederate defenses without sufficient force to seize the junction failed. His subsequent efforts continued revealing the tension between the reality of Laurel Hill's invulnerability (it had artillery and an open field of fire) and his desire for the army's leadership to judge him an effective commander. Warren had been appalled by the mauling his troops had suffered a few days earlier at Saunders Field in The Wilderness, as well as by Grant's tactics. Warren's reports had a lecturing dimension that diminished his credibility with Meade and Grant, even persuading them that future failed attacks against Laurel Hill resulted from Warren's lack of commitment rather than an extremely strong enemy position. Warren needed someone else, like Humphreys, to verify and communicate to Grant the strength of the Confederate defenses at Laurel Hill. Warren also needed someone to craft a different approach for this location than direct assaults.

Lee Decides to Withdraw the Artillery from the Salient

Three factors contributed to Lee's decision to withdraw the artillery from the salient—his worst decision of the fighting. First, while previously reliable sources had provided Lee conflicting reports of Grant's intentions, they all indicated a Federal flanking movement. Grant had almost succeeded in stealing a vital march to Spotsylvania Court House on the night of May 7–8. That could not be allowed to happen again, so Lee was forced to consider troop dispositions that allowed a quick response. Second, the recurrent rains had turned the limited number of farm roads and trails into muddy traps that would greatly impede troop movements. The artillery's mobility was paramount; the weapons and troops were a vital part of any attempt to stop the anticipated Federal movement. Yet the artillery would have an especially difficult journey in these conditions. Finally, bad luck factored into Lee's choice. Just as Lee's removal of the artillery was a lucky break for Grant, Grant's plan to assault the precise location of that withdrawal ensured that his decision to remove the artillery was a huge mistake. Only the Confederates' vigorous and costly counterattacks to hold the apex of the salient while a new defensive line was constructed prevented the destruction of Ewell's Second Corps.

Afterward

The initial fighting in Grant's Overland Campaign would necessitate command changes in both armies. Grant had considered the problem of Burnside's command for some time, and on May 24, he ordered Burnside and his corps placed under Meade to improve the coordination of the troop movements. Burnside thought the idea was excellent. He graciously accepted being under Meade, who at one time had been his subordinate.[18]

Though not as obvious a change as Burnside becoming Meade's subordinate, the seemingly good working relationship between Meade and Grant was deteriorating. On May 19, the day after the failed attempt on Lee's revised defensive line at the base of the salient, Meade wrote to his wife of his discontent. He informed her that if there were any honorable way to retire from what he considered a "false" position, he would take it.[19]

None of the four corps commanders that crossed the Rapidan with Grant at the beginning of May would make it to Lee's surrender at Appomattox Court House on April 9, 1865. Sedgwick had been killed at Spotsylvania. Following the Battle of the Crater during the Siege of Petersburg, Burnside would be relieved of command in August 1864. Experiencing lingering effects from his wound at Gettysburg, Hancock would step down from command of the Second Corps in November 1864. Finally, Warren would be relieved of command of the Fifth Corps at the Battle of Five Oaks on April 1, 1865, just over a week from the war's end at Appomattox Court House.

Changes for the Confederates were a bit more substantive. Anderson would stay in charge of the First Corps until Longstreet returned to the army in October, when Richmond and Petersburg were under siege. When Hill became too ill to command, Early temporarily led the Third Corps from May 8 to May 21. When Ewell was relieved due to illness in late May, Early replaced him. Gordon was promoted to the rank of major general and put in command of Early's old First Division. In late 1864, Gordon would replace Early as corps commander of the Second Corps, and he would go on have a prominent role at Appomattox. Lee's own health problems would come into play at North Anna, where he was ill and confined to his tent. Although he wanted to strike at Grant, Lee was not as effective in his tent as he was at the front, and the Confederates landed no real blow to the Federals.[20]

At North Anna and Cold Harbor Grant would repeat the offensive strategy and operational struggles he had initiated in his Overland Campaign at The Wilderness and Spotsylvania Court House. In leaving Spotsylvania Court House, Grant would again find his attempts to get between Lee and Richmond thwarted. With a more direct route down the Telegraph Road, Lee continued to get ahead of the Federal army and established a good defensive position at the North Anna River. Federal troops once again encountered miscues in their movements south. Hancock's move to draw Lee out had only separated him from the rest of the army. When Wright's Corps tried to head south, the road was "blocked ahead by General Burnside's column." Finally, Warren complained that he had had his orders changed three times in one night, sending him in different directions. With his own army split into three sectors and Lee maintaining a strong defensive position at North Anna, Grant wisely decided not to attempt a direct attack. He instead repeated his turning movement, this time moving south to Cold Harbor.[21]

The Federals missed an opportunity on June 2 at Cold Harbor. As at Spotsylvania Court House, a decision was made to have Hancock march his troops twelve miles overnight for an attack the next day. Concerned about these troops' fatigue and thus their ability to support the planned attack, Grant delayed the operations till the next day. Once again, Lee's army quickly established strong defensive works. Grant ordered a major assault through the darkness and fog at 4:30 a.m. on June 3, a decision reminiscent of his decision to dispatch an entire corps at the salient less than a month earlier. This time, the Rebels were ready and wielding artillery—they had had thirty-six hours to entrench.

This ill-conceived attack on the Confederate defenses cost Grant dearly. There is inherent uncertainty in the number of losses in Civil War battles and even more uncertainty about the figures for specific parts of battles. One estimate places the initial assault losses at 7,000. More recent works estimate 3,500 Union casualties in the initial morning charge and a total of 4,500 for the day. Adding to the tragedy, it would take four days to reach a temporary truce to remove the wounded from the field. It took two days for the Federals to initiate a proposal to Lee to remove the wounded. Lee desired that a "flag of truce in the usual way" be used, and Grant accepted this arrangement for the evening of June 6. The subtleties in the correspondence between the two forced Grant to admit what every soldier in the opposing armies already knew—that he had been defeated. Only a few of the recovered casualties survived, while thousands had needlessly died. Years later, as he was succumbing to his battle with cancer, Grant wrote that he had always regretted making the last assault at Cold Harbor: "At Cold Harbor no advantage whatever was gained to compensate for the heavy loss we sustained." He made no mention of waiting so long to ask for a truce to remove the wounded.[22]

Writing to his wife after the attack, Meade talked about how severe the thirty days of "protracted" fighting had been. He felt the campaign so far had vindicated his "judgment" not to attack at Williamsport after Gettysburg and at Mine Run. "I think Grant has had his eyes opened, and is willing to admit now that Virginia and Lee's army is not Tennessee and Bragg's army," he stated.[23]

The Battles of The Wilderness and Spotsylvania Court House were the opening acts of the Overland Campaign drama, and they set the common theme for the fighting. With Lee's army as his objective, Grant would move south and initiate a battle that ended in stalemate. Grant would again maneuver south, forcing Lee to follow, and the result was in the battles just mentioned. These engagements finally ended in the Siege of Richmond and Petersburg. The Union accomplishments in the East would be harder and more time consuming to attain than those in the West. Grant had said that Lee was no Joe Johnston, and Meade had said Lee was no Bragg. Finally, getting to the Siege of Richmond and Petersburg, Lee would demonstrate how much more difficult it would be to achieve progress while fighting in Virginia than in the West. While the Siege of Vicksburg in the Western Theater lasted just under two months, in the East the Siege of Richmond/Petersburg would last ten months. Leading up to the Siege of Vicksburg, Grant's Union casualties in the Battles of Port Gibson, Raymond, Jackson, Champion Hill, and Big Black River the previous May numbered fewer than 4,500, a mere 8 percent of the 55,000 casualties it would cost to get to the Siege of Petersburg/Richmond.

The two main figures of the campaign, Grant and Lee, would at different times make a simple, straightforward comment characterizing the determination of the North and the desperation of the South. Grant wrote his often-quoted remark that he would continue to "fight it out" even if it took all summer defined the steadfastness of the Northern cause. Even with all the fighting that remained, he would not deviate from this resolve.

Lee's comment defining the desperation of the South came on June 3. Confederate postmaster general John Reagan had ridden out from Richmond to see the battle at Cold Harbor. While Grant's forces were vigorously attacking, Reagan asked Lee, "General, if he breaks your line, what reserve have you?" Lee replied, "Not a regiment. And that has been my condition ever since the fighting commenced on the Rappahannock." (Note that the Rapidan River, where Grant's forces crossed on May 4, runs into the Rappahannock River a few miles upriver of Fredericksburg.)

The critical decisions made in the spring of 1864 resulted in the Battles of The Wilderness and Spotsylvania Court House. The names of Saunders Field, Widow Tapp Farm, Laurel Hill, Mule Shoe, and Bloody Angle would be added to the Civil War lexicon.

APPENDIX I

BATTLEFIELD GUIDE TO THE CRITICAL DECISIONS AT THE WILDERNESS AND SPOTSYLVANIA COURT HOUSE

As part of the overall series format, an extensive battlefield tour guide has been incorporated into the appendix. This in-depth content serves a dual purpose: allowing readers unable to visit the battlefield to gain a better feeling for the actions that occurred here and giving those who can visit a deeper understanding and appreciation of those events.

These two battlefields only have exhibit shelters, so the tour begins at Chancellorsville Visitor Center, approximately 12 miles west of Fredericksburg at 9001 Plank Road, Spotsylvania Courthouse, VA 22553. Check the National Park Service's website for current hours and contact information. Although the center focuses on the Battle of Chancellorsville, it provides patrons access to books about the Battles of The Wilderness and Spotsylvania and to staff who are knowledgeable about what occurred here. The visitor center can also provide current conditions on the battlefield's trails and roads, as well as information concerning interpretive talks presented at various sites. Every half hour an interesting twenty-two-minute film on the Battle of Chancellorsville is aired. As an added benefit, this film provides an excellent backdrop for the emotions of those who had fought here the previous year as

Appendix I

they once again moved into the area to confront one another. Check to get brochures for the Gordon Flank Attack Trail, Wilderness Crossing Trail, and Spotsylvania History Trail walking tours, as there may not be any at the stops.

No matter how much people read and study the Civil War, they can only gain insight and feeling for the terrain where the decisions and fighting occurred by walking the battlefields. This appendix provides an in-depth battlefield tour that will place you as close as possible to where the critical decisions were made or implemented. Where appropriate, primary source material such as the *Official Records* (*OR*) provides insight into historical events and the people who participated in and witnessed them. Gray text frames set off these passages from the rest of the information, making it easy to skip or come back to them as your tour schedule allows.

Ideally, the tour would progress chronologically, but the battles' events did not take place with regard to a tour. It would be impractical to drive back and forth across a battlefield to maintain chronological order; thus the stops are organized geographically. For example, at Tour Stop 3, the initial fighting on the first day and the last fighting on the second day at the Battle of The Wilderness occurred in the same area. The stop is therefore broken down into parts *a* and *b*. In some instances, like at Lee's headquarters at the Widow Tapp Farm, where various decisions were reached, action resulting from other decisions occurred at the junction of the Orange Plank and Brock Roads at roughly the same time. These decisions are addressed in two separate stops.

Driving and walking instructions immediately follow the background discussion. These will guide you through the battlefields with simple, GPS-style directions. Walking instructions to guide you around the battlefields appear in bold type. The discussion of each stop addresses Relevance to a critical decision(s), orientation and initial troop deployment, relevant military operations, and, if appropriate, additional sites of interest nearby.

At times, traffic can be very heavy and sometimes difficult to see due to seasonal foliage. Use caution at all times.

Any battlefield is the site for fluid and evolving series of events. Therefore, when looking at the maps in this appendix, keep in mind that they represent a brief snapshot in time of many unit movements. While you read, they are meant to provide a frame of reference to help you better visualize the events that occurred at each stop.

Maps in books typically have North pointing up, but at times the actual terrain dictates another orientation. Always check for North as you orient any map to its corresponding location. This is also important when viewing maps on

Park Service interpretive markers—for example, the marker at the Chewning Farm stop. Using battlefield interpretive plaques or other prominent items, each stop provides instructions for where to stand and how to orient yourself with the terrain and the appropriate map. The Battlefield America Civil War map series map number 106, *Fredericksburg and Spotsylvania National Military Park*, used in conjunction with the National Park Service's *Fredericksburg and Spotsylvania* map provides an overall perspective of the locations.

Of the eighteen critical decisions, the four predating the battle were made far from the actual fighting here, and they are briefly summarized along with the other decisions. Grant likely decided to move south from The Wilderness at his headquarters at Tour Stop 2. Even if the specific location was known for Grant's decision to move his army south for the second time and continue fighting and his decision to let Sheridan chase "Jeb" Stuart, visiting these places would provide no additional insight. The actual locations had no bearing on the decision-making process. The critical decisions are summarized as follows:

1. Lincoln Decides Grant Will Lead the Union Armies

After receiving reliable information that Grant had no political aspirations and fully supported his presidency, Lincoln nominated Grant for the revived grade of lieutenant general. The nomination was quickly approved, and Grant, now the highest-ranking officer in the Union army, would bring his style of fighting east and lay the foundation for the defeat of the Confederacy.

2. Grant Decides on the Key Commanders of the Army of the Potomac

The Army of the Potomac was the largest Union army, but Grant felt that it had not fought up to its capabilities, and he decided to reshape it. Located in the East and closest to the two capitals, the army was subject to the most scrutiny. It had come under increased criticism as a result of Maj. Gen. George Meade's inactivity following his army's victory at Gettysburg the previous year. To the surprise of many, Grant decided to retain Meade as commander. Even more surprisingly, Grant brought Maj. Gen. Phil Sheridan east with him to take charge of the Army of the Potomac's cavalry. The difference in these three men's personalities and fighting styles would create problems in the upcoming campaign.

3. Grant Decides to Locate with the Army of the Potomac

While former general-in-chief Maj. Gen. Henry Halleck had stayed in Washington, Grant located his headquarters with General Meade and the Army of the Potomac. This unprecedented decision would allow Grant to immediately implement his plans for waging war and reshape the Army of the Potomac's fighting character.

4. Grant Decides on an Offensive Operation against Lee

Grant needed to determine specifically how he would "get at Lee." For this massive undertaking, Grant moved against Lee's right flank and tried to maneuver him into more open ground where Union troops could most effectively utilize their large advantage in manpower and materiel. The major factor in his decision was logistics—the ability to not only undertake the campaign but sustain it. This decision would keep Grant in close proximity to the Army of Northern Virginia for the remainder of the war.

5. Lee Decides to Take the Initiative

Hoping to duplicate his previous year's success at Chancellorsville, Lee let the Federal army cross the Rapidan River unopposed. Confederates instead launched an assault in The Wilderness, where the terrain would negate some of the Union army's numerical advantage. Lee would attack with his two closest corps, Ewell's and Hill's, holding Grant in place until Longstreet could come up and deliver a decisive blow. This fighting would be the first in Grant's Overland Campaign, and it would set the stage for the most brutal fighting the war had seen.

6. Lee Decides to Leave His Troops in Place

As night fell on May 5, the two divisions of Lieut. Gen. Ambrose. P. Hill's corps along the Orange Plank Road, those under Maj. Gens. Henry Heth and Cadmus Wilcox, fell exhausted where they lay. The divisions' lines were so jumbled from the fighting that "a skirmish line" could drive both of them back.[1] To re-form the units, the men would have to be withdrawn from the area so dearly taken while the enemy was only yards away. Expecting Longstreet's Corps to be up before dawn and assuming that Heth's and Wilcox's men would not be engaged the next morning, Lee decided to leave the men in place. Unfortunately for the Confederates, Longstreet would be late, and Hancock's 5:00 a.m. attack would almost destroy Hill's Corps.

7. Lee Decides to Keep the Initiative

After his successful flank attack along the Orange Plank Road on the afternoon of May 6, Longstreet was wounded and taken from the field. Rather than pull back to a better position, Lee decided to press the brief advantage the flank attack had created. But with the Confederate troops in confusion, it took hours to re-form them and resume the attack. Meanwhile, Hancock's troops improved their position along the Brock Road. When the attack was finally renewed late in the afternoon, it would be one of the Army of Northern Virginia's last major corps-size offensives of the war. The attack fell short of breaking the Union line and only increased the number of casualties.

8. Ewell Decides to Delay a Critical Flank Attack

One of the rising stars in the Army of Northern Virginia was Brig. Gen. John B. Gordon, who commanded a brigade in Ewell's Corps. On the night of May 5, Gordon was ordered to move his troops from the south side of the Orange Turnpike to the extreme left of the Confederate line. On the morning of May 6, Gordon's reconnaissance revealed that he was positioned beyond the Union flank and proposed that he attack. While Gordon personally verified the situation, Maj. Gen. Jubal Early, his immediate commander, incorrectly believed that Burnside's Ninth Corps was in the area. When Gordon presented his plan that morning, Early persuaded Ewell not to authorize the attack. However, Early had done nothing to verify his own information. The flank attack was finally authorized late in the evening. The initial offensive was quite successful and caused a great deal of alarm in the Union high command. Since Ewell had authorized the attack so late, darkness limited what might have been accomplished. Had Gordon's plan been implemented earlier and properly supported, the Union right flank could have been severely damaged.

9. Grant Decides to Move South

As the fighting died down on the evening of May 6, the battle had been fought to a draw. Neither side had achieved what it had hoped. Grant had been unable to bring his soldiers' numerical advantage to bear, and the flaws in the Union army's new organizational structure were showing. Grant felt that he could not gain an advantage by continuing the fighting in this location. In all its previous engagements with Lee's forces in Virginia, the Union army had pulled back to regroup and refit. Lincoln's earlier comment "[I can] not spare this man [Grant] because he fights" and his reasoning for choosing Grant to lead the Union army would now come to the fore. Of the various options available to him, Grant broke from the eastern armies' past after a major battle in Virginia, and ordered the troops south. There, they would attempt to maneuver Lee to a location more favorable to the Union. Grant would now begin employing the tenacious fighting style that he wanted to instill in the Army of the Potomac.

10. Anderson Decides to Push On

With the wounding of Major General Longstreet, Maj. Gen. Richard H. Anderson assumed command of Longstreet's Corps. Anticipating Grant's possible movement to Spotsylvania Court House, Lee ordered a military road built to facilitate Confederate troops' movement to meet the potential threat. On the evening of May 7, Lee ordered Anderson to quietly move his troops along this new road by 3:00 a.m. Unknown to the Confederates, Union troops under Maj. Gen. Gouverneur Warren had been ordered to start marching down the Brock Road to Spotsylvania Court House at 9:00 p.m. Anderson moved out about 10:00 p.m. to try and find a place to rest his men for the night, but finding nothing suitable, he decided to push on. The criticality of this decision would not be evident until the next morning. The Confederate troops arrived at Spotsylvania Court House just before Warren's Union troops and thereby avoided disaster for Lee and his army.

11. Fitzhugh Lee Decides to Stay and Fight

A vital part of Anderson's troop movements through the night of May 7 was Maj. Gen. Fitzhugh Lee's decision to use his division of Confederate cavalry in a delaying action against the Union troops along the Brock Road. While Wade Hampton's cavalry threatened the right flank of the Union advance via the Catharpin Road, which intersected the Brock Road near Todd's Tavern, Fitzhugh Lee's cavalry dismounted and set up defensive barricades across the Brock Road. As the Union pressure mounted, Fitzhugh Lee moved farther south along the Brock Road and continued to delay the Union advance. The cavalry fighting ended in the evening, and the Union troops pulled back instead of pressing the attack. When the Federals renewed their effort to move south in the very early morning, Fitzhugh Lee's continued tenacious delaying action made it impossible for them to meet Grant's schedule for the journey to Spotsylvania Court House. Fitzhugh Lee afforded the Confederates just enough time to win the race to Spotsylvania.

12. Grant Decides to Let Sheridan Pursue Stuart

Despite Sheridan's glowing report of his success in engaging Fitzhugh Lee's cavalry, he had failed to clear the Brock Road for the Union infantry to move south. Meade arrived at Todd's Tavern late that night and found Sheridan's cavalry there. The next day, the friction between the two over the role of cavalry exploded into a heated argument. During the dispute, Sheridan stated that if he were let loose he could whip Stuart and his Southern cavalry. Grant's headquarters were colocated with Meade's. The two men conferred just after the argument, and when Meade related Sheridan's statement, it caught Grant's attention. Grant liked Sheridan's aggressive style, and he seems to have impulsively decided to grant Sheridan's wish. The Union cavalry had performed poorly up to this time. However, Grant decided to let Sheridan go his own way and leave the rest of the Union army blind. Grant would be forced to rely on infantry and his own guesses regarding Confederate positions and strength. It would prove to be a costly decision.

13. Warren Decides to Redeem His Reputation

As in The Wilderness, Warren's Fifth Corps would open the initial infantry fighting at Spotsylvania Court House. Having struggled along the Brock Road all night to reach Spotsylvania Court House, Warren and his lead troops reached the northern edge of Spindle Field on the morning of May 8. He thought the Confederates he saw were a rear-guard cavalry covering a Confederate retreat. Warren's reluctance to attack as quickly as Meade and Grant had wanted at Saunders Field on May 5 had tarnished his reputation. He sensed an opportunity to regain some of his standing by adopting Grant's style of vigorously pressing ahead without taking time for dispositions. Warren decided to send his troops across Spindle Field as they arrived along the Brock Road. The recently arrived Anderson's Corps, not the rear-guard cavalry, brutally repulsed the repeated attacks. In the subsequent days of battle, the mauling of Warren's troops at Saunders Field and now Spindle Field would greatly impede him in executing his orders to attack.

14. Grant Decides How to Use His Corps

Grant had spent his entire Civil War career in the West, and, while he knew the officers of the regular army who had served in the Mexican War, he was a "stranger to most of the Army of the Potomac." As Grant strove to get the most out of the three corps of the Army of the Potomac that reported to Meade and the one corps of Burnside's Ninth Army, he needed to assess how these corps would function in his style of fighting. By the time the Union army moved to Spotsylvania Court House, Grant would rely more and more on Hancock's Second Corps as his offensive weapon of choice against Lee. The other corps were utilized more to try and pin the Confederates in place. Notably, Grant sent Hancock to the Federal right flank on May 9 and 10 to try and attack Lee near the Po River, then recalled him to lead the major assault against the salient on May 12. Grant later moved Hancock once again to lead an attack against Lee's new defensive line on May 18. At the end of the fighting on May 20, he sent Hancock's Corps out as bait to try and draw Lee out of his defensive position.[2]

15. Upton Leads a New Type of Attack

Col. Emory Upton was an aggressive brigade commander in Major General Sedgwick's (now Wright's) Sixth Corps positioned opposite the western side of the Mule Shoe. Upton's attack on the evening of May 10 had two unique elements: the decision not to attack Confederate earthworks in the traditional manner, and the preparation that went into executing this decision. Typically, the manner of attack was to fire one or more volleys at the enemy, continue to fire while rushing the objective, engage in hand-to-hand combat on a wide front, and try to exploit any opportunity that resulted. For Upton's maneuver, the attackers would not fire until they had reached the enemy lines. Selected unites implemented this decision. They were put under an aggressive commander and informed of what was expected before they attacked at a unique section of the enemy's line. The location of the assault allowed the Union troops to conceal their position and set forth with fewer than two hundred yards of open ground to cover. Upton and his men launched the attack just after 6:00 p.m. and successfully breached the Confederate line. The attack was not properly supported, the breach was plugged, and Upton and his troops had to retreat. But the results of the assault proved that this new tactic could break the Confederate defenses.

16. Lee Decides to Withdraw the Artillery from the Salient

The defensive deployment of Ewell's Corps followed a ridgeline that formed the Mule Shoe, a north-facing salient in the shape of an inverted *U*. Although this formation had defensive drawbacks, Lee had artillery placed in the salient to greatly increase the position's defensive capabilities. On the evening of May 11, Lee received conflicting reports of Grant's intentions. Lee erroneously concluded that Grant would move east toward Fredericksburg. To counter this presumed maneuver, the Confederates would need to be prepared to move quickly. Yet the rainy conditions would make a rapid response difficult. Thus Lee withdrew the artillery from the salient and positioned it where it could be quickly shifted to support any Confederate countermove. Lee had completely misread Grant's intentions, and his decision weakened the position that would be Grant's focus in the next morning's attack. The artillery was eventually ordered back, but it was too late. It was captured as a result. The artillery in the salient would have had a devastating effect on the attacking Union troops and dramatically impacted the results of the fighting.

17. Grant Decides to Commit a Full Corps

A variety of factors contributed to the failure of Grant's May 10 attack, but

Colonel Upton's tactics was not one of them. Upton had broken through the Confederate lines with his approach, but lack of support had limited his success. Seeing what a brigade had accomplished, Grant decided to commit a whole corps to an attack using the same method. Heavy rain on May 11 delayed the assault on the Mule Shoe until dawn on May 12. The surprise attack, coupled with wet ammunition and nonexistent artillery support, produced some of the most intense hand-to-hand combat of the war. Union troops ultimately breached the Confederate line. Almost twenty thousand Federals flooded the half-mile-wide salient, resulting in their confusion as they pressed the Confederates back. A fierce Rebel countercharge under Lee's direct supervision pushed the Union troops back to the earthworks, where the fighting continued well into the night. After twenty hours of combat, the Confederates moved back at 3:00 a.m. the next morning to a new defensive line at the base of the Mule Shoe. Grant's decision to launch a full corps at the Confederate lines had been costly to both sides. But for an incredibly fierce resistance, the lack of Confederate artillery would most likely have destroyed Lee's army.

18. Grant Decides to Move South from Spotsylvania Court House

On the morning of May 13, Brigadier General Wright, now commanding the Sixth Corps, reported to Meade that he had occupied the Mule Shoe. The carnage Wright found was appalling. At first, the Federals thought that Lee's army had retreated, but they shortly realized that Lee had pulled Ewell's Corps out of the salient and moved it to the defensive works frantically constructed during the previous day's fighting. After a week of fighting and incurring heavy casualties, the two armies faced each other in about the same situation—they had again reached a stalemate. The Confederates still held Laurel Hill in strength, and the new defensive line at the base of the Mule Shoe was formidable. Once again, Grant remained resolute and ordered the troops south around the Confederate right flank. The Army of the Potomac had never had such relentless leadership. Despite Grant's failure to destroy Lee's army, the fighting would continue. Except for Cold Harbor, in the future Grant would use maneuvers more often than costly massive frontal assaults against Confederate fieldworks.

To begin the tour: From the Chancellorsville Visitor Center, proceed west out of the parking lot, and turn left onto Bullock Road (do not follow the Chancellorsville Battlefield tour to the right). Then turn right onto Virginia Route 3. Travel 4.3 miles to the stoplight at the intersection of Virginia Route 3 (Plank Road) and Virginia Route 20 (Constitution Highway). This will be after the stoplight at Brock Road. As you approach the intersection, move to

Appendix I

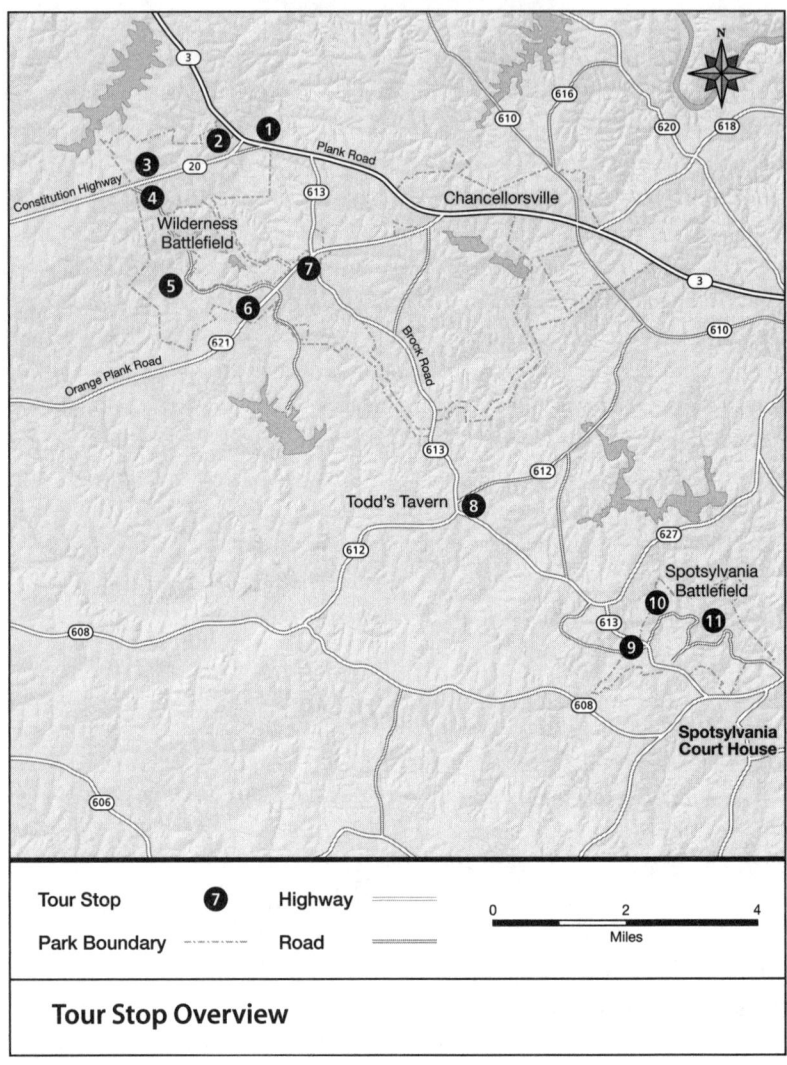

Tour Stop Overview

the left turn lane. Make a U-turn via the intersection eastbound on Virginia Route 3 and travel 0.3 mile up the slight rise to Lyons Lane on the right. Pull into this road and park near the markers. Note: there is limited space to park here, and a walk down to Wilderness Run at the bottom of the hill is covered later in an optional part of the tour.

Tour Stop 1: Wilderness Tavern

Tour Stop Relevance to Critical Decisions

After Lincoln nominated Grant for promotion to lieutenant general, Grant would be directing all Union army operations. With Lee's army as his objective, Grant made the critical decision to move by his left flank against Lee and his Army of Northern Virginia. That decision resulted in the largest Federal army moving here. This location sets the stage for the initial fighting in Grant's 1864 Overland Campaign.

Tour Stop Orientation and Troop Development

After reading the two interpretive markers, stand by the corner of the split-rail fence around the ruins closest to the markers. In 1864, the gravel road you see before you was the Orange Turnpike. It intersected the Germanna Road near Wilderness Run at the bottom of the hill, then extended up to the tavern and on to Fredericksburg. At the time of the battle, the Wilderness Tavern was deserted and overrun with weeds. The actual two-story Wilderness Tavern would have been located in the middle of the current divided highway.[3]

At the end of this road where it connects with Virginia Route 20, Grant established his headquarters. Slightly to your left but out of view is the Lacy House, where Warren made his headquarters. To your right, Virginia 3 leads to the Germanna Ford, where Warren crossed the Rapidan River.

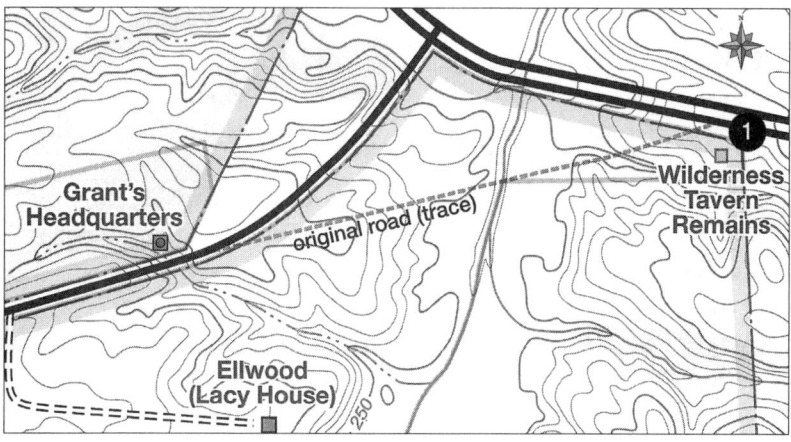

Appendix I

Remains of an outbuilding of the Wilderness Tavern. Courtesy of the author.

Given the time required to move the extensive supply wagons across the Rapidan, this location was selected for the first day's halt for the Fifth Corps. Chancellorsville was the stopping point for the Sixth Corps.[4]

ORDERS.] HEADQUARTERS ARMY OF THE POTOMAC, May 2, 1864.
1. The army will move on Wednesday, the 4th of May, 1864.
2. On the day previous, Tuesday, the 3d of May, Major-General Sheridan, commanding Cavalry Corps. . . . At midnight of the 3d of May, the Third Cavalry Division [Wilson's] . . . will move to Germanna Ford. . . . It will then move to Parker's Store, on the Orange Court-House plank road, or that vicinity, sending out strong reconnaissances on the Orange pike [Virginia Route 3] and plank roads [Virginia 621] and the Catharpin and Pamunkey roads, until they feel the enemy, and at least as far as Robertson's Tavern, the New Hope Church, and Almond's or Robertson's. All intelligence concerning the enemy will be communicated with promptitude to headquarters and to the corps and division commanders of the nearest infantry troops.
3. Major-General Warren, commanding Fifth Corps, will send two divisions at midnight of the 3d instant . . . to the crossing at Germanna Ford. . . . After crossing, [troops] will move to the vicinity of

View looking west along the original Orange Turnpike, with the intersection of the Germanna Road at the bottom of the hill. Courtesy of the author.

the Old Wilderness Tavern, on the Orange Court-House pike. The corps will move the following day past the head of Catharpin Run, crossing the Orange Court-House plank road at Parker's Store.

CIRCULAR.] HEADQUARTERS FIFTH ARMY CORPS,
May 3, 1864.
GENERAL: The First Division [Brig. Gen. Charles Griffin's], followed by the Third [Brig. Gen. Joseph Bartlett's], will move at midnight, crossing the Mountain Run at the double bridge; thence direct to Stevensburg. . . .

. . . The troops will cross the bridge at Germanna Ford as fast as possible, move out and eat their breakfasts on the other side, and then continue the march to Old Wilderness Tavern, taking up position there as fast as arriving, the First Division moving up the turnpike, toward Mine Run, about 1 mile. . . .
By command of Major-General Warren:
FRED. T. LOCKE,
Assistant Adjutant-General.

The chimney preserved here was that of an outbuilding of the Old Wilderness Tavern. The actual building would have been located where the highway

Appendix I

View from this vicinity down the Orange Turnpike in the 1860s. Grant's headquarters would have been to the right near the end of this part of the road. Library of Congress.

is today.[5] The above photograph shows how open this area was at the time of the battle.

The Union cavalry's Third Division under its inexperienced commander Brig. Gen. James Wilson reached this location by 8:35 a.m. in advance of the infantry. The division then continued on to Parker's Store on the Orange Plank Road via a wartime road just west of here. This road ran near the Lacy House and down toward the Chewning Farm (Tour Stop 5), then west of the Widow Tapp Farm (Tour Stop 6) to the Orange Plank Road near the intersection with present-day Route 611. Wilson reported from Parker's Store just after 2:00 p.m. that Confederate troops were well down the road toward Mine Run, over six miles away from this location. No soldiers were nearer than seven miles from Parker's Store. Wilson's evening report reads as follows:[6]

> HDQRS. 3D DIV., CAV. CORPS, ARMY OF THE POTOMAC,
> Parker's Store, May 4, 1864—7.40 p.m.
> Lieutenant-Colonel FORSYTH,
> Chief of Staff:
> I have executed all orders so far. Be good enough to send instructions for to-morrow. My patrols have been to the Catharpin road. Did not see Gregg, and only 2 of the enemy; also to within 1 mile

> of Mine Run, on Orange pike, skirmishing with small detachments of the enemy. Patrol to Robertson's Tavern not yet heard from. Am strongly posted, and shall be ready to move at 3.30 a.m. Will send word as soon as I hear from Robertson's again.
>
> J. H. WILSON,
> Brigadier-General.

In his report to Meade's chief of staff Humphreys on May 4, 1864 (no time stated), Sheridan provided additional assurances that there was no enemy threat from the west.

If you had been standing here on the afternoon of May 4, 1864, and looking down the dirt road with the highway at your back, you would have seen Brig. Gen. Charles Griffin's First Division, the lead division of Warren's Fifth Corps, on the far side of the ridgeline in front of you and deploying about one and a half miles down the Orange Turnpike (Virginia Route 20 / Constitution Highway). At the end of a twenty-mile hike, Griffin and his men would camp at the eastern side of Saunders Field (Tour Stop 3). Brig. Gen. Romeyn Ayres's First Brigade was situated on the right of the road, Brig. Gen. Joseph Bartlett's Third Brigade on the left of the road, and Col. Jacob Switzer's men in the rear. Pickets were loosely deployed on the presumption that no enemy troops were in the area. The men had been marching since midnight, and they had crossed the Germanna Ford about five miles northwest from here along the Germanna Highway (Virginia Route 3).[7]

Maj. Gen. John Sedgwick's Sixth Corps had also crossed the Rapidan and was camped along the Germanna Road for about three miles from the ford.[8]

By 2:00 p.m., the Federals had completed their crossing at Germanna Ford and a pleased Grant sent the following message to Washington:[9]

> GERMANNA FORD, May 4, 1864.
> (Received 1.50 p.m.)
> Maj. Gen. H. W. HALLECK,
>
> Chief of Staff:
> The crossing of the Rapidan effected. Forty-eight hours now will demonstrate whether the enemy intends giving battle this side of Richmond. Telegraph Butler that we have crossed the Rapidan.
>
> U. S. GRANT,
> Lieutenant-General.

Appendix I

To the east, Brig. Gen. David Gregg's Second Division Cavalry had preceded Maj. Gen. Winfield Hancock's Second Army Corps across the Rapidan River at Ely's Ford and moved farther south. Having started at midnight and completed the crossing of the Rapidan, Hancock's Corps went into camp near the present-day Chancellorsville Visitor Center. They camped where the graves of the men hastily buried after the previous year's fighting were now becoming exposed. "The dead were all around us," wrote one soldier of that night's encampment. Another told the "ghastly and awe-inspiring tale" of how the wounded had died in the fires.[10]

The Union signal station on Stony Mountain reported the Confederate reaction to Grant's crossing the Rapidan at 3:00 p.m. Located just southeast of Culpeper, Virginia, the station provided a good view of the area and relayed the news that the Confederates were moving:[11]

> STONY MOUNTAIN SIGNAL STATION,
> May 4, 1864—3 p.m.
> Captain FISHER:
> Enemy moving infantry and trains toward Verdierville. Two brigades gone from this front. Camps on Clark's Mountain breaking up. Battery still in position behind Dr. Morton's house, and infantry pickets on the river.
> TAYLOR.

With the remaining daylight, Warren's Fifth Corps could possibly have continued on for another five miles to Parker's Store on the Orange Turnpike. But the "very narrow and sometimes obscure" road to that location after a twenty-mile hike would have taxed the men and exposed the column's right flank and the trains' security. With Parker's Store as Warren's objective for the next day, and with fighting anticipated, fresh troops would be needed.[12]

Wilson's lack of field cavalry experience and the resulting lax Union cavalry patrols negated any Federal advantage in crossing the Rapidan unopposed. As the fighting erupted the next day, troops would be located in this area. Burnside's men would eventually march here on their way to try and exploit the gap between Lee's two attacking lines.

To Tour Stop 2

Noting traffic, pull back onto Virginia Route 3 heading east and quickly get in the left-hand lane. There is a crossover in 950 feet; make a U-turn, and head west on Virginia Route 3 (Plank Road). Note: if traffic is heavy, con-

The battlefield entrance and Grant's headquarters. Courtesy of the author.

tinue east on the Plank Road for 0.75 mile to the Brock Road intersection, where a traffic light allows for a U-turn to head west on Virginia Route 3 (Plank Road). Proceed back to the stoplight at the intersection of Virginia Route 3 and Virginia Route 20. Get in the left turn lane and prepare to turn left onto Virginia Route 20 at the stoplight. From the intersection, proceed 0.4 mile to the turnout on the right side of the road with the signs "You Are Passing Through Wilderness Battlefield" and "Grant's Headquarters." Pull into this parking area. This is the first stop on the National Park Service's tour. The roadside marker that you see from the direction you just traveled commemorates Lafayette's 1781 travels here. Take the trail a short distance to the markers for the headquarters.

Tour Stop 2: Grant's/Meade's Headquarters

Tour Stop Relevance to Critical Decision

This stop shows that Grant's decision to locate with Meade and the Army of the Potomac should be taken literally. It sets the stage for Grant's personal involvement in managing the upcoming campaign, and it shows how his presence in the field would change the character of the Army of the Potomac. While in their winter quarters on the other side of the Rapidan, Grant's and Meade's headquarters had been separated by six miles. Now that distance would be two hundred yards.[13]

Appendix I

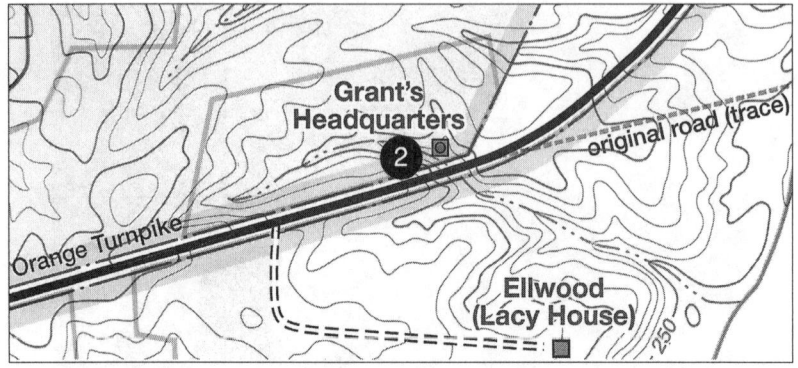

Tour Stop Orientation and Troop Development

On the morning of May 5, 1864, Grant was at Germanna Ford waiting to see Burnside's Corps cross the Rapidan when he received word from Meade that Lee's forces had been spotted.[14]

> HEADQUARTERS ARMY OF THE POTOMAC,
> Old Wilderness Tavern, May 5, 1864.
> (Received 7.30 a.m.)
>
> Lieutenant-General Grant:
> The enemy have appeared in force on the Orange pike, and are now reported forming line of battle in front of Griffin's division, Fifth Corps. I have directed General Warren to attack them at once with his whole force. Until this movement of the enemy is developed, the march of the corps must be suspended. I have, therefore, sent word to Hancock not to advance beyond Todd's Tavern for the present. I think the enemy is trying to delay our movement, and will not give battle, but of this we shall soon see. For the present I will stop here, and have stopped our trains.
> GEO. G. MEADE.

Grant responded to Meade from Germanna Ford, stating that he was waiting for Burnside to get to the ford and would leave as soon as he saw him. It was Grant's first order indicating that the charter of the army would be different than it had been previously. He ended the communication thusly: "If any opportunity presents itself for pitching into a part of Lee's army, do so without giving time for disposition." Growing impatient to get to the scene of the action, Grant left Germanna Ford about 8:45 a.m., rode to the intersection of Germanna and Orange Plank (at the bottom of the hill at Tour Stop 1), and found Meade by the roadside. Dismounting, Grant began discussing the situation with Meade. Grant's staff member Lieut. Col. Horace Porter wrote of the situation and described this knoll:[15]

> It had become evident that the enemy intended to give battle in the heart of [T]he Wilderness, and it was decided to establish the headquarters of both generals near the place where they were holding their present conference, at the junction of the two important roads....
>
> A few hundred yards to the west, and in the northwest angle formed by the two intersecting roads, was a knoll from which the old trees had been cut, and upon which was a second growth of scraggy pine, scrub-oak, and other timber. The knoll was high enough to afford a view some little distance, but the outlook was limited in all directions by the almost impenetrable forest with its interlacing trees and tangled undergrowth.

Following the canceled attack at Mine Run in late 1863, both armies went into winter quarters. Meade established his base near Brandy Station along the Orange and Alexandria Railroad, about eighteen miles northwest from here. Lee made his headquarters near Orange Court House, fewer than twenty miles away. When Grant arrived in the field on April 26, he set up his headquarters at Culpeper Court House, farther down the Orange and Alexandria Railroad from Brandy Station and about six miles from Meade's headquarters. With the initiation of the Overland Campaign, Grant and Meade would establish their headquarters here on May 5, 1864. Except for a few occasions when he rode out to personally observe the situation, Grant would remain in this area during the next few days of fighting while he received reports, issued orders, and studied maps.[16]

It was here at headquarters on the first day of fighting that Grant demonstrated his calm demeanor during a battle. After the initial preparations had been made, Grant sat on a stump, smoked a cigar, and began to whittle on a stick. As the campaign progressed and intensified over the next few weeks and Grant became more involved in the implementation and direction of the fighting, he would not be whittling any more sticks.[17]

The close and almost continuous interaction of the two men would help foster the pretense that Meade ran the Army of the Potomac. In his postwar writings, Grant stated that he had tried to make Meade's position "as nearly as possible what it would have been had [he] been in Washington." Grant also noted that he would establish his headquarters with Meade when possible to avoid the need to issue direct orders. But even before the fighting started, Grant had circumvented Meade's chain of command by ordering Brig. Gen. James Ricketts's Third Division of Maj. Gen. John Sedgwick's Sixth Corps to a position. Shortly thereafter, Ricketts received Meade's orders to go to a different location. Meade's courier Thomas Hyde suggested that since Meade's orders had come after Grant's, he should follow Meade's orders. Learning of the situation, Meade agreed with Hyde's actions but ordered him to go back and tell Ricketts to obey Grant's orders. With the courteousness typical of their correspondence, Meade informed Grant of the situation:[18]

> HEADQUARTERS ARMY OF THE POTOMAC,
> May 5, 1864—9.20 a.m.
>
> Lieutenant-General GRANT:
> I ordered General Ricketts to hold the roads leading from the enemy's line to our right flank. I am informed you have ordered him forward as one of Burnside's divisions has arrived. I would suggest Burnside's division relieving Ricketts' on the roads, also a small party of cavalry I have in front of Ricketts. Ricketts having received my order after yours is awaiting your action on this suggestion.
> GEO. G. MEADE,
> Major-General.

Although Grant and Meade would continue to have their headquarters basically colocated during the campaign, this is the only headquarters location within the parks. At the time of the battle there was a second growth of trees on this knoll. The Germanna and Orange Turnpike (Virginia Routes 3 and 20) and the Union troops moving on them would have been visible, but the fighting that would occur at Saunders Field (Tour Stop 3) would not have been.[19]

Continue back to the pullout.

Additional Nearby Site of Interest

Warren's Headquarters and Wilderness Crossing Trail

From the pullout, you can see to the west the "Ellwood Manor, General Warren's Headquarters" sign.

Noting traffic, pull back onto the road, and then pull in at the sign. If the entrance to Ellwood is closed, there is still parking space in the area within about a half-mile walk to the house.

After reading the interpretive markers, go to the front of the house and face the front yard. To your left is a marker for the trail.
Follow the blue blaze marks, which will lead you down the hill to the old Parker's Store Road, Wilderness Run, and the Orange Turnpike and Germanna Plank Road intersection. You will also be able to walk back up to the first stop of the tour.

If they are open during your tour, the museum and displays in the Lacy House are worthwhile visits.

To Tour Stop 3

Noting traffic, continue west on Virginia Route 20 for 1.25 miles to the Wilderness Battlefield Exhibit Shelter on the right, and pull into the parking lot. This shelter contains various displays of the battle. Saunders Field was one of the few open areas on the battlefield. The initial fighting took place here on Saunders Field at midday on May 5, 1864. After almost two days of conflict, the last fighting would also occur here; Lee tried to turn Grant's right flank commanded by Maj. Gen. John Sedgwick. Although they are out of chronological sequence, these engagements are presented as one stop.

Tour Stop 3a: Saunders Field, Union Perspective

This stop will present the Union perspective of the fighting that occurred here, mostly on May 5, 1864. The Confederate perspective will be covered in reference to Tour Stop 4.

Tour Stop Relevance to Critical Decisions

Grant's decision to move on Lee's right flank and Lee's decision to initiate an attack first collided here on the morning of May 5. This tour stop demonstrates the essentially offensive-minded philosophy that the two had. One underlying aspect of Lincoln's desire for Grant to lead the Federal armies was the need to instill in them a more aggressive approach to fighting. Out west, Grant had proved his abilities in this regard. Although some Union officers would be resistive, the change in culture would begin now. As for the Confederates, since Lee had become the leader of this army at the end of May 1862, aggressive tactics could be found at all levels in the army.

Appendix I

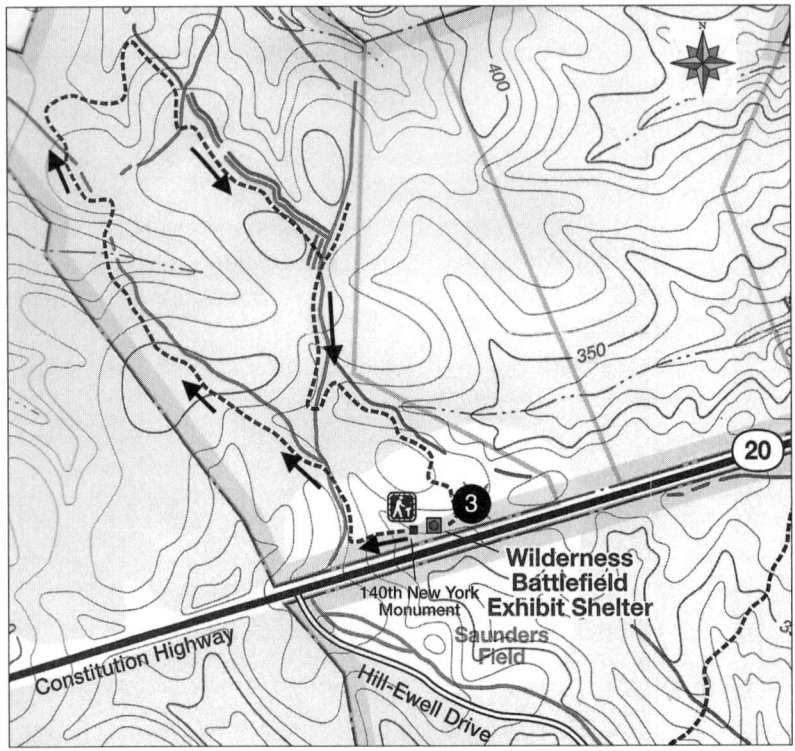

Tour Stop Orientation and Troop Development

After reviewing the information in the shelter and the various interpretive markers, go to the two markers on the south side of the shelter—"Saunders Field" and "Capture of Winslow's Battery." Looking to the east to the ridgeline on both sides of the highway, you see the position Griffin's First Division of Warren's Fifth Corps occupied on the night of May 4–5. Believing that Wilson's cavalry was monitoring the Orange Turnpike, a strong Federal picket line was not posted to the west. Pursuant to orders, on the early morning of May 5, Warren's Corps was starting to move toward Parker's Store, with Griffin's division the rear guard that was to wait until Sedgwick's Corps arrived. It would be a difficult going, as the trail went through some of the worst terrain in The Wilderness. As Warren's men started to move out, they were surprised to see Ewell's troops approaching on the opposite side of this field.[20]

A view of Saunders Field looking toward Ewell's position in the 1860s. Library of Congress.

The communications sequence that follows provides insight into the beginning of the change in fighting culture that Grant wanted to instill in the Army of the Potomac. These communications also convey the beginning of Warren's decline in the eyes of his commanders. Lieut. Col. William Throop was posted on the south side of the road. His official report of Bartlett's Third Brigade in Griffin's division shows just how surprised the Federal forces were:[21]

Report of Lieut. Col. William A. Throop, First Michigan Infantry

When my picket-line was established there was no enemy in our front. About an hour after daylight on the morning of the 5th a strong column of the enemy's infantry, preceded by cavalry, was discovered coming down the road from the vicinity of Robertson's

A similar present-day view of Saunders Field looking across the Orange Turnpike to the Confederate position. Courtesy of the author.

> Tavern. . . . Fifteen minutes before the appearance of the enemy, I had received written orders to withdraw my pickets and rejoin the brigade on the march, and was in the act of assembling the regiment when the enemy was reported advancing.

This information was quickly reported up the chain of command.[22]

> HEADQUARTERS FIFTH CORPS,
> May 5, 1864—6 a.m.
> Major-General HUMPHREYS,
> Chief of Staff:
> GENERAL: General Griffin has just sent in word that a force of the enemy has been reported to him coming down the turnpike. The foundation of the report is not given. Until it is more definitely ascertained no change will take place in the movements ordered. Such demonstrations are to be expected, and show the necessity for keeping well closed and prepared to face toward Mine Run and meet an attack at a moment's notice.
> Very respectfully, your obedient servant,
> G. K. WARREN,
> Major-General, Commanding.
> P. S.—I will remain at my old headquarters until 7 a.m.
> G. K. W.
>
> ———
>
> 6.20 A.M.
> General Bartlett sends in word that the enemy has a line of infantry with skirmishers out advancing. We shall soon know more. I have arranged for General Griffin to hold the pike until the Sixth Corps comes up, at all events.

Warren then sent word to Griffin to find out what enemy force he would be confronting:[23]

> HEADQUARTERS FIFTH ARMY CORPS,
> May 5, 1864—6.20 a.m.
> Brigadier-General GRIFFIN,
>
> Commanding First Division:
> GENERAL: The major-general commanding directs you to push a force out at once against the enemy and see what force he has.
>
> Very respectfully, your obedient servant,
> FRED. T. LOCKE,

As previously mentioned, Meade notified Grant, who was at Germanna Ford, of the situation. Stopping the Federal advance was exactly what Lee wanted from his decision to attack the Federals in The Wilderness.[24]

In keeping with the more aggressive style that Grant wanted the army to adopt, he enthusiastically endorsed Meade's actions, thereby playing into Lee's desires.[25]

> HEADQUARTERS ARMIES OF THE UNITED STATES,
> Germanna Ford, May 5, 1864—8.24 a.m.
>
> General MEADE:
> Your note giving movement of enemy and your dispositions received. Burnside's advance is now crossing the river. I will have Ricketts' division relieved and advanced at once, and urge Burnside's crossing. As soon as I can see Burnside I will go forward. If any opportunity presents itself for pitching into a part of Lee's army, do so without giving time for disposition.
>
> U. S. GRANT,
> Lieutenant-General.

This correspondence reveals an underlying difference in Warren's and his commanders' approach to the situation, as wells as Warren's tendency to appear to lecture his superiors. By 6:30 a.m., Warren wanted to get a better understanding of the situation and to wait for Sedgwick's Sixth Corps to come up before initiating any action. Meade would have been happy to fight Lee anywhere but the Mine Run defenses. Grant wanted to attack immediately—and almost impulsively—not caring what the enemy dispositions might be.

Warren's troop movements that morning to Parker's Store had left his corps scattered; Brig. Gen. Samuel Crawford's Third Division was nearing the Chewning Farm (Tour Stop 5) two miles away, Brig. Gen. James Wadsworth's Fourth Division was struggling in the thickets of Wilderness Run southwest of here, and Brig. Gen. John Robinson was near Wilderness Tavern. Only Griffin's First Division was actually facing the enemy at first. While the other divisions were re-forming, Griffin had his men begin to construct fieldworks on the east side of Saunders Field.

Meade ordered Sedgwick to move from his original route to Wilderness Tavern to Warren's exposed right flank. Part of Meade's aggressive stance might have stemmed from his belief as late as 9:00 a.m. that Lee had only left a division to fool the Federals. Instead, Meade believed, Lee was trying to form a position near North Anna, which was almost thirty miles away and where fighting would occur at the end of the month. Meade summarized his opinion to Grant:[26]

> HEADQUARTERS ARMY OF THE POTOMAC,
> May 5, 1864—9 a.m.
>
> Lieutenant-General GRANT:
>
> ... Warren is making his disposition to attack, and Sedgwick to support him. Nothing immediate from the front. I think, still, Lee is simply making a demonstration to gain time. I shall, if such is the case, punish him. If he is disposed to fight this side of Mine Run at once, he shall be accommodated.
>
> GEO. G. MEADE,
> Major-General.

Meade dramatically altered his view at 10:15 a.m., when he received word from Crawford at the Chewning Farm. According to Crawford, the Rebels were in force along the Plank Road, moving eastward, and trying to flank him.[27] Despite the desire to move quickly against the Confederates, it would take some time before Warren's Corps was consolidated and able to attack. Wadsworth was ordered to suspend his movement to Parker's Store, link up with Griffin's left flank, which would have been at the southern end of Saunders Field, and attack the enemy. Crawford held a good position at the Chewning Farm, but he, too, was ordered back, and he reluctantly complied. Complicating the situation was Brig. Gen. Romeyn Ayres, Griffin's First Brigade commander, who was positioned on the north side of the road. Ayres felt that until Sedgwick's troops could come up, any attack would expose his flank to attack from the Confederates, whose line extended past his.[28]

The Federal assault on the far ridge of Saunders Field began about noon, almost six hours after the sighting of the Confederates. At this time, Sedgwick's troops had not yet arrived to provide protection for Ayres on the Union right flank.

Face west.

Ayres's first wave consisted of his six US Regular regiments and the 140th New York. They crossed this field on your right and then proceeded up the slope to the Confederate position. As the soldiers moved out from the protection of the ridgeline behind you, they were immediately fired on. The Confederate lines extended north past the Federals' line of advance, and as Ayres had feared, firing from that direction came from Brig. Gens. James Walker's and Leroy Stafford's brigades of Maj. Gen. Edward Johnson's division of Ewell's Corps. The Second, Eleventh, Twelfth, Fourteenth, and Seventeenth US Regulars of Ayres's brigade veered to the right to meet this flanking fire. But in so doing, they exposed their left flank to enemy fire from another of Johnson's brigades, Brig. Gen. George Steuart's. The 140th New York continued straight on closer to the road, and a gap was created. Ayres sent his second wave, the 146th New York and the 91st and 155th Pennsylvania, into the forming gap against Steuart's brigade.

On the north side of the road, the 140th New York made the deepest penetration into the Confederate lines, and the only marker in Saunders Field commemorates its effort. As they moved up to the enemy position, the Confederates slowly pulled back into the woods, and the Federals followed. The Federals then encountered "a raking fire on the right flank, which caused the regiment to melt away like snow."[29]

Of the second wave, the 146th New York moved to support the gap created between the two parts of the first wave, and the 91st and 155th Pennsylvania moved to help the regulars. The men ran into the same flanking fire that their predecessor had, and Ayres's brigade was forced to retire.

On the south side of the road, Griffin's Third Brigade, commanded by Brig. Gen. Joseph Bartlett, advanced when Ayres's men did. Bartlett's First Michigan advanced on the Confederate skirmishers, driving them back. The 18th Massachusetts, 83rd Pennsylvania, and 44th New York pushed out, and the 118th Pennsylvania and 20th Maine followed them closely. Running forward under fire, Bartlett's men reached the far side of the field. Temporarily halted by the Confederate fire and their entangled formation, they were able to regroup and push on past present-day Hill-Ewell Drive (Tour Stop 4). However, this success was only temporary. The Confederates were rallying, and with Ayres's brigade stopped on the north side of the road and unable to give support, both flanks were exposed. "Every man saw the danger, and without waiting for orders to fall back, broke for the rear on the double quick."[30]

Farther south and on Griffin's left flank, Brig. Gen. James Wadsworth's Fourth Division, which had started the day's movement to Parker's Store, had also attacked the Confederate line near Higgerson's Field. It engaged Brig. Gen. John Gordon's brigade of Early's division and Brig. Gen. Daniel's brigade of Rodes' division. After moving through dense undergrowth, Wadsworth would also have initial success, but when Confederate reinforcements arrived, they too were pushed back.

As Warren's subordinates predicted, their exposed flank would prove fatal. Warren wrote of the attack in his journal:[31]

> **Journal of Maj. Gen. Gouverneur K. Warren, USA, Commanding Fifth Army Corps**
>
> May 5.— . . . Generals Meade and Grant arrived and determined to attack the force on the road near Griffin. Wadsworth was immediately gotten in line to left of Griffin with one brigade of Crawford, Robinson in support. We attacked with this force impetuously, carried the enemy's line, but being outflanked by a whole division of the enemy was compelled to fall back to our first position, leaving two guns on the road between the lines which had been advanced to take advantage of the first success. The horses were shot and the guns removed between our lines. The attack failed because Wright's division, of the Sixth Corps, was unable on account of the woods to get up on our right flank and meet the division (Johnson's) that flanked us. Wright became engaged sometime afterward. We lost heavily in this attack, and the thick woods caused much confusion in our lines. The enemy did not pursue us in the least. We had encountered the whole of Ewell's corps.

In the days ahead, the results of entering the fight without proper preparation and coordination would impact both Warren's perspective on how the campaign should not be waged and the tenor of his responses to orders.

Brig. Gen. Charles Griffin, whose First Division had been ordered in first, was so frustrated by the attack's mismanagement that he went to army headquarters at 2:45 p.m. There, he loudly announced that he had had to retreat after driving Ewell back for three-quarters of a mile because he had gotten no support on his flanks. Griffin also implied that both Wright, who had not come up, and Warren should be censured. Grant's aide Rawlins thought Griffin should be arrested, and Grant "seemed of the same mind." Surprisingly, it was the typically short-tempered Meade who defused the situation by telling Grant "It's only his [Griffin's] way of talking."[32]

Wright's assistant adjutant general described the situation in his report:[33]

Maj. Henry R. Dalton, Assistant Adjutant General, USA, of Operations of First Division

The division was ordered to go into position parallel to the plank road and advance to connect with the Fifth Corps on left, which corps had begun to feel the lines of the enemy, the formation of the division being from left to right—Second Brigade, Brig. Gen. (then Col.) E. Upton; First Brigade.... The skirmish line was moved with the greatest difficulty on account of the thick and tangled underbrush, which necessarily impeded the progress of a line, and often breaking it completely.

In his report, Upton described the same problem of moving his troops:[34]

Report of Brig. Gen. Emory Upton, USA, Commanding Second Brigade

About 11 a.m. orders were received to advance to the support of the Fifth Corps, then engaged with the enemy on the Orange Court-House pike 2 miles from Wilderness Tavern. The advance was made by the right of wings, it being impossible to march in line of battle on account of the dense pine and nearly impenetrable thickets which met us on every hand.

What headquarters completely missed regarding this incident was the underlying reason Wright had not come up in time. Headquarters did not inquired why he had been so late nor appreciated the implications of this delay for future army operations. Wright's troops had about a mile and a half to travel down the Culpeper Mine Road (Tour Stop 3b). This march normally would have taken about an hour, but in The Wilderness, it took almost four hours. The thick foliage in this area would dramatically affect troop movements and so needed to be accounted for. Without this understanding, Federal troop movements were rarely as timely as initially planned.[35]

The Federals had pulled back from this area by the time the Sixth Corps troops came up, and the Confederates had a picket line along the eastern side of Saunders Field. Once formed, Wright's troops advanced, pushing Confederate skirmishers back along their front and reestablishing the Union's forward position at the eastern edge of Saunders Field. The operations of Wright's corps would be as unsupported as those of Warren's men. Most of the fighting occurred north of Saunders Field, with the two sides advancing and retreating with the ebb and flow of the fighting there. Upton's brigade, the left flank of Wright's attacking force, was able to link up with elements of the Fifth Corps. Upton was then able to press his men forward to the east side of Saunders Field. After looking across the field in contemplation of an attack, Upton "refused to budge, saying it was madness." A brigade of Robinson's attempted a charge across the open ground, but the troops were repulsed by artillery.[36]

As part of Grant's overall plan for Hancock's assault on Lee on the Orange Plank Road the morning of May 6, Warren and Sedgwick were to move in support of Hancock's movement on the Orange Plank Road. Ewell's men struck before Sedgwick, positioned in the wooded area north of Saunders Field, could move at 5:00 a.m. After an hour of intense fighting, the two sides held the same ground that they had the night before. At 5:30, Warren reported that his "troops [were] disposed for the assault." He received and complied with 6:00 a.m. order to push his pickets well out to the front. However, Warren and his subordinates were concerned about the strength of the Confederates in their front, and they cautioned the men to wait until all preparations were in place.[37]

MAY 6, 1864—6.25 a.m.

Major-General HUMPHREYS:

General Griffin has moved up close to the enemy's position and driven him to his lines, and I am getting in artillery to open on the enemy. I think it best to not make the final assault until the preparations are made. We are driving them rapidly on the left and prisoners are coming in. General Burnside's column ought soon to be in position to intercept the retreat of the enemy's right. His [the enemy's] artillery has ceased for some time.

Respectfully,
G. K. WARREN.

Meade's quick reply ignored Warren's concerns and ordered an attack.[38]

> MAY 6, 1864—7.15 a.m.
> COMMANDING OFFICER FIFTH CORPS:
> General Hancock has taken rifle-pits and two flags. Longstreet has come up on his left. The major-general commanding considers it of the utmost importance that your attack should be pressed with the utmost vigor. Spare ammunition and use the bayonet.
> A. A. HUMPHREYS,
> Major-General and Chief of Staff

But here at Saunders Field, only the skirmishers pressed forward. Finally, feeling that Warren's troops could be better employed supporting operations on Hancock's right, where Wadsworth of Warren's Corps was already engaged, the following order was issued:[39]

> MAY 6, 1864—9.30 a.m.
> COMMANDING OFFICER FIFTH CORPS:
> The major-general commanding directs that you suspend your operations on the right, and send some force to prevent the enemy from pushing past your left, near your headquarters. They have driven in Cutler [of Wadsworth's Corps] in disorder and are following him.
> A. A. HUMPHREYS,
> Major-General and Chief of Staff

The fighting at Saunders Field was over, but the dissatisfaction with Warren was only growing.

Tour Stop 3b: Gordon's Flank Attack

Walk to the "Gordon Flank Attack Trail" marker.

Tour Stop Relevance to Critical Decision

After being moved to Ewell's left flank on the morning of the second day of fighting, Gordon found that his position extended past the Federal line. Gordon alerted Ewell, his corps commander, that he now had the opportunity to attack the exposed Federal flank. Ewell seemed in favor of the idea, but

Appendix I

Gordon's division commander, Early, objected to the plan. Ewell's critical decision to defer to Early and delay the proposed attack until almost dark cost Lee an opportunity that could have changed the battle.

Tour Stop Orientation and Troop Development

On the evening of May 6, the last of the major fighting in The Wilderness occurred near this location, with an attack on the Federal right flank. The roughly two-mile loop of "Gordon's Flank Attack Trail" takes about an hour to walk and provides an excellent perspective of these actions and an appreciation for the terrain. You will see earthworks for both sides along the way.

After reviewing the "Gordon Flank Attack Trail" marker, proceed on the trail past the 140th New York Monument to the second walking tour stop on the trail map. This stop is at the top of the ridge on the west side of the field. Note that Map 3a/3b at the beginning of this stop shows this trail.

These are the remains of field fortifications that Ewell's troops had constructed the previous day, and they were likely the first works constructed for the battle. They continue north along the trail and south of the Orange Turnpike. On the trail, follow the blue markers. You will see remains of both sides' fieldworks as you proceed along the trail. At marker 9, the trail crosses the maintenance road and the Culpeper Mine Road—stay on the blue-marked trail. Double blazes mark where the trail changes direction.

Walk north along the trail to the National Park Service "The Culpeper Mine Road" marker.

This is the Culpepper Mine Road, and it has also been called the Spotswood Road and Flat Run Road. In 1864, it connected to the Germanna Road, and Sedgwick's Corps used it at midday on May 5 while moving to take position on Warren's right. The Confederate lines extended north of here.

Continue walking to Park Service markers "Fighting of the Evening of May 5" and "Morning of May 6."

Each stop has an interpretive marker of events taking place in the area. Just to the west of here on the evening of May 5, Gordon's troops would have been marching from the Confederate right flank to the left. Fighting would occur in this area, with neither side gaining any advantage.

Continue walking to the Park Service markers "John Gordon Proposes a Flank Attack."

Gordon's troops arrived in this area late on the night of May 5. His line extended past the Federal lines manned by Brig. Gen. Alexander Shaler of

Wright's First Division just a short distance to the east. Early on the morning of May 6, Gordon received information from scouts that only cavalry picket lines protected the Federal right flank.[40]

> MAY 6, 1864.
> Lieutenant-Colonel Pendleton:
> I am having some scouting done from the position of my picket-line in the direction of the stone pike. Will report anything I learn. A reliable scouting party sent out this morning down Flat Run to the plank road leading to Germanna Ford reports that road only protected by a cavalry picket-line. They saw but one regiment of infantry. They say they went 3 miles down the river. They saw ambulances, &c., moving from direction of Germanna Ford to old pike.
>
> Very respectfully,
> J. B. GORDON,
> Brigadier-General.

Gordon personally rode to the area with some of the scouts and found "the reports correct in every particular." He was in an excellent position to strike Grant's flank, and he alerted Ewell of this opportunity through Early, his commander.[41]

> **Report of Lieut. Gen. Richard S. Ewell, CSA, Commanding Second Army Corps**
>
> About 9 a.m. I got word from General Gordon, through General Early in person, that his scouts reported the enemy's right exposed, and he urged turning it, but his views were opposed by General Early, who thought the attempt unsafe. This necessitated a personal examination, which was made as soon as other duties permitted, but in consequence of this delay and other unavoidable causes the movement was not begun until nearly sunset.

Early incorrectly felt that Burnside's Corps was in Sedgwick's rear and could threaten the proposed movement. With no reserves, a counterattack could be disastrous, and Ewell deferred to Early's opinion. Gordon made additional requests for permission to attack but was denied.[42]

Appendix I

On the afternoon of May 6, Brig. Gen. Robert Johnson's brigade, which had marched from near Richmond, arrived to reinforce Ewell. These troops moved behind Gordon's position here and were positioned on his left flank, extending the Confederate lines well past the Federal flank.

Finally, late in the afternoon, Early sent orders for Gordon to make the attack with support from Johnson's brigade.

Continue walking to Park Service marker "Gordon's Flank Attack."

This stop is located near the Federals' exposed right flank and where Gordon's men would assault on the evening of May 6. Johnston and Gordon had pulled their troops out of their defensive positions and moved quietly to the area immediately north of this marker. Gordon's six Georgia regiments were aligned perpendicular to the Federal flank here, and Johnson's four North Carolina regiments were stationed behind and slightly to the left of Gordon. It was intended that, as these two brigades pressed down from the north, Pegram's men, now commanded by Hoffman after Pegram's wounding, would press forward from their position in front of the Federal lines as the flank attack exposed their position. Additional units would then join the attack as more of the enemy line was exposed. These efforts had little effect on the impact of the offensive.

Brig. Gens. Alexander Shaler's and Thurman Seymour's mixed brigades were on the Federal right. Shaler's Sixty-Fifth New York Regiment faced north toward Gordon's hidden troops. The Sixty-Seventh New York was immediately to his left and facing west toward the Confederate earthworks.

Gordon's assault started near sundown and completely surprised the Union troops. Shaler's Sixty Fifth New York fell back, exposing the Sixty Seventh New York and then Seymour's Sixth Maryland[43]:

Report of Col. John W. Horn, Sixth Maryland Infantry

Near sundown, however, the enemy having massed heavily on our right, charged and drove in the regiments of General Shaler's command on my right in the utmost confusion, the enemy pressing on their flank and rear. To prevent the capture of my whole command, I ordered my command to fall back, which was done, but they soon became mixed up with other troops, and panic and confusion ensued. General Shaler did all that man could do to rally his troops, being captured by the enemy while so engaged... About 10 p.m. the enemy again made an attack upon the line held by General Upton in our front, but were easily driven back.

Johnston's troops were supposed to follow Gordon's and add weight to the attack, but essential information was not communicated to him. He knew nothing of the enemy's position or the ground he was to cover, nor of the plans for the attack. As so often was the case in the dense Wilderness foliage, coordination of troop movements was almost impossible, and Johnston's troops drifted toward the Germanna Road and provided little support in the attack.[44]

Gordon's initial success had caused so much confusion and panic among the Federals that staff officers of the Sixth Corps reported to Meade's headquarters. Meeting Humphreys, Meade's chief of staff, they informed him that their lines were broken and rolled up, and that the enemy was advancing down the Germanna Road. Those advancing troops were probably some of Johnston's men who were lost. The Sixth Corps' officers added that Generals Sedgwick and Wright had probably been captured. Humphreys sent staff offices down the road, but they found no enemy troops and sent word to Meade.[45]

Continue walking the Park Service trail to the next marker "Gordon's Flank Attack Falters."

As Gordon's troops reached this point, night was falling, and the momentum of the initial success was starting to wane. Seymour's and Shaler's troops had retreated in panic. In trying to rally their retreating commands, Generals Shaler and Seymour were taken prisoner about this time. The Confederates now met the troops of Brig. Gen. Thomas Neill's Third Brigade, who formed more of a resistance. Adding to the confusion, Pegram's men advanced and fired into both the Confederates and the Federals. Neill's men were forced to retreat.[46]

Continue walking the Park Service trail to the next marker, "The Federals Fall Back."

Gordon approached this point at nearly ten o'clock at night. The Federals had had time to regroup. The 14th New Jersey, 106th New York, and 10th Vermont Regiments of Brig. Gen. William Morris's brigade had been rushed to this point when the attack first started. They formed on the right of Brig. Gen. David Russell's brigade, which had been located in this area of the original Federal line. Units of Upton's brigade were also brought in to shore up this position. Three Confederate brigades of Hoffman's had moved near here, but with the darkness and confused battle lines, they did little to influence the fighting. As the Federals were able to hold their line here, the main fighting at The Wilderness was over. Col. William Ball, commanding the 122nd Regiment, which was the last of Seymour's brigade to pull back, reported on the action:[47]

Report of Col. William H. Ball, 122nd Ohio Infantry

At sunset a feint was made upon our front and a vigorous assault upon our right flank. The regiments on my right gave way one after another. When my regiment was ordered to retreat, there was not a man in the intrenchments on my right or left. So quick were the movements of the enemy that when I first discovered them in our rear, they were in rear of the center of my regiment, scattering the second line with all speed. We were then in the midst of extensive woods and thick undergrowth. The turning of our right flank was the result of gross negligence on the part of some general officer, I know not whom. The retreat was necessarily disorderly, there being barely time for possible escape.

Early's impact on Ewell and his decision to delay Gordon's flank attack until almost sunset can be judged by the attack's limited success—the men started late and were almost completely unsupported. In his official report, Gordon summarized what had been accomplished with only fifty casualties in his brigade and addressed the might-have-beens:[48]

Report of Brig. Gen. John B. Gordon, CSA, Commanding Gordon's Brigade and Early's Division, of Operations May 5–14

My personal observation satisfied me that one hour more of daylight now would have insured the capture of a considerable portion of the Sixth Army Corps. . . . The enemy's killed, according to the count kept by the officer commanding pioneer corps, amounted to nearly 400, among them one brigade commander. Several hundred prisoners were captured, among these two brigade commanders—Generals Seymour and Shaler . . .

I must be permitted in this connection to express the opinion that had the movement been made at an earlier hour and properly supported, each brigade being brought into action as its front was

> cleared, it would have resulted in a decided disaster to the whole right wing of General Grant's army, if not in its entire disorganization. The loss in my brigade amounted to about 50.

Early, never a supporter of the attack, made no effort to verify the situation. After the war, he stated that it was fortunate that darkness had allowed for the initial surprise and success, but if it had been earlier, the Federals would have seen the weakness of Gordon's attack and brought up troops against him.[49]

In his report to Halleck in Washington, Grant expressed a different view from Early:[50]

> HEADQUARTERS,
> Wilderness, May 7, 1864—10 a.m.
> (Received by mail from Alexandria, Va., 10 p.m. 12th.)
> Maj. Gen. H. W. HALLECK,
> Chief of Staff:
> ... Milroy's old brigade was attacked and gave way in the greatest confusion, almost without resistance, carrying good troops with them. Had there been daylight the enemy could have injured us very much in the confusion that prevailed: they, however, instead of getting through the break, attacked General Wright's division, of Sedgwick's corps, and were beaten back.
> U. S. GRANT,
> Lieutenant-General.

One of Warren's staff officers, Morris Schaff, referred to the possibility of coordinated attacks of Longstreet and Gordon: "I say, had Gordon struck at that hour, nothing, I think, could have saved the Army of the Potomac."[51]

From here, finish the walking tour by following the trail back to the Wilderness Battlefield Exhibit Shelter.

To Tour Stop 4

Noting traffic, leave the Wilderness Exhibit Shelter parking lot, and turn right. Proceed west on Virginia Route 20 for 950 feet, and turn left onto Hill-Ewell Drive. Note: As this turn is at the crest of the hill, oncoming traffic

Appendix I

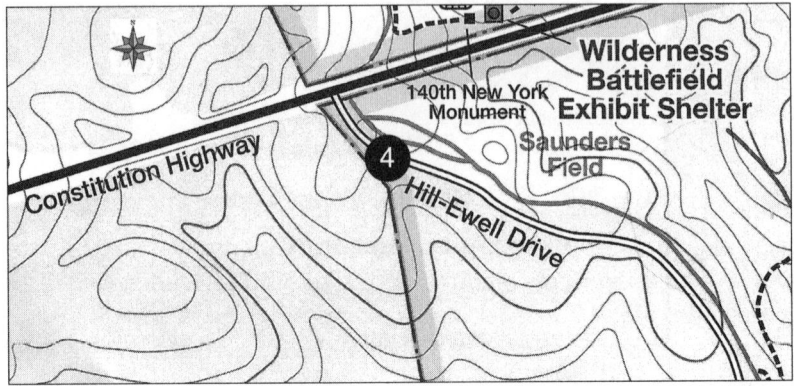

can be difficult to see. Use extra caution. Originally constructed in the 1930s, Hill-Ewell Drive follows the Confederate earthworks built between the Orange Turnpike and Orange Plank Roads.[52] Pull into the parking spaces on the right side of the road just past the National Parks Service's "Saunders Field" sign. A compass marker and dedication monument plaque are located about one hundred feet back toward the intersection. Proceed to the two wayside markers, "A Wild, Wicked Roar" and "The Confederate Line."

Tour Stop 4: Saunders Field, Confederate Perspective

Tour Stop Relevance to Critical Decisions

This stop presents the Confederate perspective of the fighting at Saunders Field on May 5, 1864, that resulted from Lee's decision to attack the Union army as it moved through The Wilderness. With Grant's move across the Rapidan, Lee's Army of Northern Virginia had left its winter quarters on May 4 and camped near Verdiersville, about seven miles down the Orange Turnpike, where the Orange Turnpike and Plank Roads were close together. Other Confederate troops were as close as Locust Grove, only three miles from this location.[53]

Tour Stop Orientation and Troop Development

At the time of the fighting, Brig. Gen. John Jones's brigade would have been here, and these works would have been much more substantial.

The forward line was improved after the fighting on the morning of May 5. The first line was either the original line or a reserve line for the forward line.[54]

Confederate entrenchments at Saunders Field. Library of Congress.

These markers also represent the site of a temporary Soldier's Cemetery No. 1, where just over one hundred Union soldiers were buried immediately after the war. Their remains were later reinterred at the Fredericksburg National Cemetery.[55]

Without walking on the remains of these earthworks, move around them to better understand the different fields of fire the Confederates would have been able to inflict on the advancing Federals.

Return to the markers, and face the open field.

Lee intended to utilize Ewell's Second Corps, moving down the Orange Turnpike behind and to your left, and Hill's Third Corps, moving down the Orange Plank, to hold Grant's army in The Wilderness until Longstreet could come up and strike a decisive blow. Lee would ride with Hill's Corps and make his headquarters later in the day at the Widow Tapp Farm, Tour Stop 6. He did not want Ewell or Hill to bring on a general engagement, only to keep Grant in position for now. In addition, as the two corps moved along their respective roads, the gap between them would increase, reducing their ability to support each other if necessary.

Early on the morning of May 5, these two corps moved out. Ewell's lead division, Maj. Gen. Edward "Allegheny" Johnson's, approached this location about the time that Warren's men across Saunders Field were moving out.

Appendix I

This is when the Federals became aware of the approaching Confederates. —Brig. Gen. John Jones's brigade was starting to deploy in this area. Next came Brig. Gen. George "Maryland" Steuart's brigade, which deployed on the north side of the road. Brig. Gen. Leroy Stafford's brigade followed, as did Brig. Gen. James Walker and his famed Stonewall Brigade, which took up position on the left of the developing Confederate position.

Once in place, Johnson's division stretched from here just south of the turnpike on the edge of Saunders Field to the Culpepper Mine Road, the "Gordon's Flank Attack Trial" stop 3 of Tour Stop 3b.[56]

Rodes's division came up next, with Battle's five brigades positioned on both sides of the turnpike, just behind the Confederate front line here. Daniel's and Doles's brigades were being moved into positions just south of here, with Daniels near Higgerson's Farm. Doles's men were on Daniel's left flank, but a gap had been created between his left flank and Jones's right flank just south of here. Rodes's last brigade, Ramseur's, had been left behind with Heth's division of Hill's Corps near Orange Court House.[57]

Early's division was still coming up the turnpike when Warren's troops advanced on this position.

The Federals attacked this position just after noon, and a second line followed at 12:30 p.m. By noon, ten thousand Confederate troops had massed here to oppose Warren, and another forty-five hundred of Early's men were coming up the road in reserve. Ayres's initial attack would strike just north of the road and hit Steuart's brigade. As the Confederates fell back and the Federals pursued, Ayres's flanks became exposed, allowing the Confederates to turn the tide. Adding to Ayres's problems, Stafford's brigade on Steuart's left was on a ridge that overlooked its flank. Stafford's men added their weight to the fight, and the Federals were forced to retreat.[58]

At this location on the south side of the road, Steuart's First North Carolina brigade was positioned nearest the road. The six Virginia regiments of Jones's brigade followed. Jones's brigade was well supported on its left by Steuart's brigade, but on its right it was vulnerable. Bartlett's advance would strike in this area, pushing first the Twenty-Fifth Virginia Regiment deployed in the field in front of you and then all of Jones's brigade back into Brig. Gen. Cullen Battle's brigade of Brig. Gen. Robert Rodes's division. Due to the force of the Union attack, the instructions not to bring on a general engagement, and the lack of reserves, the Confederates started to fall back. On Bartlett's left was Brig. Gen. Lysander Cutler's First Brigade of Brig. Gen. James Wadsworth's Fourth Division. It struck the Confederate lines near the Higgerson Farm about a half mile southeast of here and drove the Confederates back almost a mile.[59]

> **Report of Brig. Gen. Lysander Cutler, USA, Commanding Fourth Division**
>
> HDQRS. FOURTH DIVISION, FIFTH ARMY CORPS,
> August 13, 1864.
>
> COLONEL: I have the honor to submit the following report of the operations of this division during the campaign commencing May 4, and ending July 30, 1864:
>
> FIRST EPOCH.
>
> ... This division was formed in line of battle on the right of Crawford [whose First Brigade was moving up from Chewning Farm toward the Higgerson Farm], and ordered to push forward and find the enemy and attack him. The line was formed and moved forward at 12 m. Cutler on the right, Stone in the center, and Rice on the left, next Crawford. The enemy was soon found and attacked. He was driven nearly a mile by Cutler's brigade, capturing 289 prisoners and three battle-flags. Rice lost nearly all of his skirmish line as prisoners, and a large number of men and officers killed and wounded. Stone's brigade gave way soon after meeting the enemy, thus letting the enemy through our line. The First Brigade (Cutler's) continued to drive the enemy until it was ascertained that the troops on both flanks had left, and that the enemy was closing in his rear, when he was obliged to fight his way back, losing very heavily in killed and wounded.

Again, this success in penetrating the Confederate defenses would expose units' flanks. When Brig. Gen. George Doles's brigade of Rodes's division had arrived earlier, it had taken up a position about one hundred yards south of and parallel to the Orange Turnpike. Bartlett's push back to Battle's position had exposed Doles's left flank as it was exposing his own. Cutler's advance also threatened Doles's right flank. With Doles and Battle barely holding their own, Daniel's brigade of Rodes's division and Gordon's brigade of Early's division were ordered up:[60]

Appendix I

Report of Brig. Gen. John B. Gordon, CSA, Commanding Gordon's Brigade and Early's Division, of Operations May 5–14

HEADQUARTERS GORDON'S DIVISION,
July 5, 1864.

MAJOR: I beg to submit the following report of the operations of my brigade from May 5 to 14, 1864:

On the morning of May 5 I was ordered by Major-General Early to move along the old stone pike from Locust Grove in the direction of the Wilderness Tavern. When within 3 or 4 miles of the latter point I discovered the Confederate troops, who had preceded me, and had engaged the enemy in my front, rapidly retreating, and was informed by Major-General Early and Lieutenant-General Ewell that the enemy was driving back our line in confusion, and received orders to form my brigade at once on the right of the pike, for the purpose of checking the enemy's advance and saving the artillery, which at that time was moving back along the pike under the enemy's fire. I moved my brigade by the right flank and formed at right angles to the road with as much expedition as the nature of the ground and the fire from the enemy's artillery and advancing infantry would admit. Some of my men were killed and wounded before the first regiment was placed in position. As soon as the formation was completed I ordered the brigade forward. The advance was made with such spirit that the enemy was broken and scattered along the front of my brigade, but still held his ground or continued his advance on my right and left. For the protection and relief of my flanks I left a thin line (Thirty-first and Thirty-eighth Georgia Regiments) to protect my front, and changed front to the right with three regiments (Thirteenth, Sixtieth, and Sixty-first Georgia), and moved directly upon the flank of the line on my right, capturing several hundred prisoners, among them one entire regiment, with its officers and colors. At the same time I caused the regiment on the left (Twenty-sixth Georgia) to make a similar movement to the left, which was also successful. By this time portions of Battle's brigade rallied, and with other troops of Rodes' division came forward and assisted in driving the enemy back and establishing the line, which was afterward held.

With Confederate reinforcements up, the Federals would be forced to retreat with heavy losses. The Thirty-First and Thirty-Eighth Georgia Regiments of Gordon's brigade and Dole's Brigade pushed forward in the woods south of Saunders Field close to 400 yards from Hill-Ewell drive.[61]

Sedgwick's Corps was moving down the Culpeper Mine Road to support Warren, and Wadsworth's troops that had been on his left would be pulled back and sent to aid operations on the Plank Road. Unlike Warren, who neither used his reserve of Robinson's Second Division nor adequately engaged Crawford's Third Division, Ewell involved his whole corps. He had no reserves, and there could be no assistance from Hill on the Orange Plank Road.

When Warren ordered Wadsworth's troops south, Ewell moved Early's Corps, less Gordon, to the north side of the road on Johnson's left, extending his lines in that direction. Fighting in the afternoon was with Wright's First Division of Sedgwick's Corps north of Saunders Field, and it would involve these troops of Early's Corps as well as "Allegany" Johnson's troops. Wright was unable to penetrate the Confederate defenses in this area, only adding to the casualty count. By four o'clock in the afternoon, fighting on this end of the battlefield ceased. Ewell had successfully achieved Lee's desire not to bring on a general engagement, and he sent his aide Maj. Campbell Brown to tell Lee that he could hold his ground "with ease against any force, so far developed."[62]

Additional Nearby Sites of Interest

On your way to the next stop, you will pass the Higgerson Farm, which covers some of the fighting in the area where the Federal advance pushed through and was then forced back.

To Tour Stop 5

Proceed 1.9 miles to the pullout on the right for the Chewning Farm marker.

Tour Stop 5: The Chewning Farm

Tour Stop Relevance to Critical Decisions

The Battle of The Wilderness resulting from Lee's critical decision to attack produced many more localized tactical decisions. The Chewning Farm represents a range of possible outcomes that would have depended on the timing and enthusiasm of the execution of orders. Most notable would be Burnside's inability to reach and secure this area on May 6.

Appendix I

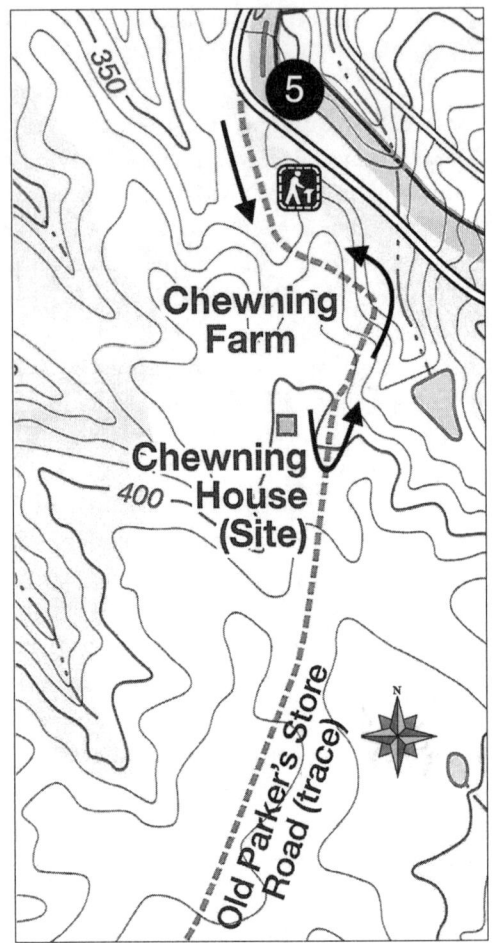

Tour Stop Orientation and Troop Development

After reviewing the interpretive markers, at the pull out, take the short trail for about 250 yards through the farm field to the two interpretive markers at the top of the hill. Of the two interpretive markers, take note of map on the "Key Terrain" marker and how this location is behind the Tapp Field. Both armies would occupy this area at different times during the fighting. The Chewning Farm is one of the few clearings in The Wilderness, and in 1864 it had views of Warren's headquarters at the Lacy House and of the Wilderness Tavern down to the Orange Plank Road. Maj. Washington Roebling,

Warren's aide-de-camp, was with Wadsworth's troops as they moved into this area on the morning of May 5 as part of Warren's overall movement to Parker's Store. He recounted his impressions of the terrain: [63]

> This field was a commanding plateau overlooking the ground to the north and west and connecting with the plank road by two good roads, one leading to Parker's store, and the other to a point a mile east of the store. It became evident at once that it would be of the utmost importance to hold that field, as its possession would divide Lee's army in two parts if he attempted operations down the plank road, in pike at the same time; and if he attempted to pass us intact the 2nd Corps further down the plank road he would fall upon his rear; again it was the best fighting ground in the whole neighborhood.

As the battle progressed, both sides would become more aware of this location and its implications. Situated in the gap between his two corps, it presented a concern for Lee that had to be monitored. If he could control the field, his two corps could be linked. For Grant, the field presented an opportunity to get behind Lee's troops.

Pursuant to orders, at 6:00 a.m. on May 5 Warren started his troops toward Parker's Store along the road that existed at that time and ran by this location. Crawford's Third Division was the first to leave its encampment of the night, and the soldiers arrived at this location in the morning. Crawford informed headquarters that he saw the enemy moving along the Orange Plank Road.[64]

> ONE MILE FROM PARKER'S STORE, May 5, 1864—8 a.m.
> Colonel LOCKE, Assistant Adjutant-General:
> I have advanced to within a mile of Parker's Store. There is brisk skirmishing at the Store between our own and the enemy's cavalry. The general's order is received, and I am halted in a good position.
> S. W. CRAWFORD,
> Brigadier-General, Comdg. Third Division.
> [First indorsement.]

Appendix I

> Major-General HUMPHREYS,
> Chief of Staff:
> The above just received, 9 a.m. General Warren is examining Griffin's front.
>
> Very respectfully, your obedient servant,
> FRED. T. LOCKE,
> Assistant Adjutant-General.
>
> [Second indorsement.]
> MAY 5.
> Dispatch from Crawford received. I have sent to Wilson, who, I hope, will himself find out the movement of the enemy.
> GEO. G. MEADE.

This was the first time that headquarters was informed of the enemy troops moving on this road, and of the possibility that Lee could get between Hancock, who was then near Todd's Tavern, and the rest of the army, which was closer to Wilderness Tavern.

With Warren being forced to attack Ewell along the turnpike on May 5, he wanted to cover his left flank, first with Wadsworth's Fourth Division and then with Crawford's Third Division. In a series of messages, Warren ordered Crawford to move to support Wadsworth. Roebling stated the importance of this location:[65]

> MAY 5, 1864.
>
> General WARREN:
> It is of vital importance to hold the field where General Crawford is. Our whole line of battle is turned if the enemy get possession of it. There is a gap of half a mile between Wadsworth and Crawford. He cannot hold the line against an attack.
>
> W. A. ROEBLING,
> Major and Aide-de-Camp.

But with the mounting pressure on his front, Warren felt he needed Crawford to support Wadsworth:[66]

> MAY 5, 1864—11.50 a.m.
> General CRAWFORD:
> You must connect with General Wadsworth, and cover and protect his left as he advances.
> G. K. WARREN,
> Major-General.

Crawford complied and began to move his men north from here to make the ordered connection. This was a missed opportunity for the Federals. With Crawford at this location, he would likely have caused major disruptions to Confederate troop movements.

When Wilcox's division of Hill's Corps came up the Orange Plank Road, Lee, concerned about this gap, ordered Wilcox to move through the woods and connect with Ewell. Lead elements of Wilcox's men arrived here as the last of Crawford's men were leaving, and they captured some twenty to thirty of the fleeing Federals. Wilcox left the brigades of Brigadier Generals McGowan and Scales here, and he reported to Lee that Federals could be seen at the Wilderness Tavern from this location. Wilcox pushed on with his remaining force and soon met up with Gordon, who was on Ewell's extreme right flank. With Hancock pressing Hill hard on the Orange Plank Road, Lee had sent a courier to Wilcox with orders for him to return to the plank road "with all possible speed." As they were moving back, across this field, they could see the enemy traveling toward the Orange Plank Road.[67]

Although this location was devoid of troops for the rest of the day, it likely occupied the thoughts of both commanders. At dawn on May 6, Hancock's assault had been moved from 4:30 to 5:00 a.m., when he struck Hill's disorganized troops and pushed them back along the Orange Plank Road. Just as his troops were approaching the Tapp Farm, Longstreet's Corps was coming up to stop the rout. Men of Hill's Corps that were not still engaging the enemy were stationed near the Tapp Farm, and once they recovered from the early morning fighting, they were sent here.

Grant had approved Hancock's delay because "of the representation that Burnside could not get up in time." Burnside had been ordered to attack the left flank of Lee's army on the Orange Plank Road. In a recurring theme, both the impact of the thick foliage in The Wilderness and the failure to account for Burnside's slowness would lead to an overly optimistic assessment of troop movements.[68]

Appendix I

> HEADQUARTERS ARMY OF THE POTOMAC,
> May 6, 1864. (Received 8 a.m.)
> General HANCOCK:
> General Wadsworth, with 5,000 on Birney's right is directed to, take your orders. Two of Burnside's divisions have advanced nearly to Parker's Store, and are ordered to attack to their left, which will he your front. They ought to be engaged now, and will relieve you. Our only reserve is Burnside's third division, yet here, and I don't want to send it, if possible.
> GEO. G. MEADE,
> Major-General

Burnside had started early in the morning, and he arrived at the Wilderness Tavern at daybreak. The men continued forward, and after marching about a mile, they stopped for breakfast. Moving along the road with two divisions, they were approaching Parker's Store at about the same time as Hill's men. Hill and his staff had ridden just ahead of his troops, and as they dismounted while moving to the house, they saw Federals approaching the far side of the field. Hill directed his staff as follows: "Mount, walk your horses, and don't look back." When the men were out of sight, Hill directed an aide to go back to headquarters for reinforcements. Anderson's division was just coming up to the Tapp Farm, and a brigade was sent to Hill, who then engaged the enemy and stopped Burnside's advance.[69]

These Union troops were Burnside's lead division, commanded by Brig. Gen. Robert Potter. Men of the Sixth New Hampshire had seen a horseman, likely someone in Hill's party, riding slowly away from the cabin on a road leading southeast. These troops halted near the cabin, and they shortly saw Rebel troops moving toward them. Forced to retreat, the Federals lost many prisoners. This missed opportunity was the Union troops' last chance to control this important location. Col. William Palmer, Hill's chief of staff, reported this exchange:[70]

> As I passed a group of prisoners an officer asked, "Were you not at the house a short time ago?" I told him, "Yes." He abused his officers and said, "I wanted to fire on you, but my colonel said you were farmers riding from the house."

The rest of Potter's troops came up and engaged in battle. Just as Potter was getting ready to charge the enemy position, he received orders to move to the left to support Hancock on the Orange Plank Road.[71]

> BURNSIDE'S HEADQUARTERS,
> May 6, 1864—10 a.m. (Received 10.50 a.m.)
> Lieutenant-General GRANT:
> Burnside has gained 1½ miles to his left to connect with Wadsworth, and now moves at once toward Hancock's firing, with Potter's division deployed, supported by a brigade. I should think Hancock's firing a mile away.
> C. B. COMSTOCK,
> Lieutenant-Colonel, &c

Just north of here, Maj. Gen. Stephen Ramseur's brigade had attacked and turned Burnside's advance toward this location. With Burnside's troops turning south to help support Hancock, Lee had succeeded in linking his two corps. As a result, this location would remain in Confederate control until the move to Spotsylvania Court House.[72]

To Tour Stop 6

From the Chewning Farm, proceed 1.5 miles to the intersection with the Orange Plank Road. Note 1: A quarter of a mile before the intersection, you will see a pullout for stop 6 of the National Park Service Wilderness Battlefield Tour on the right. This is on the far side of the Widow Tapp Field, and it has two interpretative markers. Note 2: Just before the intersection, you will see a pullout on the right side of the road. This features interpretative markers giving an overview of the fighting on May 6, 1864. Turn right onto Orange Plank Road, and go 0.5 mile to the "Battlefield Tour" arrow pointing to the Widow Tapp Farm pullout on the right.

Tour Stop 6: Widow Tapp Farm

Tour Stop Relevance to Critical Decisions

When Lee's army moved out on the morning of May 5 on the Orange Turnpike and Orange Plank Roads, he accompanied Hill and his Third Corps to this location, where they established their headquarters. The fighting on

Appendix I

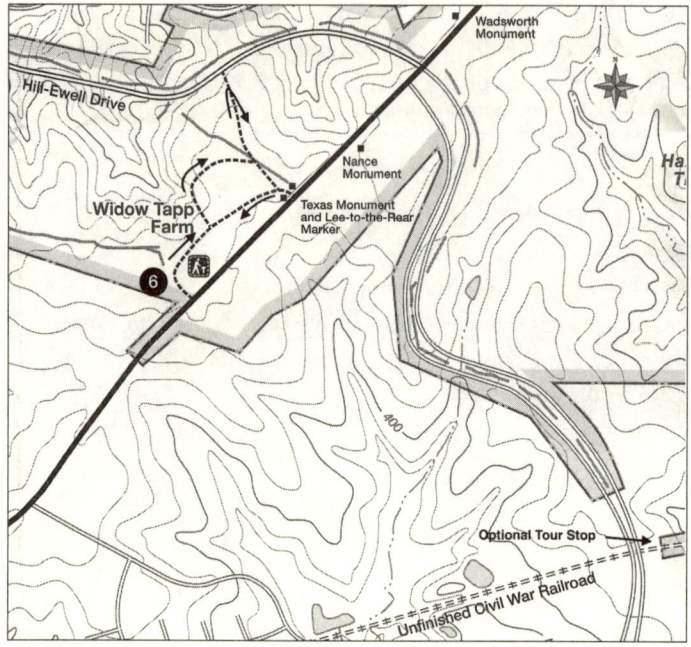

this road would be "fought immediately under the eye of the Commanding-General," and would inspire the troops in their unequal contest. It was from here that Lee made his critical decisions to leave Heth's and Wilcox's troops in place on the night of May 5, and to assault the Federal lines on the afternoon of May 6. It was also here that Lee first tried to personally lead his men in a desperate charge to stem a Federal advance.[73]

Tour Stop Orientation and Troop Development

From the parking lot, take the trail to the left of the "Wilderness Campaign" interpretive marker to the cannon. At the time of the battle, Poague's battalion of sixteen guns were arrayed in the open field between one and two hundred yards in front of you from left to right. The right side was near the road. In fighting on May 5, the previous day, Lee had failed to capture the critical junction, and Grant had failed to dislodge Lee. They both would try to rectify that situation.[74]

Face the field with your right shoulder toward the Orange Plank Road.

View from Poague's artillery position across the Widow Tapp Field. Courtesy of the author.

Lee, riding with Hill, arrived here in advance of the troops. Stuart and the generals' staff were also with them. Dismounted and resting, they were surprised to see enemy skirmishers across this field to the left. With virtually no support, the Confederates were extremely vulnerable. The Federal skirmishers were too far off to accurately identify Lee and his staff, and not knowing the Rebel strength, they moved away and back into the woods. This close call demonstrated how The Wilderness could conceal troop movements by either side.[75]

After Heth's division headed out along the Orange Plank Road for the Brock Road intersection to take up a line of battle, Wilcox's division was initially directed toward the Orange Turnpike to link up with Ewell and close the gap between the two corps. Wilcox made it to the Chewning Farm, left McGowan's and Scales's brigades there, and moved on with Lane and Brig. Gen. Edward Thomas. Just as he made contact with Gordon, then positioned on Ewell's extreme right, Wilcox received orders to move back to support Heth, who was under attack. Fighting on the Orange Plank Road would commence with an attack by Getty's division of Sedgwick's Sixth Corps, and Hancock's Second Corps quickly joined the action.[76]

With Hill's Corps so heavily pressed, Lee revised his plans for Longstreet, ordering him to change his direction of march and to come to Hill's aid on the Orange Plank Road. Lee felt that Longstreet would be up before daylight the next day to relieve Hill.[77]

> MAY 5, 1864—7 p.m.
> Lieutenant-General EWELL,
> Commanding, &c.:
> ... General Longstreet and General Anderson are expected up early
> ...
> C. MARSHALL,
> Lieutenant-Colonel and Aide-de-Camp

Lee sent the following report on the day's fighting to Richmond:[78]

> HEADQUARTERS ARMY OF NORTHERN VIRGINIA,
> May 5, 1864—11 p.m. (Received 6th.)
> Hon. SECRETARY OF WAR:
> ... A strong attack was made upon Ewell, who repulsed it, capturing many prisoners and four pieces of artillery. The enemy subsequently concentrated upon General Hill, who, with Heth's and Wilcox's divisions, successfully resisted repeated and desperate assaults. ...
> R. E. LEE.

The vulnerable position of the Confederate line needed to be rearranged, and its condition caused Wilcox and Heth great concern. With his headquarters so close to Lee's, Wilcox went to see him about rectifying the situation. Before Wilcox could state his concerns, Lee informed him that Hill's remaining division, commanded by Anderson's, and Longstreet's Corps would be up. Lee also told him, "[The two divisions] that had been so actively engaged will be relived before day." With this assurance, Wilcox made no suggestion about the lines to Lee. Heth's concern for the lines later prompted him to talk to Hill. But after several conversations, Hill emphatically stated that the men were no to be disturbed.[79]

The decision to leave Heth's and Wilcox's troops in place was based on the belief that Longstreet would be up to relieve them, and that they therefore would not become engaged.

At 5:00 a.m. on May 6, nearly half of the Federal army attacked. Hancock's Corps and three brigades of Getty's division of Sedgwick's Corps moved on the Confederate front from the Brock Road. Wadsworth's division plus another brigade of Warren's Corps from the woods, just over one-half mile from here to the northwest, hit Hill's left flank north of the Orange Plank Road.[80]

The combined forces overpowered the ill-prepared Confederates and pushed them from their positions. Wilcox rode back to Lee and reported the conditions of his command. But the Federal advance was not without problems. Soldiers became tangled as they moved "through woods so dense that at the best no body of troops could possibly preserve their alignment." An officer of the 141st Pennsylvania of Birney's Third Division reported, "During the entire morning's operations there had been neither a general nor staff officer along this portion of the line. It seemed to have been left without a commander, and each regiment acted independently of the others."[81]

Wadsworth's advance southward collided with the right flank of Birney's troops and crowded his division toward the left. After an hour's severe fighting, Hill's divisions under Wilcox and Heth were driven for a mile and a half back to the Confederate headquarters here at Tapp Field.[82]

The Confederates fell far enough back to the rear to allow Poague's guns, which the enemy could not see, to be used. As the last of the Confederates were retreating, these sixteen guns, located on the western part of the field, were ordered to open fire. The artillery fire was both unexpected and effective since the terrain had limited its use the day before.[83]

As the assault temporarily slowed, Lee and Hill personally helped to rally the troops. The situation was so desperate that Lee sent his adjutant, Lieut. Col. Walter Taylor, to Parker's Store to prepare the trains for a movement to the rear. While the Confederate line was collapsing back to this location, Longstreet arrived:[84]

> Like a fine lady at a party, Longstreet was often late in his arrival at the ball, but he always made a sensation and that of delight, when he got in. . . .

Less than a brigade had left the road when Longstreet in person arrived. He was informed where General Lee would be found—within one hundred and fifty yards.[85] Longstreet's lead troops were Maj. Gen. Charles Field's and Brig. Gen. Joseph Kershaw's divisions. They were immediately put into action as they came up, Fields to the left of the road into the Tapp Farm clearing and Kershaw's to the right. Brig. Gen. John Gregg's Texas Brigade was the first unit of Field's men to enter the fray, followed closely by Brig. Gen. Henry Benning's Georgians and Col. William Perry's Alabamians. In the ensuing excitement as Gregg's men moved through Poague's artillery, Lee was ready to personally lead the counterattack. Accounts of this incident, typically referred to as "Lee to the rear," differ. Basically, however, the troops made the general turn back with the assurance that they would go forward.

Gregg's, Benning's, and Perry's brigades pressed on and halted the Federal advance, and several documents indicate that Lee's presence inspired them. Gregg's unit suffered a 50 percent casualty rate, and some of the companies in the other brigades "were entirely obliterated." A monument on the Tapp Field walk commemorates this desperate moment. Traveling along this 0.6-mile loop trail, shown on the map for this stop, is recommended to get a better appreciation of the rapidly evolving fighting in this area on the morning of May 6. The spur trail takes you back to Hill-Ewell Drive and the markers you passed just before the Brock Road intersection.[86]

As the rest of Longstreet's Corps came up, they were placed into the battle. Hill's last division under Anderson came up about 8:00 a.m. and was put under Longstreet's direct command and held in reserve. As Longstreet was funneling his troops into battle, Hill and the men of Wilcox's and Heth's unengaged divisions moved toward the Chewning Farm.[87]

Along and just south of the Orange Plank Road, Kershaw's division, now led by Henagan, moved to confront the Federal tide. The terrain south of the road had more dense undergrowth and scrubby pine, making movements here more difficult than in the open field to the north. As Henagan's troops moved forward, the men of Heth's division streamed back. Birney's and Getty's Federals followed closely. Humphreys's brigade was ordered to Henagan's left and Bryan's brigade to his right. By this time, the Federal line had become disorganized, and it was running low on ammunition. Each regiment seemed to act independently of the others, and the troops were pushed back as far as the present day location of the Vermont Monument at Tour Stop 7.[88]

The artillery and the countercharge had their intended effect. Reports from troops in Wadsworth's leading units along the Plank Road were consistent:[89]

Report of Brig. Gen. Lysander Cutler, USA, Commanding Fourth Division.

At 4:30 a.m. on the 6th we moved forward, attacked the enemy, and drove him across the plank road, where a junction was made with the Second Corps. The division was then formed in four lines, the left resting on the plank road. These lines were, by order of General Wadsworth, closed in mass to avoid the artillery fire of the enemy. While in this position it was furiously attacked by infantry and artillery, driven back, and badly scattered, a large portion of them taking the route over which they had marched the night before.

Report of Col. Richard Coulter, Eleventh Pennsylvania Infantry, Commanding Second Brigade

Shortly after daylight, 6th instant, advance was resumed, Twelfth Massachusetts as skirmishers, General Cutler's brigade, Fourth Division, on left, . . . Enemy's skirmishers were driven with small loss, and plank road soon gained, when Hancock's line was met advancing, and direction was changed to the right. Moved now along both sides of plank road about one-quarter mile, under brisk fire, when farther advance was checked by strong force of enemy, supported by artillery

By 8:00 a.m. Longstreet's counterattack had ground to a halt, and after several hours of intense fighting, neither side had made much progress. About 10:00 a.m., the Army of Northern Virginia's chief engineer, Major General Smith, reported to Longstreet that his reconnaissance had found a way for the Confederates to use an unfinished railroad cut to move unseen on Hancock's left flank. Need had dictated the arriving units' hasty deployment, resulting in their intermixture. Using available troops on the south side of the Orange Plank Road, Longstreet ordered his chief of staff Lieut. Col. Moxley Sorrel to conduct Brigadier Generals Anderson and Wofford (of Field's and Kershaw's division, respectively, and both of his corps), as well as Brig. Gen. "Billy" Mahone (of Anderson's division of Hill's Corps), on the route Smith had recommended. Longstreet would support the flank attack with a simultaneous frontal one.[90]

Since the fighting and effects of the flanking movement occurred nearer to the Brock Road, the details of this attack's impact are covered in the discussion of the next stop, Tour Stop 7.

Just after noon, following Sorrel's successful flanking attack and the subsequent routing of Federals on both sides of the Orange Plank Road, Longstreet went forward to ascertain how to press the advantage. He rode down the Orange Plank Road from the Tapp Farm with Brig. Gen. Micah Jenkins of Field's division, along with staff and couriers. The group was about thirty yards in front of Fields when men of Mahone's brigade mistook them for Union troops and fired on them. Jenkins was killed and Longstreet seriously wounded. Before being removed from the road, Longstreet turned command over to Field and urged him to press the attack before the Federals could rally. In addition, Union Brigadier General Wadsworth was mortally wounded in Longstreet's flank attack. Wadsworth's memorial is on the left just as you pass the Hill-Ewell Drive intersection. There is no marker where Longstreet was wounded, but an interpretive marker is located at the next stop.[91]

This is the situation Lee faced when he had to make the critical decision to either try and maintain the initiative or withdraw, make a strong defensive position, and let Grant come to him. From his location here at the Tapp Field, Lee had firsthand knowledge of both the attacks' success and the Confederate troops' disorder. He ultimately decided to keep the initiative. However, Lee needed to rectify the lines before he could move against the Federals, who had fallen back to their works along the Brock Road. As with all operations in The Wilderness, the thick foliage would be a liability in troop movement. Small trees that had been hit by bullets and thus bent over made the already dense undergrowth even worse, creating an almost impassable barrier. It would take almost four hours for Lee to get the troops realigned and ready to assault the ever-improving Federal defenses along the Brock Road.[92]

When the Rebel lines were finally rectified, Longstreet's two divisions moved out. Field's unit traveled north and along the Orange Plank Road, and Kershaw's men set out south of the road. At Gettysburg, the open field that Lee's frontal assault had to move across had allowed Meade's troops to disrupt the assault formations. Here, however, it was the thick foliage that disorganized the units as they advanced.

This assault would hit the Federal position about 4:15 p.m., and it is covered in material concerning Tour Stop 7.

In his last dispatch of the day, sent to Richmond from this location at 8:00 p.m., Lee summarized the actions of May 6 and listed the fallen general officers:[93]

> HEADQUARTERS ARMY OF NORTHERN VIRGINIA,
> Via Orange Court-House, May 6, 1864—8 p.m.
> (Received Richmond, 4.45 p.m. 7th.)
> SECRETARY OF WAR:
> Early this morning as the divisions of General Hill, engaged yesterday, were being relieved, the enemy advanced and created some confusion. The ground lost was recovered as soon as the fresh troops got into position and the enemy driven back to his original line.

> Afterward we turned the left of his front line and drove it from the field, leaving a large number of dead and wounded in our hands, among them General Wadsworth. A subsequent attack forced the enemy into his intrenched lines on the Brock road, extending from Wilderness Tavern, on the right, to Trigg's Mill. Every advance on his part, thanks to a merciful God, has been repulsed. Our loss in killed is not large, but we have many wounded; most of them slightly, artillery being little used on either side. I grieve to announce that Lieutenant-General Longstreet was severely wounded and General Jenkins killed. General Pegram was badly wounded yesterday. General Stafford, it is hoped, will recover. [He did not.]
>
> R. E. LEE.

To Tour Stop 7

Leaving the Widow Tapp Farm pullout, turn left, and proceed 1.25 miles to the pullout on the right just before the Brock Road (Virginia Route 613) intersection. A side stop is available to the unfinished railroad cut used in Longstreet's flank attack. The Fawn Lake community is located just before the Hill-Ewell intersection. Turn right onto Longstreet Drive, and tell the guard that you are going to the Civil War–era unfinished railroad cut. Proceed almost 1 mile to just past the athletic fields on the right. There, a left turn will take you to the opposite-direction lane of Longstreet Drive and to a small parking lot. After parking in the lot, walk back out to the road. Part of the unfinished railroad is just south of the parking lot. Return to the Orange Plank Road, turn right, and proceed to next stop. As you proceed east on the Orange Plank Road just past Hill-Ewell Drive, you will see the monument to Wadsworth on your left near the place where he was mortally wounded. There is a very small pullout here, and as traffic can be heavy at times, this stop is tricky. Continue east on Orange Plank Road just over 0.1 mile and past the "Longstreet Wounding" sign on the right. Pull into the small parking space. This is the location of both Longstreet's wounding and the flank attack's position. Noting traffic, continue east a short distance to the parking lot on your right just before the intersection of the Orange Plank and Brock Roads.

Appendix I

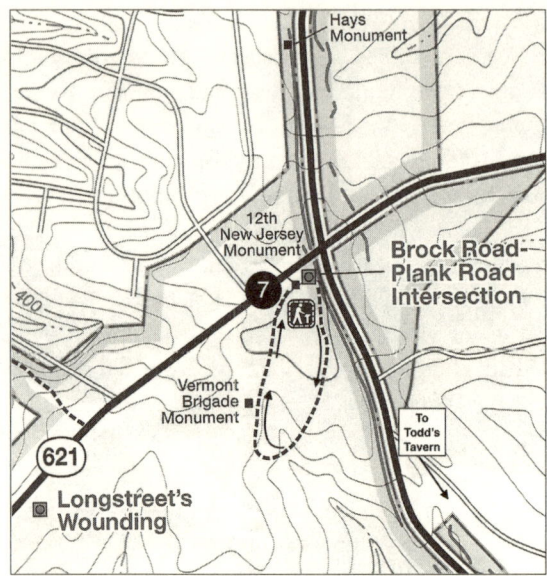

Tour Stop 7: Orange Plank and Brock Road Intersection

Tour Stop Relevance to Critical Decisions

This location is related to three critical decisions: Lee's to attack Grant, Lee's to leave the men in place, and, finally, Lee's to press the attack after Longstreet's wounding on the afternoon of May 6. Just as Grant could have split Lee's forces at the Chewning Farm, Lee could have split the Army of the Potomac by gaining control of this intersection.[94]

Tour Stop Orientation and Troop Development

From the parking area, proceed to the Wilderness Compass Rose located at the southwest corner of the intersection. As you walk there, you will pass the trail to the Vermont Monument, which is the trail by which you will return to the parking lot. Note that the same distance is marked on the compass from here to Washington and to Richmond. The second part of Lee's decision to move against Grant's army as it negotiated The Wilderness occurred here on May 5, 1864. Looking back down the Orange Plank Road, Heth's lead division of Hill's Corps marched to this critical intersection. Looking roughly north along the Brock Road, Getty, a member of Sedgwick's Corps, marched to this location. Looking roughly south, lead elements of Hancock's division were at Todd's Tavern, four miles away, and starting to march here.

From his headquarters at the Widow Tapp Farm, Lee had ordered Heth to move down the Orange Plank Road and take the intersection without bringing on a general engagement. At 8:00 a.m. Brig. Gen. Samuel Crawford of Warren's Corps, who was at the Chewning Farm, reported to Warren's headquarters, "There is brisk skirmishing at the Store between our own and the enemy's cavalry." Headquarters received a second report at 10:15 in which Crawford stated that the enemy was moving to get on his flank. Brig. Gen. George Getty, Sedgwick's Second Division commander, received orders from Meade to quickly move to the junction and support the cavalry. When he got here, Getty found the Federal cavalry retreating in the face of infantry skirmishers from Brig. Gen. John Cooke's brigade of Heth's division. Getty's First Brigade, commanded by Brig. Gen. Frank Wheaton, was brought up "double-quick" and pushed the enemy back. "The rebel dead and wounded were found within 30 yards of the cross-roads, so nearly had they obtained possession of it."[95]

Face down the Orange Plank Road that you just arrived on.

Getty notified Sedgwick of Hill's presence, then began to deploy his brigades in two lines along the Brock Road. Wheaton's men were on both sides of the Orange Plank Road, Brig. Gen. Henry Eustis's Fourth Brigade was on the right, and Brig. Gen. Lewis Grant's Second (Vermont) Brigade was on the left. For the next few hours, only skirmishing occurred. Hancock received orders to move up the Brock Road from Todd's Tavern at 11:40 a.m., and to be prepared to move out on the Orange Plank Road. The head of Hancock's column reached Getty's left at 3:30 p.m. Hancock informed Meade that he would attack with two of his divisions and Getty's when they got into position. Once again, a corps commander was ordered to attack without proper understanding of the situation and suitable preparation.[96]

HEADQUARTERS ARMY OF THE POTOMAC,
May 5, 1864—3.15 p.m.
Major-General HANCOCK,
Commanding Second Corps:
The commanding general directs that Getty attack at once, and that you support him with your whole corps, one division on his right and one division on his left, the others in reserve; or such other disposition as you may think proper, but the attack up the plank road must be made at once.
A. A. HUMPHREYS,
Major-General and Chief of Staff

Appendix I

The narrow and heavily wooded sides of Brock Road had slowed Hancock's progress, and the head of Hancock's Corps, Maj. Gen. David Birney commanding the Third Division, was just arriving at 3:30—later than Hancock's whole corps was assumed to be in position and ready to attack. Mott's Fourth Division was behind Birney's, and Gibbon's Second and Barlow's First Divisions were still two miles away struggling up the Brock Road. Getty received orders from Meade to attack at once without waiting for Hancock. "This order was reiterated by Lieut. Col. Lyman, of General Meade's staff, in person." Getty launched his attack at 4:15 p.m. with his own troops on both sides of the Orange Plank Road.[97]

While the Federals were moving troops into position along the Brock Road and making field fortifications, Heth's lead division of Hill's Corps was moving into place about a mile down the Orange Plank Road and taking advantage of the terrain as much as possible. Heth placed Brig. Gen. John Cooke's brigade across the road, with Brigadier General Kirkland directly behind, Davis's brigade (commanded by Col. John Stone in Davis's absence) north of the road to Cooke's left, and Brig. Gen. Henry Walker's brigade south of the road on Cooke's right.[98]

If Getty's, Birney's, and Mott's troops could make a coordinated attack, Heth's 6,500 troops would initially face perhaps 17,000 Federals. Maj. Gen. Cadmus Wilcox's division had been sent to link up with Ewell's right flank near the turnpike, thereby closing the gap between the two Confederate corps. Brig. Gen. Richard Anderson's division, which had left Orange Court House earlier in the day, was far down the road.[99]

Getty's attack south of the road saw Grant's Second Brigade moving out through the dense underbrush and soon encountering the enemy. Brig. Gen, Lewis Grant said "The rebels had the advantage of position, inasmuch as their line was partially protected by a slight swell of ground, while ours was on nearly level ground." The musketry fire was so intense that neither side could advance. With Birney's division of the Second Corps coming up the Brock Road, three regiments from his First Brigade were sent in to break the impasse. Their limited initial success gave way to heavy losses, but they held their line until dark. Finally, Brig. Gen. Gresham Mott sent the two brigades of his Fourth Division forward. Col. Robert McAllister's First Brigade went in behind the Vermonters, while Col. William Brewster's Second Brigade (the Excelsior Brigade) went in but drifted into McAllister's troops. Both units were subject to heavy fire. The brigades' pulled back to their original position behind the breastworks they had constructed along the Brock Road upon arriving. Three lines of works can still be seen along the Brock Road, two to the west and one to the east.[100]

Report of Col. Robert McAllister, Eleventh New Jersey Infantry, Commanding First Brigade, Fourth Division

On receiving the enemy's fire, to my great astonishment, the line began to give way on the left. It is said first the Excelsior Brigade, then my left regiment—First Massachusetts Volunteers—and regiment after regiment, like a rolling wave, fell back, and all efforts to rally them short of the breast-works were in vain. To assign a cause for this panic is impossible, unless it was from the fact that a large number of troops were about to leave the service. I think this had much to do with it.

McAllister's concern that expiring troop enlistments were "a cause of the panic" was validated two days later when Lee cautioned Ewell about misleading Federal troop movements:[101]

HEADQUARTERS,
May 7, 1864—10.30 a.m.

Lieut. Gen. R. S. EWELL,
Commanding, &c.:

GENERAL: General Lee directs me to say that the Richmond Dispatch, of yesterday, contains extracts from a Northern paper which state that the United States Government has acceded to the demand of the Pennsylvania troops to be discharged at the expiration of three years from the date of their muster into the State service, instead of the United States service, and that 5,000 men will thus be lost to Grant's army. It is said that the time of their discharge is to-day, but the general does not know certainly. Some two-years' men were captured at Chancellorsville last year whose term of service expired a few days after the battle, and it may be that the three-years men enlisted at the same time. The general thinks it best to bear this in mind, to avoid being misled by movements to the rear. I inclose [sic]the latest from General Stuart.

Very respectfully, your obedient servant,

C. MARSHALL,
Lieutenant-Colonel and Aide-de-Camp.

Despite the retreat of Mott's men and the mounting Confederate pressure, the One-hundred-twenty-fourth New York Regiment of Birney's First Brigade was able to move forward and hold their line.

On the north side of the road, Brig. Gen. Frank Wheaton's First Brigade advanced, and Brig. Gen. Henry Eustis' Fourth Brigade advanced on his right flank. When Wheaton's men had advanced about an eighth of a mile, they met Cooke's Confederates. Wheaton's first line continued to fire until its ammunition was exhausted, at which time it was replaced by a second line, but they too could not budge the Confederates. They remained in this position, sustaining heavy losses, until they were relieved at 6:00 P. M. by a Brig. Gen. Alexander Hays's Second Brigade of Birney's Division of the Second Corp.[102]

Eustis' troops fared no better. The thick foliage obscured the Confederates, and when they got close to Stones' advanced sharpshooters, they received a withering fire. The sharpshooters withdrew to their lines, but the advancing Federals had been stopped, and could go no further. In compliance with orders to support Getty's efforts, Birney sent Brig. Gen. Alexander Hays Second Brigade north of the road. It too struggled with the terrain, and when it became engaged, it took severe losses, including Hays who was killed about one-quarter of a mile north of here and is marked with a monument. Making no progress and taking casualties, they moved back.

Located at the Widow Tapp Farm (Tour Stop 6), Lee had initially sent Wilcox's division north to link up with Ewell's right flank and plug the gap between the two corps. When Heth's division, located near the Brock Road, was attacked, Lee was forced to abandon his effort to link the two units. He recalled Wilcox's troops and ordered them to the Orange Plank Road to support Heth's heavily pressed division. Brig. Gens. Samuel McGowan's and Alfred Scales's brigades were the first to arrive. McGowan moved basically along the road, and Scales traveled farther to the south. The next brigade in was Brig. Gen. Edward Thomas's, which went to the north side of the road in support of Stone's brigade, which faced nearly parallel to the road. Wilcox's last brigade, Brig. Gen. James Lane's, went to the extreme right and pushed forward. With the addition of Wilcox's troops, the Confederates were able to make some progress toward the Brock Road. Meade's aide-de-camp was with Hancock near this intersection, and he reported as follows:[103]

HEADQUARTERS SECOND CORPS,
May 5, 1864—5.05 p.m.
Major-General MEADE,
Headquarters Army of the Potomac:

> GENERAL: There is a general attack as per diagram. It holds in some places, but is forced back to the Brock road on the left. Gibbon is just coming up to go in, and Barlow is to try a diversion on the left; a prisoner of Archer's (Tennessee) division says he was told that Longstreet was to-day on their right.
> Respectfully,
> THEO. LYMAN,
> Lieutenant-Colonel, Volunteer Aide-de-Camp.

Gibbon's Second and Third Brigades were sent to the vicinity of this intersection where you now stand. Barlow's Second and Fourth Brigades moved against the Confederate right flank just down the road. These troops went in with those Federal units already in place, and the intense fighting continued.[104]

> MAY 5, 1864—5.50 p.m.
> Major-General MEADE,
> Headquarters Army of the Potomac:
> GENERAL: We barely hold our own; on the right the pressure is heavy. General Hancock thinks he can hold the plank and Brock roads, in front of which he is, but he can't advance.
> THEO. LYMAN,
> Lieutenant-Colonel, Volunteer Aide-de-Camp.
> Fresh troops would be most advisable.

Fighting continued until dark, with the Federals pushing the Confederates back from the Brock Road area. A 7:00 p.m. communication to Ewell from Lee's headquarters described repeated enemy assaults and ongoing fighting but also stated, "We hold our own as yet."[105]

Although some Federal troops had moved back to their defenses along the Brock Road, many remained where they were when the fighting finally subsided. The hard-pressed Confederates remained in place. A common theme in the writings of Northern and Southern participants in the fighting is the impact the terrain and dense foliage had on the operations. Phrases such as "The thicket was so dense that neither [enemy] could see the other" and "engaging the enemy's infantry in tangled brush" were typical of these soldiers' accounts. The limited visibility made it almost impossible to see adversaries until they fired their weapons, and this situation gave a decided edge to those holding defensive positions. Moreover, managing troop movements in these conditions was nearly hopeless. Units often became separated from one another, and they got turned in the wrong direction. Here, where the fighting had moved back and forth over the same ground, lines became hopelessly entangled. Heth told Hill that a skirmish line could drive his two divisions with the lines as they now were.[106]

Here, as at Saunders Field, the Federals had not been able to mount a coordinated attack, and the fighting had ended in a draw. Just north of here, Wadsworth's division had been sent to hit Hill's exposed left flank. However, its late start, the difficult terrain, and the Fifth Alabama's spirited demonstration allowed darkness to halt the movement. Major General Humphreys, Meade's chief of staff, lamented, "An hour more of daylight, and he [Hill] would have been driven from the field." The Confederate command found itself in this situation while deciding whether to withdraw the men or leave them in place in anticipation of Longstreet's relieving them. The decision to leave the soldiers in place would play out in the morning.[107]

The May 6 fighting for this intersection would have three distinct phases: Hancock's early morning assault; Longstreet's arrival and counterattack stopping Hancock, followed by a Confederate flank attack; and Lee's last assault to take the Brock Road.

On the evening of May 5, Grant formulated his strategy for the next morning. Hancock and Wadsworth would strike Hill at 4:30 a.m.—hopefully before Longstreet came up. Burnside was to move to the gap (the Chewning Farm) between Ewell's and Hill's two corps in the field and move against Hill's rear. Warren and Sedgwick were to attack along their fronts and prevent Ewell from supporting Hill. Meade felt that the troops could not get into position by 4:30 a.m. and suggested a start time of 6:00 a.m. Concerned that Lee might take the initiative, Grant decided on 5:00 a.m.[108]

HEADQUARTERS ARMY OF THE POTOMAC,
May 5, 1864—11.30 p.m. (Received 12 midnight.)
Major-General WARREN,
Commanding Fifth Corps:
The attack ordered for to-morrow will be made at 5 a.m., instead of 4.30. You will make all your arrangements accordingly and attack punctually at that hour.
By command:
A. A. HUMPHREYS,

> Major-General and Chief of Staff.
> (Same to Generals Sedgwick and Hancock.)
>
> HEADQUARTERS ARMIES OF THE UNITED STATES,
> Near Wilderness Tavern, May 5, 1864—8 p.m.
> Major-General BURNSIDE:
> Lieutenant-General Grant desires that you start your two divisions at 2 a.m. to-morrow, punctually, for this place. You will put them in position between the Germanna plank road and the road leading from this place to Parker's Store, so as to close the gap between Warren and Hancock, connecting both. You will move from this position on the enemy beyond at 4.30 a.m., the time at which the Army of the Potomac moves.
> C. B. COMSTOCK,
> Lieutenant-Colonel and Aide-de-Camp.
> If you think there is no enemy in Willcox's front, bring him also.
> C. B. C.

Grant's intended coordinated assault on Hill's two battered divisions of his Third Corps began at 5:00 a.m. Hancock had arranged his three corps to give depth to the attack, and they struck on both sides of the Orange Plank Road. To the north, Wadsworth's division was bearing down on Hill's left flank. But Burnside's Corps would be late and not in a position to deliver a fatal blow. First south of the road and then north, Hill's troops were pushed back. Isolated pockets of resistance emerged, but Hancock's juggernaut could not be stopped. On Hill's left flank, Wadsworth struggled through the difficult terrain and some rare artillery fire to press from the north, eventually pushing the Confederates to the Orange Plank Road and linking up with Hancock's troops. Hill's position that could not have stopped a skirmish line, as Heth said, had no chance against this force.[109]

> HDQRS. SECOND CORPS, ARMY OF THE POTOMAC,
> May 6, 1864—5.40 a.m.
> [General A. A. HUMPHREYS:]
> GENERAL: We have driven the enemy from their position, and are keeping up the plank road, connected with Wadsworth, taking

> quite a number of prisoners. My attack is being made with three divisions on both sides of the plank road.
>
> WINF'D S. HANCOCK,
> Major-General of Volunteers.

As the Federals pressed on, some of the momentum of the assault faded. With Wadsworth's men coming in from the north and Hancock's from the east, the attackers became more bunched up and less effective. Once again, the terrain made it impossible to coordinate troop movements, and the intense firing had left the forward units low on ammunition. About 7:00 a.m., Brig. Gen. Alexander Webb's First Brigade of Gibbon's division was ordered north of the Orange Plank Road to support Getty's troops. When Webb advanced through the dense undergrowth, he completely missed Getty and instead ran into Confederates. The enemy eventually forced him to retreat. At 8:00 a.m. a brigade of Stevenson's division of Burnside's Corps arrived and was put into action, but again to no benefit.[110]

Hancock now faced a dilemma about how to support the assault here. The Federals knew they were engaging two of Longstreet's three divisions, but they did not know where Pickett's division was located. The unit was actually near Richmond.[111]

> **Reports of Maj. Gen. Winfield S. Hancock, USA, Commanding Second Army Corps**
>
> The enemy was now making some demonstrations on my extreme left, which led me to apprehend an attack in that direction and gave me some uneasiness, but I was notified at 8.15 a.m., by a dispatch from General Humphreys, that General Sheridan, with one division of cavalry, had been directed to attack the enemy on the Brock road. It was supposed that Longstreet's corps was marching on that road toward my left. . . . The impression that Longstreet was executing the flank movement, concerning which I had been cautioned during the night, was strengthened by a report that infantry was moving on the Brock road from the direction of Todd's Tavern about 2 miles from my left.

After Hancock issued conflicting orders, Col. John Brooke's brigade of Barlow's division was sent down the Brock Road, and "Leasure's brigade,

Ninth Corps, and Eustis' brigade, Sixth Corps, were held in readiness to support Barlow."[112]

Hancock's problems only increased. At 9:45 a.m., he received a message from headquarters that he "make immediate disposition to check" the enemy movement on Warren's left (Hancock's far right), and that there were no troops to spare. In response, Hancock directed Birney to send troops to address this issue. Two brigades were sent and were no longer available to fight at this location. By ten o'clock, the troops had reached a stalemate here, and Warren and Sedgwick received orders to suspend their attacks and prepare for an attack on Hancock's right.[113]

Chronologically, Longstreet's flank attack occurred next. It was implemented about three-quarters of a mile west along the Plank Road (past the Wadsworth Monument), closer to Tour Stop 6 and to the south. This assault's impact is covered below.

By late morning, the troops on Hancock's left flank were predominantly those of Birney's, Mott's, and Getty's divisions. The area had been relatively quiet, and the Federal troops were trying to get some rest. Brig. Gen. Robert McAllister's brigade of Mott's division was on the left, and he used the lull to make a reconnaissance of his troops' location. McAllister saw the railroad cut and the enemy in it, and he sent word of his discovery to Mott. But it was too late.[114]

The Confederate flank attack began about 11:30 a.m. As Sorrel was hitting McAllister's flank, units of Kershaw's division began to attack in front. The surprised Federals were then forced to retreat. Colonel Franks's brigade of Barlow's division was on Mott's left, and both the brigade and Mott were overrun. Birney, who was in charge of the troops in the area, advised the troops south of the Orange Plank Road to withdraw to the Brock Road. The Federal troops re-formed in two lines along the road on the same ground they had occupied in the morning before their assault. After the war, Hancock told Longstreet that he had rolled him up like a wet blanket.[115]

On the north side of the Orange Plank Road, the Federals did not fare any better. During his morning attack, Wadsworth had pushed down from the north on Hill's haphazard line, driving the men back to the Widow Tapp Farm. There, Longstreet's troops had come into action just in time to avert disaster. They now occupied the area north of the Orange Plank Road that Wilcox's and Heth's soldiers had held. As his troops tried to form a line along the Orange Plank Road to face Sorrel's flank attack, they were subjected to Confederate artillery firing down the road on their flank. Sharpshooters also took a heavy toll on these troops. In an effort to rally his men toward the front, Wadsworth spurred his horse over some defensive works and was mortally wounded. (As previously mentioned, a marker with a small pullout is located along the north side of the Orange Plank Road just before Hill-Ewell Drive.) With the pressure mounting, Webb, who commanded the troops in the immediate area, ordered his men to "break like partridges through the woods for the Brock Road." Both sides had incurred heavy casualties, and the Federals were essentially in the same position as when they had started. Hancock informed Meade that, due to Longstreet's "heavy attack," he had been forced back to the Brock Road and was reestablishing his line of battle.[116]

The initiative had shifted to the Confederates, but at 3:00 p.m., Hancock received orders concerning an attempt to possibly regain it. He was to rest the troops until 6:00 p.m. and then make a "vigorous attack." Hancock used the time to strengthen his defensive works and position the reinforcements he received, which included a mixture of units. Seven different brigades from all three of Meade's army corps would be present on the north side of the Orange Plank Road. Getty's troops were positioned here at the intersection with three brigades as reserves back along the Orange Plank Road. Birney's and Mott's divisions were situated south of this junction, and Barlow's brigades, which were temporarily under Gibbon, were now on the Union far left flank.[117]

At 4:15 p.m. Lee implemented his critical decision to assault Hancock's position here along the Brock Road. The attack was not as coordinated as hoped. Field stated that "the almost impenetrable growth" and brush prevented some of his men from reaching this point. Protected by breastworks, the Federals stopped the advance and held Confederates all along the line in check for half an hour. The only portion of the line the Rebels would finally breach was in Mott's front just south of the Orange Plank Road; the breastworks caught fire here. Some of the Federal troops in the immediate area retreated. Jenkins's brigade of South Carolinians, now commanded by Colonel Bratton, briefly breached the works, but the troops did so in an area near where reserves and a strong contingency of Federal artillery were located.[118]

Report of Col. James R. Hagood, First South Carolina Infantry

Under a destructive fire I attained the enemy's works and drove him from them. He retired to a second line, keeping up a terrific fusillade, assisted by several pieces of artillery. . . . I abandoned this position only when the troops on my left gave way (there were none on my right during any part of the advance) and the enemy threatened to cut me off.

Brig. Gen. John Brooke of Barlow's division had been ordered up along the Brock Road, and he arrived as Mott's men were retreating.

> ### Report of Brig. Gen. John R. Brooke, USA, Commanding Fourth Brigade
>
> ... about 5 p.m. to move with that part of my brigade in the second line to the assistance of Mott's division. ... I reached General Mott's line in time to see it leave the works (which were on fire in many places) and the enemy plant their colors on them. I at once changed front to the left, and charging drove the enemy from our front. I then had the fire put out, and held the position until relieved by Owen's brigade, of the Second Division.

This was also the area where the reserves had been placed along the Orange Plank Road. A mixed unit under Rice was sent in first, and the third line under Carroll was then sent in.[119]

> ### Report of Brig. Gen. Samuel S. Carroll, USA, Commanding Third Brigade, of Operations May 3–13
>
> I massed the brigade in the third line in rear of his headquarters to the right of the plank road and remained so until about 3.30 p.m., when Longstreet's corps charged and drove a portion of the troops from the breast-works on the Brock road to the left of the plank road and planted their colors there. At this juncture General Birney in person ordered me to regain the breast-works, which I did in double-quick at the point of the bayonet, and shortly afterward resumed my position in rear of his headquarters.

Near this intersection, the artillery of Capt. Edwin Dow of the Sixth Maine Battery and Capt. Frederick M. Edgell of the First New Hampshire Battery delivered a "destructive fire" of shell, case shot, and double-shotted canister.[120]

Appendix I

> ### Report of Capt. Frederick M. Edgell, First New Hampshire Battery
>
> . . . forced back our first lines (whose works were already untenable from the burning of the dry logs composing them), and the batteries in position here at once opened fire with case and canister on the advancing enemy, who, after repeated attempts to carry the works, were driven back with great loss, and the line reoccupied by our troops.

Take the Vermont Monument Trail loop a few yards south of the compass ending back at the parking lot.

The trail is about 0.4 mile long. The remains of the fieldworks that the Federal troops used are still visible along the Brock Road today. Several interpretive markers are situated along the trail, and they describe the difficulty of advancing through the terrain. The trail will take you to the Vermont Monument honoring the brigade commanded by Col. Lewis Grant. The top of this monument is the outline of the Camel's Hump mountain in Vermont.

Lee's frontal assault to deliver a decisive blow to Hancock's line had failed. Afterward, Hancock reported to headquarters:[121]

Vermont Monument marking the fighting location of Col. Lewis Grant's Second Brigade of Getty's Second Division. Courtesy of the author.

> HEADQUARTERS SECOND ARMY CORPS,
> May 6, 1864—5.25 p.m.
>
> Major-General MEADE:
>
> At 4.15 p.m. the enemy made a very determined assault upon my lines, covering a great part of the front. The attack was strongest from the left up to the plank road. The enemy was finally and completely repulsed at 5 o'clock. . . . I wish now to know whether to make the assault you mentioned. . . . The enemy's attack was continuous along my line and exceedingly vigorous. Toward the close one brigade of the enemy (Anderson's brigade) took my first line of rifle-pits from a portion of the Excelsior Brigade, but it was finally retaken by Colonel Carroll. The attack and the repulse was of the handsomest kind. Please send me your orders.
>
> Your obedient servant,
>
> WINF'D S. HANCOCK,
> Major-General.

Meade replied to this report:[122]

> HEADQUARTERS ARMY OF THE POTOMAC,
> May 6, 1864—5.45 p.m. (Received 6.10 p.m.)
>
> Major-General HANCOCK,
> Commanding Second Corps:
>
> Your dispatch is received. The major-general commanding directs that you do not attack to-day. Remain as you are for the present.
>
> A. A. HUMPHREYS,

The fighting in The Wilderness would end with Gordon's flank attack (Tour Stop 3b) later in the evening, and Grant would decide to continue to fight by moving to Spotsylvania Court House.

To Tour Stop 8

Leaving the turnout parking lot, turn right (south) onto Virginia Route 613 (Brock Road). Proceed 4.6 miles to Todd's Tavern and the Virginia Civil War Trails sign for the pullout on the right-hand side of the road.

Appendix I

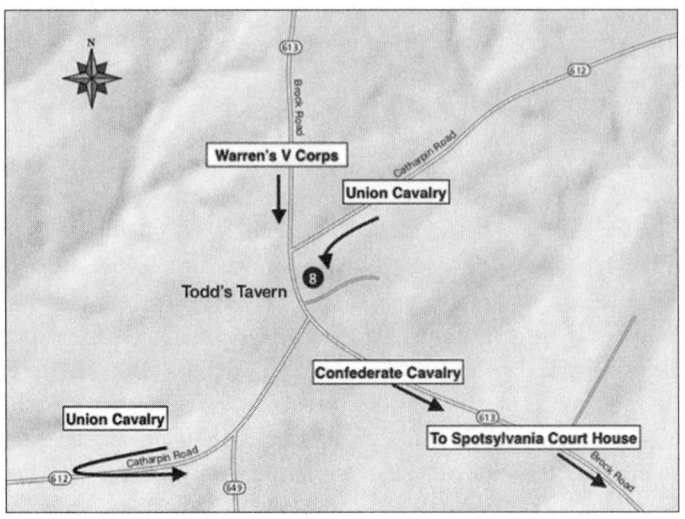

Tour Stop 8: Todd's Tavern

Tour Stop Relevance to Critical Decisions

The intersection of the Brock and Catharpin Roads here at Todd's Tavern had both strategic and tactical significance for the fighting in The Wilderness and the transition to Spotsylvania Court House. Lee's decision to attack in The Wilderness would force Grant to change his plans for Hancock's Corps, starting with Hancock's orders changing his direction of march to move up the Brock Road here at Todd's Tavern on May 5. Later, when Grant made a critical decision to move to Spotsylvania Court House, the Brock Road that passes through this location became a vital part of his planned movement south. The need for a quick passage along this road was crucial to the movement's timing. Fitzhugh Lee's critical decision to fight a delaying action against the Federals' drive south started here. Finally, due to Sheridan's failure to clear this road, the friction between Meade and Sheridan resulted in Grant's critical decision to cut Sheridan loose from the army to pursue the Confederate cavalry.

Tour Stop Orientation and Troop Development

After reading the various interpretive markers, orient yourself by looking at the "Todd's Tavern" marker. In 1864, you would have been looking at the following:

Longstreet's initial route to The Wilderness was along the Catharpin Road and then north on the Brock Road placing him in Grant's path, where

Civil War–era photo of Todd's Tavern. Library of Congress.

he hoped to "intercept the enemy's march." With the crisis developing on Hill's front on the Orange Plank Road, Longstreet's orders were changed. He was to come via Parker's Store on the Orange Plank Road to more quickly relieve Hill.[123]

In Grant's move south, Warren's and Sedgwick's Corps were to cross the Rapidan River at Germanna Ford, while Hancock's Corps would cross at Ely's Ford and camp at Chancellorsville. Hancock's orders on May 4 were to have his Second Division move at 5:00 a.m., followed by his Third and Fourth Divisions, to Shady Grove Church, about three miles west of here along the Catharpin Road (Virginia Route 612). On the morning of May 5, he marched down the Catharpin Road to get on Lee's far right. Hancock's troops were spread from near Chancellorsville to about 1 mile west of this location on the Catharpin Road when he got word to stop:[124]

Todd's Tavern location, present day. Courtesy of the author.

> MAY 5, 1864—7.30 a.m.
> (Received 9 a.m.)
> COMMANDING OFFICER SECOND CORPS:
> The enemy are on the Orange pike about 2 miles in front of Wilderness Tavern in some force. Until the matter develops the major-general commanding desires you to halt at Todd's Tavern.
> A. A. HUMPHREYS,
> Major-General and Chief of Staff.

Lee's assault on Grant would force him to change his plans for his move south. At 11:40 a.m., Hancock received orders to move north along the Brock Road, exposing the Federal flank and rear.[125]

> HEADQUARTERS ARMY OF THE POTOMAC,
> May 5, 1864. (Received 11.40 a.m.)
> Major-General Hancock, Comdg. Second Corps:
> The major-general commanding directs that you move up the Brock road to the Orange Court-House plank road, and report your arrival at that point and be prepared to move out the plank road toward Parker's Store.
> A. A. HUMPHREYS,
> Major-General and Chief of Staff.

The Federals did not know where Longstreet's Corps actually was at this time, and they were concerned. To guard against this exposure, Brig. Gen. David Gregg's Cavalry Division, composed of brigades commanded by Brig. Gen. Henry Davis and Col. Irvin Gregg, would cross Ely's Ford in front of Hancock. Thus these troops were now the closest to support him. Sheridan ordered Gregg to this location with the intended destination of Shady Grove Church, three miles down the Catharpin Road. To support Gregg, Sheridan dispatched Brig. Gen. George Custer's brigade of Brigadier General Tolbert's First Cavalry Division (currently commanded by Brig. Gen. Wesley Merritt). But Custer encountered Brig. Gen. Thomas Rosser's Confederate cavalry at the intersection of the Brock and Furnace Roads, about a mile up the Brock Road from here. Both sides brought up their artillery and pounded away at each other. Rosser eventually pulled back, but he had immobilized the Federal cavalry in this area for the afternoon.

Musketry and artillery firing from this fighting could be heard at headquarters for Hancock's Second Corps. This development only reinforced Hancock's concern that Longstreet might be threatening him from this direction. Adding to the Federals' concern, the Twenty-sixth Michigan of Barlow's division captured "an important dispatch from General Lee." The message suggested that Lee wanted to put artillery near the Trigg Farm, close to the intersection of the Brock and Furnace Roads just north of here.[126]

Here at Todd's Tavern, Maj. Gen. Fitzhugh Lee's cavalry sparred with Gregg's, but Gregg retained control of the intersection. At the end of the day, the Federals controlled two vital intersections on the Brock Road leading to Hancock's flank. But with the continuing dilemma about Longstreet's whereabouts and Meade's concern for the supply trains, both cavalry units were ordered two to four miles back along the roads they were on.[127]

> HDQRS. CAVALRY CORPS, ARMY OF THE POTOMAC,
> Chancellorsville, Va., May 6, 1864—2.20 p.m.
>
> Brigadier-General GREGG,
> Commanding Second Division:
>
> The general commanding directs that you fall back from your present position at Todd's Tavern and relieve General Wilson's division, now occupying Piney Grove Church and Alrich's.
>
> JAS. W. FORSYTH,
> Lieutenant-Colonel and Chief of Staff.

> HDQRS. CAVALRY CORPS, ARMY OF THE POTOMAC,
> Chancellorsville, Va., May 6, 1864—2.30 p.m.
>
> Brigadier-General GREGG,
> Commanding Second Cavalry Division:
>
> GENERAL: General Custer and Colonel Devin have been ordered back to the Furnaces; General Merritt to the plank road leading from this point to Old Wilderness Tavern. General Wilson, on being relieved by you at Piney Branch Church and Alrich's as directed in previous note, will report for orders at these headquarters.

Appendix I

> Very hard infantry fighting along our whole infantry line. General Hancock's left flank has been turned by the enemy.
>
> JAS. W. FORSYTH,
> Lieutenant-Colonel and Chief of Staff

Fitzhugh Lee's cavalry followed, harassing the withdrawing Federals.[128]

By the morning of May 7, Grant had decided to move from The Wilderness. He would need a clear Brock Road to Spotsylvania Court House for the troops to travel, and the task of opening that route would fall to Sheridan's cavalry. Orders to that effect were issued for the troop movements:[129]

> ORDERS.] HEADQUARTERS ARMY OF THE POTOMAC,
> May 7, 1864—3 p.m.
>
> The following movements are ordered for to-day and to-night: ...
>
> 6. At 8.30 p.m. Major-General Warren, commanding the Fifth Corps, will move to Spotsylvania Court-House by way of the Brock road and Todd's Tavern. ...
>
> 13. Major-General Sheridan, commanding Cavalry Corps, will have a sufficient force on the approaches from the right to keep the corps commanders advised in time of the appearance of the enemy. ...
>
> S. WILLIAMS,
> Assistant Adjutant-General

With Gregg's withdrawal from his front the previous evening, Fitzhugh Lee's division of two brigades under Brig. Gen. Lunsford Lomax and Brig. Gen. William Wickham had moved into this area and constructed rough fieldworks. Maj. Gen. Wade Hampton's division was posted about two miles southwest of here along the Catharpin Road at the Corbin's Bridge across the Po River. Note that there is nothing at the bridge today and no place to pull over.

In the morning, Sheridan would use Merritt's First Division, First and Second Brigades under Custer and Devin that had encamped for the night on the Furnace Road. Sheridan would also use Gregg's Second Division with Davies's and Irvin Gregg's brigades located near Piney Branch Church. To start, Merritt was ordered to move his First and Second Brigades back to the position they had held the previous day. At the Brock Road, they encountered units of Fitzhugh Lee's cavalry which "obstinately" held the road behind barricades. The Federals pushed Fitzhugh Lee's troops back but not as far as Todd's Tavern. Meade felt more secure about his supply wagons at ten o'clock, and he authorized Sheridan to detach units for offensive operations. In response, Sheridan directed the First and Second Divisions of cavalry to drive the Confederates from Todd's Tavern. Gregg's two brigades of the Second Division, led by Col. Irvin Gregg and Brig. Gen. Henry Davies, started down the Catharpin and Piney Branch Roads respectively, and Gibbs's reserve brigade of the First Division was sent down the Furnace Road.[130]

Fitzhugh Lee was being threatened from his front and flank, so he withdrew about two miles south of Todd's Tavern and set up new defensive positions. One in front of the other, he deployed his two brigades under Wickham and Lomax across the Brock Road near the intersection of Piney Branch Road.[131]

When Col. Irwin Gregg's brigade reached here from Piney Branch, it pushed on down the Catharpin Road toward Corbin's Bridge, where it encountered the strong position of Hampton's division of Confederate cavalry. After driving the Confederate skirmishers back, Gregg's troops pulled back near Todd's Tavern. The Confederates followed, and a brisk fight ensued, but they could not dislodge Gregg. While Gregg's troops were fighting here, Merritt started out with Gibbs's and Devin's brigades to clear Fitzhugh Lee's troops from the Brock Road. Meanwhile, Davies's brigade was proceeding down Piney Branch Road. Gibbs found the Confederate cavalry dismounted and behind fieldworks, and he attacked with limited success and substantial losses. Devin's troops arrived after 4:00 p.m. Aided by the first line of Confederate works catching fire, Devin was able to drive Wickham's brigade back to the second line held by Lomax just south of the intersection of Piney Branch Road. Davies's brigade arrived at this point and added weight to the Federal assault. Lee's barricades held until dark, when Devin was ordered back to the edge of the woods, and Gibbs encamped on the battlefield.[132]

Sheridan sent a glowing report to headquarters:[133]

> HDQRS. CAVALRY CORPS, ARMY OF THE POTOMAC,
> May 7, 1864—6.15 p.m.
> Major-General HUMPHREYS,
> Chief of Staff:
> I have the honor to report that I attacked the rebel cavalry at Todd's Tavern this afternoon, and, after a sharp and hotly contested action, drove them in confusion toward Spotsylvania Court-House. Our

Appendix I

> cavalry behaved splendidly. I cannot estimate the casualties. Two brigades of General Gregg's and two of General Torbert's [Merritt's] were engaged.
> I am, very respectfully,
> P. H. SHERIDAN,

Before returning to his headquarters at Alrich, several miles from here, Sheridan sent another boastful report from Todd's Tavern at 8:00 p.m.:[134]

> HDQRS. CAVALRY CORPS, ARMY OF THE POTOMAC,
> Todd's Tavern, May 7, 1864—8 p.m.
> Major-General HUMPHREYS,
> Chief of Staff:
> GENERAL: The cavalry made a very handsome fight here this afternoon. We found the whole rebel cavalry here, Hampton's and Fitzhugh Lee's divisions, and drove them on the Spotsylvania road about 3 miles. They were very handsomely repulsed and drove on the road to Beech Grove Church. They had constructed barricades and rifle-pits, which we charged and captured. I had only four brigades engaged—Merritt's, Davies', Colonel Gregg's, and Colonel Devin's. They all behaved splendidly. I captured prisoners from Lomax's, Wickham's, Rosser's, Young's, Gordon's, and Chambliss' brigades, and killed Colonel Collins, of the Fifteenth Virginia Cavalry.
> I am, very respectfully, your obedient servant,
> P. H. SHERIDAN,
> Major-General, Commanding.

Despite what Sheridan reported, the objective of an open Brock Road to Spotsylvania Court House had not been met.

Here at Todd's Tavern, another facet of the growing antagonism between Meade and Sheridan would emerge, one that factored into Grant's critical decision to let Sheridan move independently of the army in pursuit of "Jeb" Stuart's cavalry. Meade and members of his staff arrived here after midnight. They were ahead of Warren's troops, who were leading the army's move to Spotsylvania Court House. Much to his frustration, Meade found Gregg's cavalry division camped here, and Merritt's men camped not too far down the road. Neither division commander had received any further orders. Bypassing the normal chain of command, at 1:00 a.m., Meade ordered Merritt and Gregg to "immediate" action:[135]

> HEADQUARTERS ARMY OF THE POTOMAC,
> Todd's Tavern, May 8, 1864—1 a.m.
>
> General TORBERT or MERRITT,
> First Cavalry Division:
> You will immediately move your command beyond Spotsylvania Court-House, placing one brigade for the present at the Block house, picketing the roads approaching the Court-House, and disposing the other two so as to cover the trains that will be north of the Ny [Ni] River, between that and the Orange plank road. It is of the utmost importance that not the slightest delay occur in your opening the Brock road beyond Spotsylvania Court-House, as an infantry corps is now on its way to occupy that place.
>
> GEO. G. MEADE,
> Major-General, Commanding,
>
> Brigadier-General GREGG,
> Commanding Second Cavalry Division:
> You will immediately move your division to the vicinity of Corbin's Bridge and watch all the roads approaching from Parker's Store, and as soon as General Hancock has occupied Todd's Tavern you will send a force on the Brock road to notify General H. of the approach of the enemy.
>
> GEO. G. MEADE,
> Major-General, Commanding.

At the same time, Sheridan, either unaware of the urgency of the infantry movement down the Brock Road or deliberately ignoring it, ordered Gregg and Merritt to move out on the Catharpin Road at 5:00 a.m., cross the Corbin's Bridge on the Po River, and then travel on to Shady Grove Church.[136]

Appendix I

> HDQRS. CAVALRY CORPS, ARMY OF THE POTOMAC,
> May 8, 1864—1 a.m.
>
> Brigadier-General GREGG,
> Commanding Second Cavalry Division:
>
> GENERAL: I am directed by the major-general commanding to instruct you to move with your command at 5 a.m. on the Catharpin road, crossing at Corbin's Bridge and taking up position at Shady Grove Church. General Merritt, with the First Division, will follow you on the same road, and on arriving at Shady Grove Church will take the left hand or Block house road, moving forward and taking up position at that point (via Block house). Immediately after he has passed you will move forward with your division on the same road to the crossing of Po River, where you will take up position, supporting General Merritt. General Wilson, with his division, will march from Alsop's by way of Spotsylvania Court-House and the Gate to Snell's Bridge, where he will take up position. Your ammunition train and authorized headquarters wagons will accompany you. The balance of your division train will proceed to Spotsylvania Court-House in charge of Lieutenant-Colonel Howard, chief quartermaster, Cavalry Corps, when they will be reported to you. The infantry march to Spotsylvania Court-House to-night.
>
> JAS. W. FORSYTH,
> Lieutenant-Colonel and Chief of Staff.

The fact that Hampton's cavalry held a strong position at Corbin's Bridge and Longstreet's Corps (now under Anderson) was marching behind on its race to Spotsylvania Court House indicates that Sheridan was totally unaware of the tactical situation here at Todd's Tavern.

Warren's troops would pass through this intersection during the night and reach Merritt's headquarters at 3:30 a.m. At 6:45a.m., Warren issued a report:[137]

> MAY 8, 1864—6.45 a.m.
>
> Major-General HUMPHREYS:
>
> ... The cavalry in my front here have, I think, made no advance to-day. I sent my men forward at General Merritt's intimation that, under the circumstances, I could push the enemy faster than he could.

> It is difficult to do much with the troops in an expeditious manner in these dense woods.
>
> Respectfully,
>
> G. K. WARREN,
> Major-General of Volunteers

Fitzhugh Lee's actions here and farther down the Brock Road resulted from his judicious decision to delay Grant's move to Spotsylvania Court House. These actions played a vital role in helping Lee's army arrive at the next stop just before Grant.

Return to your car and proceed to the next stop.

To Tour Stop 9

From this pullout, head south for 3.75 miles. At the "Spotsylvania Court House Battlefield" sign, turn left onto Grant Drive. Note that this is the only entrance to the park. Access from the Fredericksburg Road goes only to private property, and at the park boundary, this is a one-way road exiting the park. Pull into the parking lot by the interpretive shelter.

Tour Stop 9: Spindle Field / Laurel Hill

Tour Stop Relevance to Critical Decisions

This tour stop addresses the convergence of three critical decisions. Anderson's decision to push on through the night, coupled with Fitzhugh Lee's decision to delay the Federal infantry's southward movement down the Brock Road, allowed the Confederates to reach this location just before Warren's troops arrived. That action set the stage for Warren's decision to regain some of his lost reputation by quickly attacking the Confederates to attain the Union's immediate goal of Spotsylvania Court House.

Tour Stop Orientation and Troop Development

This stop first addresses the Federal movement to this location, then includes a walk to where the Confederates moved into position, and finally returns to the starting location to cover the action that occurred here.

The most notable feature at this stop is the monument to Maj. Gen. John Sedgwick, commander of the Sixth Corps who was killed at this spot on

Appendix I

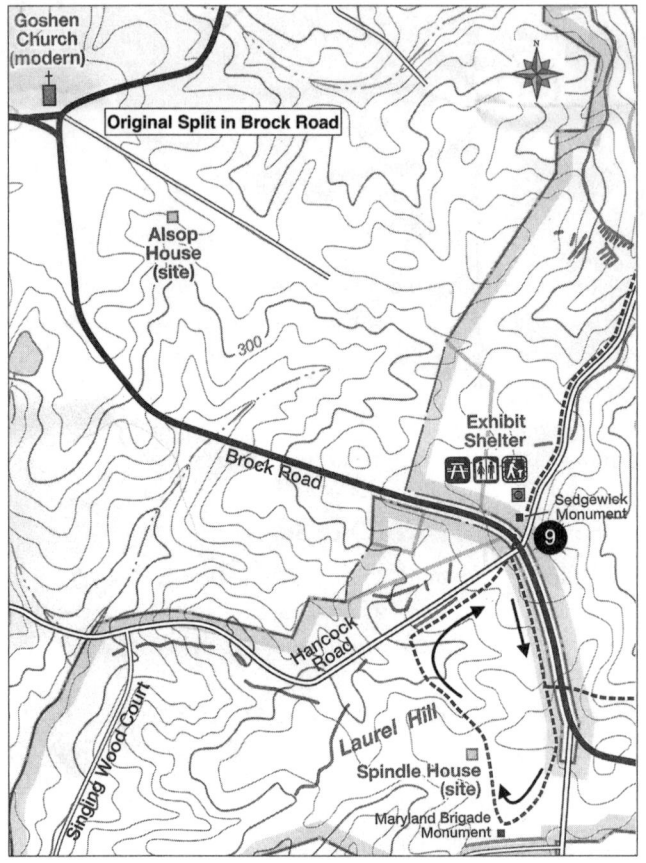

the morning of May 9, the day after the Federals' arrival here. By that time, Warren's line had shifted so that his left flank was at the junction of the Brock and Park Service Grant Drive roads. Sedgwick's right flank, occupied by the First Brigade of the Third Division and two batteries of artillery, was stationed in this area just to the east of the intersection and in front of the monument. Playfully admonishing a soldier who had dodged a sharpshooter's bullet, Sedgwick said, "They couldn't hit an elephant at this distance." After the soldier had explained himself, Sedgwick laughingly replied, "All right my man; go to your place." The next instant, a Confederate sharpshooter's bullet killed the general. Per his previous request, the command of the Sixth Corps went to Brigadier General Wright.[138]

At times, a park ranger will be at this shelter to provide information. After reviewing the information at the shelter, walk to the interpretive markers, and review them, paying particular attention to the "Laurel Hill Trail" marker. You will be taking this trail in the opposite direction to maintain the proper sequence of troop movements here. Move to the intersection near the entrance sign, and face north along the Brock Road. At the time of the battle, Grant Drive did not exist. Fitzhugh Lee's troops moved to the south side of Spindle Field as Warren's troops moved to this location.

By the morning of May 8, 1864, the two armies had traveled through the night in a race to reach Spotsylvania Court House. Warren's Fifth Corps, Grant's lead infantry force in his move south, had replaced Merritt's cavalry that was fighting against Fitzhugh Lee's cavalry delaying action.[139]

> MAY 8, 1864—5 a.m.
>
> Major-General HUMPHREYS,
> Chief of Staff:
>
> GENERAL: The head of my column reached General Merritt's headquarters at 3.30 a.m. I have word that the rear of it got off about 12.30, the delay being caused by the halt at the head of the column from obstruction. . . . General Merritt has for some time been skirmishing, and his column near me is advancing. . . . He will have to clear the way or make way a little himself for me to get to the front. Perhaps he will succeed without my help; and, if not, a little rest to my men will help them very much. I am aware of the importance of getting on to Spotsylvania Court-House as soon as may be, and should have taken the front and attacked if the cavalry had not been moving up to do it. I have one division now massed. The fighting is about a mile in front of me.
>
> G. K. WARREN,
> Major-General.
>
> ———
>
> MAY 8, 1864—6.45 a.m.
>
> Major-General HUMPHREYS:
> I started my troops forward at 6 o'clock; nearly two divisions have passed where General Merritt's headquarters is. General Robinson's division had the advance and he has been directed to spare no effort to clear the road. . . . I am now going to the immediate front to direct the advance. . . . They report the enemy chopping along their front during the night. . . . The cavalry in my front here have,

Appendix I

> I think, made no advance to-day. I sent my men forward at General Merritt's intimation that, under the circumstances, I could push the enemy faster than he could. It is difficult to do much with the troops in an expeditious manner in these dense woods.
> Respectfully,
> G. K. WARREN,
> Major-General of Volunteers.

At the intersection of the Brock and Gordon Roads that you passed on the way here, Lyle's First Brigade of Robinson's Second Division finally dislodged the Confederates from their defenses. In 1864, the Brock Road split there into an eastern branch that no longer comes through and a western branch that is the Brock Road you traveled. These two roads came together at this location. Warren's Second Division, commanded by Brig. Gen. John Robinson, took the eastern branch, with Lyle's brigade in front and Col. Richard Coulter's Second Brigade and Col. Andrew Denison's Third Brigade following. Brig. Gen. Charles Griffin's First Division took the western branch, with Brig. Gen. Romeyn Ayres's and Col. Jacob Switzer's brigades taking the road, and Brig. Gen. Joseph Bartlett's men taking a ravine to the right of the road. Lyle's brigade was the first to push on to this location:[140]

> MAY 8, 1864—8 a.m.
> Major-General HUMPHREYS:
> The opposition to us amounts to nothing as yet; we are advancing steadily, the enemy uses artillery, two pieces, on the road. General Robinson has gone ahead with a brigade, mostly in line. I follow close with columns filling the road, and artillery; if there is nothing but cavalry, we shall scarcely halt, if our troops can be made to move, but they are exceedingly hesitating, I think. General Robinson's orders are to use only the bayonet, and carry every battery the enemy shows. It is believed to be Fitzhugh Lee's cavalry.
> G. K. WARREN,
> Major-General.

This is how the Federal position was developing that morning.

View across Spindle Field to Laurel Hill. Courtesy of the author.

Turning and facing south, carefully cross Brock and Grant Drive Roads, and follow the "Laurel Hill Loop Trail" shown on the map at the beginning of this stop. Walk south along the Brock Road until you reach the intersection of Block House Road, and face north back to where you started. Note that on the walking tour, you will be going in the opposite direction of that shown on the marker in order to have the correct timing for troop dispositions.

Fitzhugh Lee's cavalry had been fighting its delaying action along the Brock Road for five hours. When Warren's infantry finally broke through, Lee's men retreated to this area, constructed "very slight rail breast works on the edge of a pine thicket," and positioned a few of Maj. James Breathed's horse artillery that had been on the Brock Road. The cavalrymen were joined by their corps commander, "Jeb" Stuart. Adding to Fitzhugh Lee's problems, Wilson's cavalry was at Spotsylvania Court House fewer than two miles down the Brock Road.[141]

Anderson's critical decision to press on through the night had brought his troops to the Block House Bridge about a mile southwest from here as the crow flies. He finally found a place to bivouac about dawn. After only a brief rest, Anderson received word that Fitzhugh Lee's cavalry was hard pressed by Union infantry. Anderson then sent his lead brigade, Kershaw's, now headed by Col. John Henagan, and Brig. Gen. Benjamin Humphreys to help. They hurried to this location and took position behind some slight rail breastworks that Fitzhugh Lee's cavalry had hastily constructed. The first available artillery to come up was Maj. John Haskell's battalion, and it too was sent here to be deployed.[142]

Appendix I

Henagan's Third South Carolina Regiment arrived here first and was posted on the east side of the Brock Road. The Second South Carolina followed and was put on the west side. Troops' timing was so close this morning that one of Henagan's men wrote, "We occupy the rail piles in time to see a column, a gallant column, moving towards us, about sixty yards away." Haskell's artillery was moving up the Block House Road behind you and to your right. As the infantry moved in, Fitzhugh Lee pulled his cavalry troops back and went to support Rosser, who was confronting Wilson's cavalry at the courthouse. Stuart would remain here and direct the immediate fighting in this area. Anderson had also sent Brig. Gen. William Wofford's and Brig. Gen. Goode Bryan's infantry brigades to help Rosser.[143]

Continue to walk clockwise on this loop trail, and in about nine hundred feet you will come to an interpretive marker, "Fighting For The Fences." Just behind it is the Maryland Brigade Monument, which represents the farthest position Warren's troops reached on May 8. As you continue walking the loop back to the starting point, you will cross the open field area near the ruins of the Spindle House. Denison's Maryland brigade crossed here just after Lyle's assault.

As you travel it, the latter part of the trail will be along Hancock Drive back toward the intersection. You will see some of the entrenchments Warren's troops constructed, as well as artillery revetments.

When back at the intersection, face Spindle Field:

View of Spindle Field from Laurel Hill, where Denison's Seventh Maryland made the farthest Union Advance on May 8, 1864. Courtesy of the author.

Robinson wanted to wait until the rest of his brigades could come forward, but Warren ordered an immediate attack. While Coulter's and Denison's brigades were coming up, Lyle's men went down the east side of the Brock Road at 8:30 a.m. in a column of his five regiments. Major Roebling of Warren's staff stated, "Up to this time we thought we were fighting cavalry." Lyle's brigade charged over the open ground and was subjected to musket and artillery fire from the Confederate position where you just were. These Federals made it to within thirty feet of the Confederate works before they were pinned down for a time. Humphreys's Confederate brigade was arriving up the Block House Road, and it moved to the right and on Lyle's left flank, forcing him back "in some confusion." An officer of the Thirty-Ninth Massachusetts felt that "Had there been any support for the brigade, . . . the enemy's line would have been broken."[144]

Fifteen minutes after Lyle's advance, Denison's brigade marched across the open field west of the Brock Road. Confederate sharpshooters opened fire as soon as the Union troops broke from cover. As they advanced, the men of the Third Brigade of Robinson's Second Division fired a volley, then stopped to reload. This action slowed the advance and "caused the ranks to become intermingled" as following units pressed forward. Robinson was wounded at this time, and Col. Richard Coulter replaced him.[145]

Report of Brig. Gen. John C. Robinson, USA, Commanding Second Division

At daylight on the morning of the 8th I overtook the advance guard of cavalry, which was engaged with the enemy. I immediately deployed two brigades, holding the third in reserve, pushed by the cavalry (commanded by Brigadier-General Merritt), and drove the light troops and artillery of the enemy from one position to another, through woods and across open fields for about 3 miles. Coming to another field I could plainly see the enemy's line in the edge of timber beyond. I here halted and reformed the division, and again advanced to the attack. The division was soon checked, and it became evident that here was the enemy's main line, but his strength was undeveloped. Knowing that my brave men would follow wherever I led the way, I placed myself at their head and led them forward to the attack. At this moment a part of Griffin's division advanced out of the woods on my right. Cheering my men on, we had arrived within 50 yards of the works when I received a musket-ball in the left knee, resulting in amputation of my leg. This unfortunate

> wound caused the result I feared, for as I was borne off the field I saw that our troops were repulsed and the attack had failed. Our loss this day was heavy, but I have never been able to learn the number of killed and wounded.

The Federals were going against the area where Stuart was physically directing the defense. He ordered the Second South Carolina to hold its fire until the Federals were in range. At that point, Stuart ordered the troops, "Give it to them good and hold this position to the last man." Colonel Denison was now wounded, and command went to Col. Charles Phelps, who would be captured. "A blast of rifle fire struck the brigade at short pistol range and the broken ranks were swept by canister from a battery on the left."[146]

Bartlett's men went in after Denison's brigade and still farther to its right, fixing bayonets as they charged. They soon discovered Henagan's troops behind their breastworks, and they were subjected to a "murderous fire" as they engaged the enemy. "We stood face to face, not over fifteen feet apart, for over half an hour," stated a member of Bartlett's Eighty-Third Regiment. But the enemy soldiers held their ground, and this attack, too, fell back with heavy losses.[147]

As the advance stalled, Griffin's last two brigades, commanded by Ayres and Sweitzer, were brought up still farther to the right. However, they were not engaged. By this time, Haskell's artillery had arrived and added its weight to the Confederates' already strong position. The Federal effort became a confused affair, and "charging the enemy's position found them in too large a force to push forward." Warren tried to rally the troops, but it was no use. Robinson's and Griffin's men and the artillery that had been supporting the Federals were pulled back to near the Alsop Farm. Robinson's Second Division was so badly damaged that it would cease to exist, the brigades being absorbed into the Third and Fourth Divisions.[148]

Warren had lost the race to this location by the narrowest of margins, and he had been unable to mount a coordinated and concentrated attack to secure the vital Brock and Block House Roads intersections. A captain in the Second Brigade of Robinson's disbanded division wrote, "It was the common impression that had the affair been properly managed at the start, the position could have been carried and held."[149]

At 10:15 a.m., Warren sent word of the morning's results to headquarters, adding that he was now advancing his Third and Fourth Divisions. Ironically, Meade could not yet accept the possibility that Longstreet's troops had gotten here before Warren's:[150]

HEADQUARTERS FIFTH ARMY CORPS,
May 8, 1864—10.15 a.m.

Major-General HUMPHREYS:

I reached the vicinity of the blacksmith shop at the intersection of the road from Piney Branch Church, the road we were on, toward Spotsylvania Court-House. I promptly attacked the enemy here with what was on hand of General Robinson's division, led by himself in person, and General Griffin's division by himself. General Robinson's troops fought with reluctance, and fell back, himself severely wounded in the knee. This exposed General Griffin's left, and part of his command fell back too, all in much confusion, refusing much of our attempts to stop them, till they got out of fire. They will soon stop, and can perhaps be assembled again to-day. General Griffin, however, held on. I sent General Crawford to his support and then General Cutler, and they have held on, and at this time are again advancing in fine style. We have taken prisoners from General Longstreet's corps.

WARREN.

[Indorsement.]
An aide who brought this says that the prisoners belonged to General McLaws' division, Longstreet's corps. I am in hopes this may be some mounted infantry given to Stuart to strengthen him. I hardly think Longstreet is yet at Spotsylvania. Sedgwick's best division joined Warren just as this dispatch was written. I have told him to call for the one here if necessary, and to attack vigorously without loss of time, but I fear the *morale of his men is impaired.*

GEO. G. MEADE.

After the initial repulse of Warren's 8:30 a.m. attack, both sides would continue to move troops to this area. In the afternoon, Meade wanted Sedgwick to join in an immediate assault:[151]

HEADQUARTERS ARMY OF THE POTOMAC,
May 8, 1864—1 p.m.

Major-General SEDGWICK:

You will proceed with your whole corps to Spotsylvania Court House and join General Warren in a prompt and vigorous attack on the enemy now concentrating there. Use every exertion to move with the utmost dispatch.

GEO. G. MEADE,
Major-General, Commanding

HEADQUARTERS ARMY OF THE POTOMAC,
May 8, 1864—1.30 p.m.

Major-General WARREN:
Sedgwick's whole corps is sent to join you in the attack on the wing. Wilson sends word he has taken prisoners from both of Longstreet's divisions. It is of the utmost importance the attack of yourself and Sedgwick should be made with vigor and without delay.

GEO. G. MEADE.

By the end of the day, five different brigades would attack the Confederate line at Laurel Hill. Throughout the remainder of the battle, Laurel Hill's defenses would be strengthened, and the infantry behind fieldworks and the artillery in this area could not be broken. E. Porter Alexander, commander of Anderson's artillery, found that the rail breastworks constructed by the cavalry were "a beautiful place to locate [the] whole old Battalion in a mass, having a fine sweep over fairly open grounds."[152]

At the end of the day, Warren sent the following reports to headquarters:[153]

HEADQUARTERS FIFTH ARMY CORPS,
May 8, 1864—9 p.m.

Major-General HUMPHREYS:
GENERAL: The result of General Crawford's demonstration was to show the enemy not intrenched there. He captured a number of prisoners, in all, to-day, some 50, perhaps. The enemy seem nervous and opened a stampeding fire after dark on our left. The enemy is cutting trees, and artillery is heard moving guns to his right.

Respectfully, yours,
G. K. WARREN,
Major-General

Confederate entrenchments at Spotsylvania Court House. Library of Congress.

> MAY 8, 1864—9 p.m.
>
> Major-General MEADE:
>
> DEAR GENERAL: I have sent my statement of all that I know of importance instead of coming myself. I am rather amused at your conclusion from my remark about having any opinion to give, as indicating a loss of something, but in truth I only meant that there is but one thing left—to fight it out. I think it is very hard to do, but the object of war, a settlement, can never be come to without doing so. I am so sleepy I can hardly write intelligibly.
>
> G. K. WARREN.

Fighting would continue here at Spindle Field until May 14, when Warren's Fifth Corps moved to the Federal left flank on the Fredericksburg Road. The artillery that Warren heard being moved onto the Confederate defenses would not be taken, and neither would the effective earthworks at Laurel Hill.

To Tour Stop 10

From the parking lot, proceed 0.7 mile to the second Spotsylvania Battlefield stop, "Upton's Road," and pull into the parking lot.

Appendix I

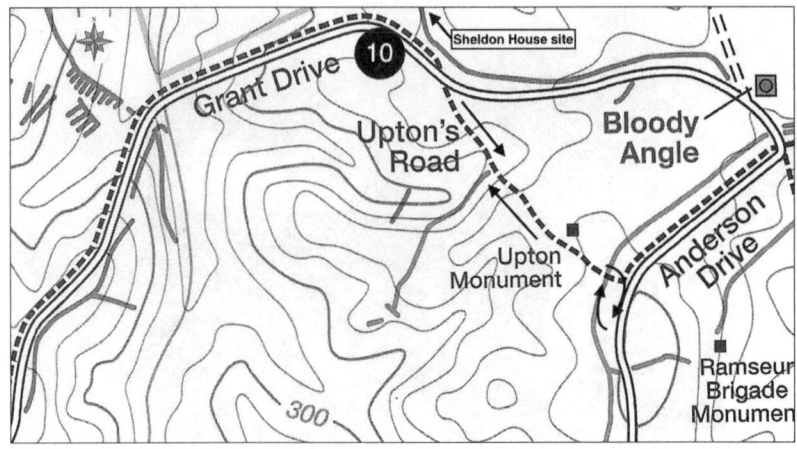

Tour Stop 10: Upton's Attack on Dole's Salient

Tour Stop Relevance to Critical Decision:

This stop covers the location and fighting concerning the decision to implement Col. Emory Upton's more innovative column formation in the May 10, 1864, assault on the Confederate entrenchments. Brig. Gen. George Doles's brigade of three Georgia regiments manned the Rebel defenses. Up to this time, the typical frontal assault, like those at Saunders Field in The Wilderness and at Laurel Hill near here, had failed.

Tour Stop Orientation and Troop Development

As the men of Sedgwick's Sixth Corps arrived from the fighting at The Wilderness, they were deployed on Warren's left. The soldiers extended the Federal position along this road, which did not exist at the time of the battle. With Sedgwick's death on May 9, Brig. Gen. Horatio Wright assumed command of the corps, and Brig. Gen. David Russel took command of Wright's old First Division, located behind the works that line the road here.

After reviewing the field markers, for initial orientation purposes, face the "Wartime Woods Road Used By Upton" marker along the trail.

As Sheridan's cavalry was off pursuing "Jeb" Stuart and therefore available to ascertain the Confederate line, Grant assumed that the Confederate right flank extended much farther northeast. Burnside's Corps was stationed far to the east on the Fredericksburg Road near the Ni River. Brig. Gen. Gresham Mott's Fourth Division of Hancock's Corps was positioned just under a mile northeast of here, making a tenuous connection between Wright and Burnside. With Mott so far removed from Hancock, he was temporarily put under Wright:[154]

> HEADQUARTERS ARMY OF THE POTOMAC,
> May 10, 1864—7 a.m.
>
> [General MOTT:]
>
> GENERAL: While in your present position you will be under the direction of Brigadier-General Wright, commanding Sixth Corps. I wish you would push out reconnoitering parties in all directions and obtain all possible information of the country and the enemy.
>
> GEO. G. MEADE

As they arrived after Anderson's initial clash with Warren's troops, Ewell's Confederates were deployed on Anderson's right. Following a ridgeline, the four brigades of Maj. Gen. Robert Rodes's division were those of Brig. Gen. Cullen Battle, Brig. Gen. Stephen Ramseur, and Brig. Gen. Junius Daniel, as well as the men of Doles's regiment. During the night of May 8–9, Maj. Gen. Edward Johnson's division would move to the end of the defenses, forming a

Start of the "Wood Road" that Upton's troops took to attack Dole's Salient. Courtesy of the author.

salient called the Mule Shoe. Gordon, now commanded Early's old division, also had troops in the Mule Shoe. Col. William Monaghan, who had replaced the wounded Hays, was positioned near the western apex of the Mule Shoe. Col. Clement Evans, now in charge of Gordon's old brigade, Brig. Gen. Johnson's reassigned brigade from Rhodes's division, and Col. John Hoffman who had replaced the wounded Pegram, were placed in reserve behind the area manned by Rodes's division.

Grant's original plan for May 10 was to have Hancock, currently on the Federal far right flank, renew his movement across the Po River and strike Lee's left flank. Then Warren and Wright would attack. Hancock's effort had started the previous evening with troops moving across the Po River, but he had pulled back for the night. Alerted to the danger, Lee had moved Mahone's and Heth's divisions to the area, forcing Hancock to abandon the plan for his assault on the morning of May 10.[155]

While Hancock was operating on the Federal far right, Warren ordered some of his units forward against Laurel Hill. The men encountered a "galling fire of bullets and canister and were compelled to return to the rifle-pits." At 9:30 a.m., Meade informed Grant that Wright had reported "the enemy on his left moving rapidly to his right" toward Hancock. Grant thought Lee might have been moving these troops from the salient and the Laurel Hill area, thus providing the Federals an opportunity to regain the initiative. Grant formulated a plan to pull Hancock's troops back from the far right, combine them with Warren's Corps, and launch an overall frontal attack on the Confederates. It would take time to get Hancock back, so the intended coordinated attack would commence at 5:00 p.m. All four corps commanders received orders to that effect at 10:00 a.m.:[156]

> HEADQUARTERS ARMY OF THE POTOMAC,
> May 10, 1864—10 a.m.
>
> Major-General HANCOCK:
> You will immediately transfer two divisions of your corps to General Warren's position, and make arrangements, in conjunction with the Fifth Corps, to make a vigorous attack on the enemy's line punctually at 5 p.m. General Warren's instructions are herewith sent to you. The remaining divisions will be so disposed as to keep up your present threatening attitude on the enemy's left, but so that it can be withdrawn promptly to your support, if necessary.
>
> GEO. G. MEADE,
> Major-General, Commanding

HEADQUARTERS ARMY OF THE POTOMAC,
May 10, 1864—10 a.m.

Major-General WARREN:

General Hancock has been ordered to throw out two divisions of his corps on your right, . . . [to support it] your corps [will make] an attack on the enemy in your front . . . at 5 p.m. this day. You will accordingly make all necessary dispositions. Major-General Hancock will by virtue of seniority have the command of the combined operations.

GEO. G. MEADE,
Major-General.

HEADQUARTERS ARMY OF THE POTOMAC,
May 10, 1864—10 a.m.

Brigadier-General WRIGHT:

You will make an attack on your front promptly at 5 p.m., using General Mott's division for this purpose. Generals Warren and Hancock will attack at the same time.

GEO. G. MEADE,
Major-General, Commanding

HEADQUARTERS,
May 10, 1864—10.30 a.m.

Major-General BURNSIDE:

A general attack will be made on the enemy at 5 p.m. to-day. Reconnoiter the enemy's position in the mean time, and if you have any possible chance of attacking their right do it with vigor and with all the force you can bring to bear. Do not neglect to make all the show you can as the best co-operative effort.

U. S. GRANT,
Lieutenant-General

Twenty-four-year-old Col. Emory Upton would lead twelve specially selected regiments for Wright's part of the attack. The men in these units would be drawn from Wright's First and Second Divisions, and Mott's division, temporarily under Wright's command, would support them.

Appendix I

This overall operation would require a degree of coordination that Grant's army had not yet achieved, and Grant's subtle reproach of Warren's performance by putting Hancock in charge of their joint force would affect the timing. Barlow's division was in some danger during its withdrawal from the Federal far right flank. Hancock was dispatched to ensure things went as smoothly as possible with the movement, and in his absence, Warren suddenly perceived an opportunity to attack Laurel Hill ahead of schedule. Brig. Gen. John Gibbon, a well-respected veteran of Hancock's Corps, had been directed to coordinate with Warren. The two generals went to the ground that would have to be covered, and Gibbon declared "that no line of battle could move through such obstacles to produce any effect." When Warren and Gibbon arrived back at Meade's headquarters, Gibbon reiterated his concern to no avail. Meade relied wholly on Warren's judgment.[157]

HEADQUARTERS ARMY OF THE POTOMAC,
May 10, 1864—3.30 p.m.

Major-General HANCOCK,
Commanding Second Corps:

The opportunity for attack immediately is reported to be so favorable by General Warren that he is ordered to attack at once, and Gibbon is directed to co-operate with him. Wright is ordered to be ready to attack at once, if necessary, or to send support to Warren in an emergency. As soon as Barlow is secure, let Birney return.
By command:

A. A. HUMPHREYS,
Major-General and Chief of Staff.

HEADQUARTERS ARMY OF THE POTOMAC,
May 10, 1864—4 p.m.

[General HANCOCK:]
I have ordered Warren to attack at once, and use Gibbon. Send Birney back if you don't want him, or, if not too late, attack with Birney and Barlow. Let me know at once what you do, and hurry Birney back as soon as possible if not wanted.

GEO. G. MEADE,
Major-General.

Warren's attack resulted in more Federal casualties and a disrupted schedule for the overall assault, which was already late. Maj. Washington Roebling stated, "This preliminary attack showed that the enemy was all set for us, and that the subsequent attack would have little chance. Nevertheless the two divisions of Gibbon and Birney made a charge about 6 o'clock and were repulsed even more disastrously."[158]

Wright had corresponded with Mott about the conflicting orders Mott had received and the difficulty of getting sufficient troops back into position after they had been ordered to connect with Burnside.[159]

> HEADQUARTERS FOURTH DIVISION,
> May 10, 1864—2.05 p.m.
> General WRIGHT:
> GENERAL: I was ordered to connect with General Burnside, on my left. My line of pickets is so extended that my troops for assault will not be more than 1,200 to 1,500 men, so that I am very weak. To call in the pickets on my left will take all the time, if not more than 5 p.m. Shall I send for them?
> G. MOTT,
> Brigadier-General.

Wright replied:[160]

> HEADQUARTERS SIXTH ARMY CORPS,
> May 10, 1864.
> Brigadier-General MOTT,
> Commanding Fourth Division, Second Army Corps:
> I regret that your skirmish line is so extended, but if you cannot withdraw a part of the left in full time, you will not attempt to do so, but advance the whole at 5 p.m., as previously ordered, following it at the proper moment by your column of attack, made as strong as your numbers will permit. Use your artillery freely whenever there is a chance, as it serves to inspirit our men and demoralize the enemy, even if it does not hurt them much. I think you can enfilade their line with it. I rely much on the effect of your attack.
> H. G. WRIGHT,
> Brigadier-General, Commanding.

Appendix I

Mott's attack was intended to strike the Confederate flank in support of Upton's simultaneous frontal assault. The statement that Wright would rely heavily on the effect of Mott's attack demonstrated the Federals' extreme lack of understanding of the Confederate works. With up to 1,500 troops, Mott would have to march over open field for almost a mile. He would hit the defenses well north of where Wright anticipated they would strike the Confederate line. Meanwhile, Upton's 5,000 troops would be attacking over only two hundred yards. Compounding Mott's problem, he was never informed that Warren's debacle had delayed the time of the overall attack from 5:00 to 6:00. Starting off at 5:00 p.m., Mott quickly came under enemy fire and was easily driven back. He basically had no impact on the fighting.

You are standing in front of what Upton called the "wood road" that extended from the Sheldon/Scott site, approximately 1,200 feet in back of the Federal trenches behind you, then through Dole's Salient, on to the Harrison House (Lee's headquarters) in the salient behind Confederate lines, and then on south to the Brock Road.[161] Earlier in the day, Lieut. Ranald Mackenzie of the US Army Corps of Engineers had performed a reconnaissance of the Sixth Corps front and found this location. Mackenzie had reviewed this position with Russell, now commanding Wright's First Division, and it was selected for an attack.[162]

Twelve regiments from different brigades of two Sixth Corps divisions were designated for the assault, and Upton was chosen to lead it. Upton was an aggressive officer who had successfully utilized a column assault at Rappahannock Station in November 1863. The same tactic would be used here. Ironically, the 121st New York and 5th Maine that Upton had led in the 1863 offensive would be on the front line of this one at Spotsylvania Court House.

Preparation for the attack was thorough. First, Mackenzie showed Upton the point of attack. Then Russell, Upton, and the twelve regimental commanders personally covered the ground. To ensure the Confederacy had as little warning of the attack as possible, men in the Sixty-Fifth New York Regiment of Col. Nelson Cross's Fourth Brigade were deployed as skirmishers. They drove the enemy pickets from the woods in front of you. Federal artillery fired on the Confederate works, targeting the part of the enemy line where Upton's attack would strike.[163]

Upton and his troops assembled at the Shelton House and then moved to this area about five o'clock. Upton's signal to attack was to have been a halt in artillery fire near five o'clock, but with delays in other areas, the firing continued, and the assault was postponed. Shortly before six o'clock, headquarters sent orders to stop the artillery and proceed with the attack.[164]

Walk a short distance down this trail until you reach the interpretive marker "Forming for the Attack."

Upton's troops moved down the trail you are now on in four lines of column of attack as shown. When they arrived at this location, then lay down.[165]

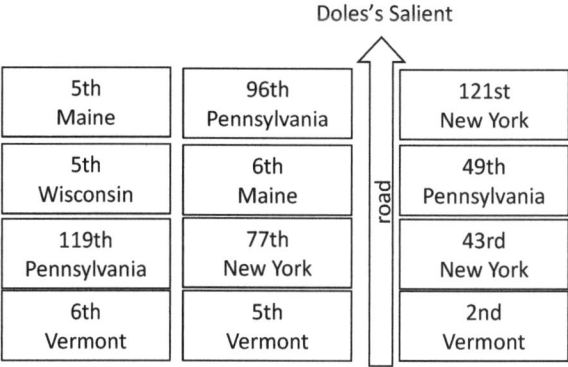

At the time of the battle, Upton stated that the distance to cover was about two hundred yards, and this location would have been the tree line.

Proceed about sixty feet on this trail to the Upton's Charge Monument.

The topography here, and at other parts of the battlefield, is composed of various rolling swales that afforded the men limited protection. To help conceal the movement, Union skirmishers had driven the Confederate pickets back to their defenses. The ridgeline you are looking at juts out enough here to form a slight salient that was manned by Doles's Georgia brigade of Ewell's division. This marker is about the height of a soldier. Given the slight rise to the location of the Confederate works and the logs forming the top of the works, it would have been difficult for Dole's men to see Upton's troops before they reached this point.

The troops were instructed to move directly on the enemy works and hold their fire until they had breached the works. To ensure this, the weapons of the first line were ordered to be loaded and capped, while those of the other lines were loaded but not capped. All lines fixed bayonets. Once the enemy line was breached, the 121st New York and 96th Pennsylvania of the first line were to face right and attack an artillery battery. Meanwhile, the 5th Maine, also of the front line, was to turn left and open an enfilading fire on the enemy. The three regiments of the second line were to halt at the works and fire straight ahead, and the third line was to lie down behind them and await orders. The fourth line was to wait at the edge of the woods (marker) as a reserve to be used where needed most. At 6:10 p.m., the troops moved to this position and then rushed the Confederate works:[166]

> At command, the lines rose, moved noiselessly to the edge of the wood, and then, with a wild cheer and faces averted, rushed for the works. Through a terrible front and flank fire the column advanced, quickly gaining the parapet. Here occurred a deadly hand-to-hand conflict. The enemy sitting in their pits with pieces upright, loaded, and with bayonets fixed, ready to impale the first who should leap over, absolutely refused to yield the ground. The first of our men who tried to surmount the works fell pierced through the head by musket-balls. Others, seeing the fate of their comrades, held their pieces at arms length and fired downward, while others, poising their pieces vertically, hurled them down upon their enemy, pinning them to the ground.

Follow the path to the Dole's Salient marker and visualize what it would have been like to fight your way through the abatis and enemy fire just to reach your objective.

After reviewing the interpretive markers, imagine how hard it would be to see the marker if it were located at the "Forming for the Attack" interpretive marker where Upton's attack started.

The openness here now sharply contrasts with what was here on May 10, 1864. Upton described the defenses where you are standing as being "of a formidable character with abatis in front and surmounted by heavy logs, underneath which were loopholes for musketry. . . . There were also traverses at intervals along the entire work." The log parapets had to be climbed over, and the advancing troops made easy targets before they could engage in the deadly hand-to-hand fighting that occurred here.[167]

View looking south from Dole's Salient to Captain Smith's Confederate battery. Courtesy of the author.

From your left to right facing the tree line, the three regiments of Doles's brigade located here were the Twelfth, Forty-Fourth, and Fourth Georgia. To Doles's right were Brig. Gen. James Walker's five regiments of the Stonewall Brigade. Daniel's North Carolina regiments were positioned to Doles's left. A second line of entrenchments had been constructed farther back behind these works where Daniel's troops were located.

Turning to your left, face south.

Just to the left of these North Carolina regiments and between the junction of Daniel's and Doles's brigades was Capt. Benjamin Smith's battery of Hardaway's artillery battalion—the objective of Upton's first two regiments as previously mentioned. The two artillery pieces that you see represent the location of Smith's battery.

Proceed south along Anderson Drive to these guns.

The Parrot artillery pieces here are representative of the Napoleon artillery present at the fighting. This position represents the deepest penetration Upton's men made in this area. Smith's battery fired a few rounds of canister into the enemy from here. Seeing the Confederates "rapidly advancing towards the enemy's line," they stopped firing and began to cheer. These troops were in fact prisoners being taken to the rear. Fearing that they would hit their own men, Smith's battery briefly stopped firing. As "one dense mass of Federal infantry" advanced, they were able to get a few shots off, but this position was soon overrun.[168]

Upton's first three lines created a breakthrough that he estimated was at least one-half mile long. The troops then fought to extend it. Regiments of the first and second lines pushed south and temporarily captured the Confederate artillery here. However, the retreating artillerists wisely took the gun rammers, making the seized pieces useless to the Federals.[169] Next, Upton's troops hit Daniel's North Carolinians that had been on Doles's left flank. Pushing the Rebels back, the Federals temporarily captured sixteen pieces of artillery before they were in turn repulsed.[170]

Return to the markers at Dole's Salient.

On Upton's northern flank, his troops pushed Walker's men back until the Confederates were finally able to "refuse" their line across their flank. Upton described the attack in his official report:[171]

Appendix I

> ### Report of Brig. Gen. Emory Upton, USA, Commanding Second Brigade.
>
> Pressing forward and expanding to the right and left, the second line of intrenchments, its line of battle, and the battery fell into our hands. The column of assault had accomplished its task. The enemy's lines were completely broken and an opening had been made for the division which was to have supported on our left, but it did not arrive.

Unaware that Mott's attack had started and ended before his troops had even begun, Upton felt that added reinforcements from Mott and the Sixth Corps would have allowed him to press all the way to the Brock Road. Written fifteen years after the battle, this statement mentions neither the darkness at that time, nor the other Confederate troops in position that would have made advancing to the Brock Road highly unlikely.[172]

At this time, Ewell, riding ahead of his staff, told Daniel's troops to hold the line because reinforcements were on the way. Concerned about the skirmishers being pushed back on this front, Ewell was already looking to send troops here:[173]

> ### Report of. Lieut. Gen. Richard S. Ewell, CSA, Commanding Second Army Corps
>
> About 4 p.m. I learned that General Doles' skirmishers were driven into his works. He was ordered to regain his skirmish line at any cost., but while preparing to do so his lines were attacked and broken.

Battle's and Ramseur's brigades of Rodes's division were quickly brought up, and they and the remaining troops of Daniel's brigade began to press the 121st New York and 96th Pennsylvania. When Ewell met Brig. Gen. Robert Johnston of Gordon's division riding to the sound of the fighting, he was ordered up also. As in The Wilderness on May 6, the crisis was such that Lee at this time "started for the breach, with the purpose of leading the troops in the effort to regain the lost ground." To the staff officers and others who urged him back, Lee replied that he would "relinquish his purpose" if they would see to it that the lines were reestablished. These forces were able to stop Upton's advance.

With the return of the artillerymen and their gun rammers, they turned the guns on the Federals and delivered artillery support to the infantry.[174]

On Upton's left, the Fifth Maine and Fifth Wisconsin had attacked the flank of the Stonewall Brigade, forcing the Second and Thirty-third Virginia Regiments to pull back toward the west angle of the Mule Shoe. As the Federals were pressing forward, Walker was able to re-form his brigade perpendicular to the fieldworks. He sent for help from Johnson's division, whose troops were on the other side of the salient. "Maryland" Steuart's brigade was in the right position to deliver a counterattack on Upton, and it was sent back across the salient a half mile to meet the Federals, arriving next to Walker's brigade. Steuart's Third North Carolina Regiment "suffered severely" when it deployed across the line of the Federal advance. The Confederates had stopped Upton's men on the southern end of the breakthrough, and Walker's and Stewart's men were able to do the same on the northern side.[175]

To hold what had been gained, Upton ordered his men on the reverse side of the breastworks. He then rode back to get his fourth line up only to find that they had already advanced. The twelve regiments of his assault had become so mixed up that Upton stated, "There was not a single unit under my current control."[176]

Finally, Gordon's old brigade under Col. Clement Evans, came up from the south side on Upton's right flank. The dead and wounded from the recent fighting hindered their movement, as did the traverses that had been constructed perpendicular to the Confederate works. Finally, though, the Twenty-Sixth Georgia got in a flanking position to ensure that the Confederates would retake the works. Surprisingly, Ewell's losses exceeded Upton's by only two to three hundred.[177]

Upton described the attack in his official report:[178]

Report of Brig. Gen. Emory Upton, USA, Commanding Second Brigade

Night had arrived. Our position was three-quarters of a mile in advance of the army, and, without prospect of support, was untenable. Meeting General Russell at the edge of the wood, he gave me the order to withdraw. I wrote the order and sent it along the line by Captain Gorton, of the One hundred and twenty-first New York Volunteers, in accordance with which, under cover of darkness, the works were evacuated, the regiments returning to their former camps. Our loss in this assault was about 1,000 in killed, wounded, and missing.

While Upton's tactic here had successfully breached a strong Confederate entrenchment, its lack of support and delayed implementation meant that it did not achieve all that it might have. Mott was unjustifiably criticized for not coming up on Upton's flank in support. His distance from this location, the strong Confederate response to his movement, and the limited number of troops at his disposal had all made it impossible for him to alter the outcome. Mott's commanders would have been more appropriately faulted for not putting him in a position to realistically effect the outcome. However, Wright did have troops closer and in greater numbers than Mott, but he did not engage them.

The hour and a half of hand-to-hand combat that occurred here was just a preview of what would happen in two days at the Mule Shoe north of here—our next stop.

Standing with the markers at your back, face east.

The closeness of the fighting around the salient can be gauged here. In the far distance across the field, you can see a brown National Park Service maintenance building. Closer in, you can see the interpretive markers at the McCoull House in the center of the salient, which will be part of the next stop. In this roughly straight line of sight, the Ramseur Brigade Monument is not visible. The monument honors the unit's May 12 advance as part of the effort to retake the west side of the salient, and it is just beyond the next ridge and to the right of this line of sight. Looking to your left, you can see the Bloody Angle of the Mule Shoe and the parking lot for the next tour stop.

Proceed back along the trail to the parking lot, and go to the last stop.

To Tour Stop 11

From the Upton trail pullout, go 0.25 mile to the third Spotsylvania Battlefield stop, and pull into the parking lot. Walk to the interpretive sign "The Mule Shoe Salient" on the trail at the north side of the parking lot.

Tour Stop 11: The Mule Shoe

Tour Stop Relevance to Critical Decisions

Three critical decisions converged here on May 12, 1864, that resulted in some of the most horrific fighting of the war. First, as Upton's May 10 attack had demonstrated potential to penetrate Confederate defenses, Grant would endeavor to repeat the tactic on a much larger scale. To implement this approach, Hancock's Corps would once again be his weapon of choice. Finally, concerned that Grant might be moving to Fredericksburg, Lee decided to remove the artillery from the salient the night before to be able to move quickly to thwart Grant's anticipated move from Spotsylvania.

To put the fighting at this location in perspective, consider the following from Gordon Rhea's book on Spotsylvania:[179]

> The Civil War has seen its share of horrors. The Bloody Lane at Antietam, the stone wall of Fredericksburg, and the Wheatfield at Gettysburg were synonymous with carnage. They pale, however, when measured against the slaughter along the short stretch of earthworks where the salient's western tip bent south.

Tour Stop Orientation and Troop Development

The ferocity and duration of the back-and-forth fighting here at the Mule Shoe means that parts of the walk will be duplicated to keep the tour in chronological order. After reviewing the overall situation on the evening of May 11, 1864, you will move from the parking lot to the historic trail and walk the part of the trail that follows the west side of the salient, called the Bloody Angle, to the apex of the salient. There, you will review Lee's removal of the artillery. Then, you will walk the trail leading to the Landrum House for a review of the Federal preparations and the initiation of the assault. After walking back the way you came, you will review the initial action against the salient, first at the apex, then on the east side, and then back to the west side toward the parking lot. Before reaching the parking lot, you will take a trail through the woods to the McCoull House, thence to the Ramseur Monument, and then back to the starting location. A map of the walk appears on the next page.

Grant's overall concept for the May 12 morning assault was for Hancock, duplicating Upton's tactic, to break the Confederate position at the apex of the Mule Shoe. Wright was to move and exploit Hancock's gains, and Warren and Burnside were to apply pressure on Lee's extreme flanks. Duplicating Upton's five-thousand-man main assault on Dole's Salient using Hancock's approximately twenty thousand troops to attack the Mule Shoe salient would require Hancock to move from his position on the Federal right flank to the center of the Federal line.

As the Confederates maneuvered to this area on May 8, it was late, and their deployment followed the ridgeline to form the position dubbed the Mule Shoe. Initially, the Confederate line extended into the open field in front of the Landrum House. By the evening of May 9, soldiers had been

Appendix I

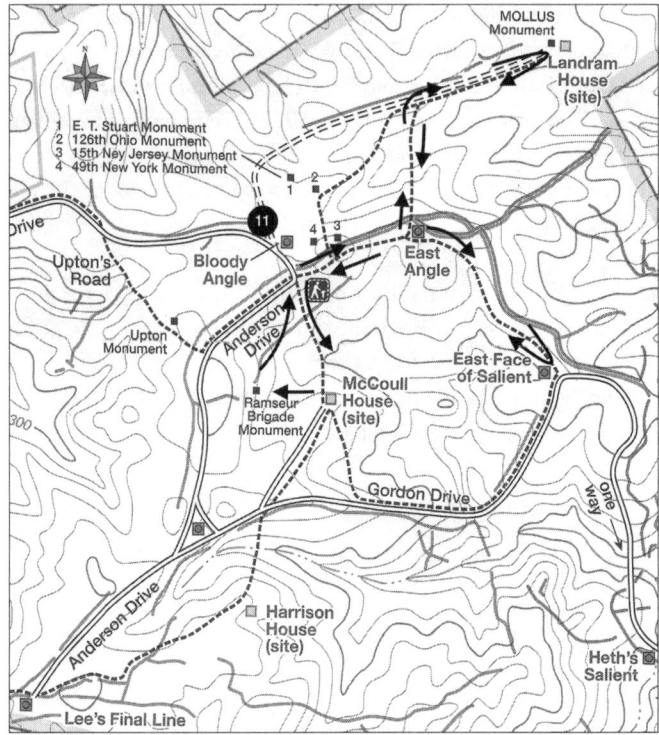

pulled back to the ridge that formed the eastern portion of the salient as we know it now. On the evening of May 10, Steuart's brigade had been pulled from the east side of the salient defenses to meet the crisis created by Upton's breakthrough, but the unit was returned after the lines had been restored.

Farther down, Early's division manned the eastern Confederate defenses facing Burnside. These stretched from south of Spotsylvania Court House along the Brock Road to south of Steuart's position with about a one-quarter mile gap between them. Lane's brigade had moved closer to Steuart's right flank by the night of May 11, but a gap still existed.

On the evening of May 11, Ewell's Corps was in place to defend the Mule Shoe. Troops in Maj. Gen. Edward Johnson's division manned the salient, followed by Maj. Gen. Robert Rodes's division on his left flank, which would place him in the area where Anderson Drive starts from the parking lot. Brig. Gen. John Gordon's division manned a reserve line back along the battlefield park's Gordon Drive just south of the McCoull House. At the time of the battle, the McCoull House could be seen from here, but the view is now obscured by the trees. Gordon had been put in a reserve position with orders from Lee "to support any portion of the line about the salient which might be attacked."[180]

Confederate troops at this location to your right would have been part of a combined Louisiana brigade. Casualties had forced the consolidation of Louisiana troops in Brig. Gen. Harry Hays's brigade in Gordon's division and Brig. Gen. Leroy Stafford's brigade in Johnson's division into a single brigade now assigned to Johnson's division. The Louisiana troops here were under the immediate command of Col. William Monaghan. To their right along the western side of these defensive works were the troops of Walker's Stonewall Brigade that had been engaged with Upton's troops the day before. The remaining part of the combined Louisiana brigade was under Col. Zebulon York, and it was situated where the Mule Shoe started to bend to the right. Starting roughly at the apex, Witcher's Brigade was in position, and Steuart's brigade on the east side of the Mule Shoe followed, completing the defenses.

Rodes's division was located to the left (south) of Monaghan's troops, completing the end of the Mule Shoe and forming a connection toward Laurel Hill. The unit's disposition was as follows: Daniel's brigade, Ramseur's, and then Battle's. What was left of Doles's brigade was behind this line. Pegram's brigade (Hoffman) in Gordon's division was behind Monaghan's, and Gordon's brigade (Evans) was along the park road. Gordon's remaining brigade under Brig. Gen. Robert Johnston was stationed south of the Harrison House facing the Mule Shoe.

To make the salient more defensible, artillery was posted with the infantry units throughout the Mule Shoe. Lee and Grant both knew that the most vulnerable portion of a salient is its apex. Thus, Lee would locate most of his artillery there, and Grant would make it the focus of his assault. Troops deployed in a salient present a defensive challenge. An attacker's incoming fire converges on the salient, while a defender's outgoing fire diverges. In the case of the Mule Shoe, what made it defensible was putting artillery throughout.

Brig. Gen. Alexander Long commanded the Confederate artillery in Ewell's Corps, and four of his five artillery battalions were involved with supporting the infantry here. At the apex of the Mule Shoe, Lieut. Col. William Nelson's three batteries, in counterclockwise order, were Kirkpatrick's, Massie's, and Milledge's. Two of Page's battalions, Montgomery's and Carter's, were stationed on Nelson's right. Page's other two battalions, Fry's and Reese's, were positioned with Witcher's and Steuart's brigades. Carrington's battery of Maj. Wilfred Cutshaw's battalion was to Nelson's immediate left. Cutshaw's remaining two batteries were posted with troops on the west side, Tanner's battery with Hoffman's brigade and Garber's battery with Daniel's Brigade. Completing the artillery dispositions in the Mule Shoe, Jones's and B. H. Smith's batteries of Hardaway's battalion were located between and slightly behind the interface between Daniel's and Ramseur's brigades. (Refer to the map on page 141.)

Appendix I

Meade's headquarters sent out a request at 7:30 a.m. on May 11 for what number of troops would be needed to hold their position and how many would be available for offensive operations:[181]

> CIRCULAR.] HEADQUARTERS
> ARMY OF THE POTOMAC,
> May 11, 1864—7.30 a.m.
> Corps commanders will ascertain what force is sufficient for holding securely their positions and what number of troops will then remain free for use in offensive movement or in extending the front of the army. It is of the utmost importance that there should be some available force for these purposes, and the commanding general desires that corps commanders will examine their positions carefully with these views and to determine what further works, if any, are necessary to reduce the number of men holding them to the lowest number possible . . .
>
> By command of Major-General Meade:
> S. WILLIAMS,
> Assistant Adjutant-General

Wright, whose corps would basically stay in position to guard against any Confederate offensive movements and to support Hancock's accomplishments, replied in this manner:[182]

> HEADQUARTERS SIXTH CORPS,
> May 11, 1864—9.15 a.m.
> Brig. Gen. S. WILLIAMS,
> Assistant Adjutant-General:
> GENERAL: In answer to your circular of this morning I respectfully report as follows: It will require a force of about 8,000 men to hold securely the rifle-pits and picket-line of the corps, leaving about 6,000 men available for offensive operations elsewhere . . .
>
> Very respectfully, your obedient servant,
> H. G. WRIGHT,
> Brigadier-General, Commanding Corps

With the maturing of Grant's plan, orders were sent to the affected corps commanders out that afternoon:[183]

> HEADQUARTERS ARMY OF THE POTOMAC,
> May 11, 1864—4 p.m.
> Major-General HANCOCK,
> Commanding Second Corps:
> GENERAL: You will move, as soon after dark as it can be done without attracting the attention of the enemy, the divisions of Birney and Barlow, with which and Mott's division you will assault the enemy's line from the left of the position now occupied by General Wright and between him and General Burnside . . . Gibbon's division . . . will be moved before daylight . . .
> GEO. G. MEADE,
> Major-General.

> HEADQUARTERS ARMY OF THE POTOMAC,
> May 11, 1864—6.30 p.m.
> COMMANDING OFFICER FIFTH CORPS:
> . . . The Second Corps will be withdrawn to-night from its position, which you will hold in addition to the position now held by you. . . . To aid in meeting an attack of the enemy or his advance upon our right, General Wright, commanding Sixth Corps, will post at Alsop's, in the vicinity of your present headquarters, a division of his corps, under the command of General Russell, that will be held ready to move wherever required. Another division of his corps, under General Wheaton, will be held ready near his own position to support wherever needed. The combined attack of Burnside and Hancock will take place at 4:30 a.m. to-morrow, at which hour your troops will be in readiness . . .
> A. A. HUMPHREYS,
> Major-General and Chief of Staff

> HEADQUARTERS ARMIES OF THE UNITED STATES,
> Near Spotsylvania Court-House, Va., May 11, 1864—4 p.m.
> Maj. Gen. A. E. BURNSIDE,
> Commanding Ninth Army Corps:

> Major-General Hancock has been ordered to move his entire corps under cover of night to join you in a vigorous attack against the enemy at 4 a.m. of to-morrow, the 12th instant. You will move against the enemy with your entire force promptly and with all possible vigor at precisely 4 o'clock to-morrow morning.... Generals Warren and Wright will hold their corps as close to the enemy as possible, to take advantage of any diversion caused by your and Hancock's attack, and will push in their whole force if any opportunity presents itself.
>
> U. S. GRANT,
> Lieutenant-General

The immediate challenge was to quietly move Hancock's Corps in the dark and rainy conditions for the early attack. Hancock met his three division commanders who were located on the Federal right, Barlow, Birney, and Gibbon, at 7:00 p.m. at his headquarters. They were informed that the Second Corps would make a major attack on the enemy's right flank (the Mule Shoe) at daylight the next day. Barlow's division, with the farthest to travel, was the lead division. Barlow said "We staggered and stumbled along in mud and intense darkness." After two hours, the unit arrived after midnight at the Brown House. Birney's division followed Barlow's, and as planned, Gibbon left last and would be placed in reserve when the units were formed near the Brown House.[184]

Walk to the east end of the parking lot on the trail between the split-rail fence and the road. This path turns to the left and takes you up the west side of the salient. Once you turn north on the path, you will see a trail marker on your right; later, you will take this trail to the McCoull House. The interpretive signs along this route will be reviewed on the walk back. Continue on the trail to the apex of the Mule Shoe, and stand at the interpretive markers reading "Fatal Mistake at the East Angle" and "Dawn Assault."

On the afternoon of May 11, Lee began to get conflicting information from his flanks about Federal troop movements. Unlike Grant, who had allowed Sheridan to go after "Jeb" Stuart, Lee still had some cavalry support. His second son, Maj. Gen. William H. F. Lee , commanded a division of Confederate cavalry and was located on the army's right flank. The younger Lee reported from Gayle near the Fredericksburg Road and the Ni River, noting that Burnside's troops seemed to be moving east on that road, which would take them to Fredericksburg:[185]

> GAYLE'S, May 11, 1864—3.30 p.m.
>
> General R. E. LEE:
>
> GENERAL: The enemy has changed his position slightly. They are withdrawing, I think, across the Ni. A wagon train which was parked about two miles from court-house has disappeared. . . . Since commencing this I think the move looks more like a general move, as they are withdrawing their flanks from the Ny [Ni] at Anderson's house. Up to this time there has been no movement toward the Telegraph road. If I can get a good chance I will open, as I suppose your object for my remaining quiet has been removed by their move.
>
> Respectfully,
>
> W. H. F. LEE,
>
> Major-general.

An hour later, he provided this additional information:

> GAYLE'S, May 11, 1864—4.30 p.m.
>
> General R. E. LEE:
>
> GENERAL: There is evidently a general move going on. Their trains are moving down the Fredericksburg road, and their columns are in motion. Sergeant Chandler has just returned from a scout and reports that they are moving their wounded to Belle Plain. Their trains were moving all night. I am moving lower down where I can better operate.
>
> Respectfully,
>
> W. H. F. LEE,
>
> Major-General.

The Telegraph Road, now US 1, ran south from Fredericksburg and east of Spotsylvania Court House. If Grant was moving in this direction, the movement could show Grant intend to move around Lee's right flank to get to Richmond. Since no movement toward this road had been reported, this eventuality was unlikely. But Belle Plain was Grant's supply base northeast of

Fredericksburg on the Potomac Creek, and if he was indeed traveling in that direction, the reported troop movement up the Fredericksburg Road would make sense.

But Wade Hampton's cavalry had also provided information about Federal movement on Lee's left flank:[186]

> POOLE'S, May 11, [1864]—3.40 p.m.
>
> General EARLY:
>
> GENERAL: A scout has just come in and reports the enemy in large force of infantry marching toward Shady Grove, and when he left, an hour ago, they were one mile this side of Todd's Tavern. I have just taken nine prisoners, who say that their army is in motion, one part moving up the river and the other down. There is evidently a movement up the river, and I am now trying to develop it. I shall remain here till I ascertain more fully what is in my front.
>
> Yours, very respectfully,
> WADE HAMPTON, Major-General.

Early's troops were moving to the east from the Confederate left flank to form the lower right defenses on May 11, and thus Hampton's information was sent to Early as the closest corps commander. A movement "up the river" would indicate that Grant was moving in the opposite direction of his other information that Lee had. In either event, Lee felt that he would have to be ready to move quickly to counter Grant's actions. The rain and mud that would make Barlow's troop movements so difficult would have the same effect on the Confederate artillery, so orders were issued to withdraw the Rebel artillery to a position that would allow it to be quickly moved. Brig. Gen. William Pendleton, Lee's chief of artillery, recounted the situation:[187]

> ### Report of Brig. Gen. William N. Pendleton, CSA, Chief of Artillery
>
> Late in the afternoon of this day [May 11] the commanding general, having reason to believe the enemy withdrawing, and intending to leave him no time to gain distance upon us, directed the general chief of artillery to have brought back from the front line before it should be entirely dark all guns so situated as to be difficult to

withdraw at night, so that everything might be ready to march at any hour.

Orders then were issued to Brig. Gen. Armistead Long, chief of artillery for Ewell's Second Corps:[188]

Report of Brig. Gen. Armistead L. Long, CSA, Chief of Artillery, Second Army Corps

Late in the afternoon [of May 11] I received orders to have all the artillery which was difficult of access removed from the lines before dark, and was informed that it was desirable that everything should be in readiness to move during the night; that the enemy was believed to be moving from our front. I immediately ordered all the artillery on Johnson's front, except two batteries of Cutshaw's battalion, to be withdrawn, as it had to pass through a wood by a narrow and difficult road, and the night bid fair to be very dark.

To counter the threat on the left, Mahone's troops had been ordered back to their previous position. This threat was later felt to be a feint, and Mahone was recalled to rejoin Early. First Corps artillery commander Brig. Gen. E. Porter Alexander noted, "The order to the chiefs of artillery, however, was not recalled, and consequently 22 guns of Page's and Cutshaw's battalion were, about sundown, withdrawn from the position about to be attacked." About thirty guns in all were withdrawn about one and a half miles behind the salient near the town. Only eight guns remained in the salient.[189]

During the night, "Allegany" Johnson noticed indications that convinced him that an attack on his front was imminent. He sent a staff member, Maj. Robert Hunter, to tell Ewell that the position could not be held without artillery. Ewell stated that Lee had positive information that the enemy was moving to turn the right flank, and that the artillery needed to be moved accordingly. When Hunter returned to Johnson's headquarters at the McCoull House, Johnson went in person to convince Ewell of the pending assault. Persuaded, Ewell sent orders for the artillery to be returned.[190]

> ### Report of Brig. Gen. Armistead L. Long, CSA,
> ### Chief of Artillery, Second Army Corps
>
> At 3.30 a.m. on the 12th I received a note from General Johnson, indorsed by General Ewell, directing me to replace immediately the artillery that had been withdrawn the evening before; that the enemy was preparing to attack. I immediately ordered Page's battalion to proceed with all haste to the assistance of General Johnson. He moved his battalion with great rapidity, but just as he reached the point to be occupied the enemy broke Johnson's line and enveloped and captured all of Page's guns except two, which were brought off by Captain Montgomery. At the same time two batteries of Cutshaw's battalion were captured. The enemy thus captured twenty guns—twelve from Page and eight from Cutshaw.

After Ewell recalled the artillery, "Alleghany" Johnson had Hunter issue a circular to each brigade commander stating that all indications pointed to an assault at daybreak, that the artillery had been ordered to return, and that troops should be prepared to "repel the enemy."[191]

This is the last of the actions that resulted from Lee's critical decision to remove the artillery from the salient the night of May 11. Proceed over the bridge protecting the earthworks along the trail in front of you, and continue up the slight rise to the road leading to the Landrum House site. You will see interpretive markers and Federal earthworks and artillery revetments along the way. After reviewing the site and interpretive markers at the Landrum House, position yourself by the markers facing the open field.

On the night of May 11, Confederate skirmishers of Witcher's and York's front-line brigades in the salient were positioned in this area at the Landrum House. This area was more open at that time. About one-quarter of a mile to the north of here, the Union skirmish line started and would wander all along the Union lines.

After midnight, Hancock's troops were assembling in a two-column formation about a half mile north of here at the Brown House. Barlow's First Division formed the Federal left flank with the first line of Miles's and Brooke's brigades and the second line of Smyth's and Brown's (Brown had replaced the relieved Frank). The two brigades of Birney's Third Division, led by Ward and Crocker, made up the front line on the right flank of the Federal assault force. The brigades led by McAllister and Brewster of Mott's Fourth Division were located behind Birney. Gibbon's three brigades, under Carroll, Owen and Webb, were the last to leave the Federal right flank. Men in these units slogged through the night in the rain and mud and arrived just before the attack was scheduled to start. They were placed in reserve behind Barlow's division.

Unlike Upton's attack two days before, this offensive reserved no time to properly reconnoiter the Confederate position. Hancock was ordered to "storm" the enemy position, but he was not even certain as to the exact point to be attacked. Barlow, who commanded Hancock's lead division in the assault, had no information on the enemy strength or position. Nor did he know what other Union troops would be involved—only that the Second Corps was to take part. Barlow's only insight into the enemy position he was to attack was a sketch that Lieutenant Colonel Merriam of the Sixteenth Massachusetts had drawn on a wall in the Brown House. Merriam had been involved with Mott's assault in the area on May 10, and Barlow said of his drawing, "[It was] the sole basis of which the dispositions of my division were made."[192]

Also, while Upton had planned extensively before his assault, the second wave of troops in this attack received no guidance as to what to do once the breakthrough occurred. The lack of knowledge, planning, and preparation for this offensive was such that Barlow ludicrously stated, "For Heaven's sake, at least, face us in the right direction, so that we shall not march away from the enemy, and have to go around the world and come up in his rear." He further characterized the Federal assault planning as haphazard but stated that it was a "brilliant attack."[193]

Due to the dense fog, Hancock delayed the start of the assault from 4:00 a.m. to 4:35 a.m., at which time his troops quietly started off from behind the Union picket line to the Confederate works 1,200 yards away.[194]

While traveling to the front of the Confederate works before you, soldiers of the 145th Pennsylvania of Brooke's brigade in Barlow's front line had gone about half a mile when they passed the Landrum House on their left. This observation provides you with insight to their position in the attack. The firing of the Confederate skirmishers in this area had given warning that something was happening, and these skirmishers soon rushed back to their lines. The few remaining Confederate artillery pieces fired on the Federals when they were in the depression ahead of you. One officer in the attacking corps observed that the fire "would have proved disastrous." But in anticipation of a night attack, he noted, the guns "had been trained on the ridge we had just crossed and so the shells passed over our heads, and did no damage." Although the abatis temporarily slowed the advance, this obstruction was quickly torn away, and the entrenchments on the east side of the salient were topped. Hand-to-hand fighting ensued. The Confederates' initial shots misfired, and rapidly putting new caps on the firearms did nothing for the damp powder.[195]

Proceed back along the path you took here. By stopping at the bottom of the trail and noting the Confederate works to your front and the terrain sloping up to the right, you will better appreciate what the Federals were looking at and how the most intense fighting would be funneled to the west side of the salient.

Barlow's troops were on your left moving toward the east side of the Confederate works, and Birney's were on your right. In the salient at the top of the hill, Witcher's Virginia brigade was stationed to the left side, and York's Louisianans were almost directly in front of you.

Continue to the top of the hill, and recross the footbridge. Note that you will be returning here after stops on the east side of the Mule Shoe. Take the left-hand part of the trail along the east side of the salient about five hundred feet to the interpretive marker "The Confederate Line." (Refer to the map for this tour stop.)

This interpretive marker explains the works in front of you. Barlow's advancing troops swayed to their left with some musketry fire and one "discharge of artillery." These soldiers then swept through Witcher's and then Steuart's brigades that were in this area. Barlow's leading brigade, Brooke's, struck the salient at its apex between Witcher's Virginians and York's Louisianans. By 4:50 a.m. the 26th Michigan and 140th Pennsylvania of Miles's brigade had planted their colors on the Confederates' outer fieldworks, likely in this area. It was about this time that the returning Confederate artillery arrived and was captured.[196]

Continue down the trail about two hundred feet to the interpretive marker "A Mass Capture."

From this marker, you can see stop 6 on the Park Service's Spotsylvania Battlefield Tour—"The East Face of the Salient"—which you will pass on your way out of the park. The initial phase of the surprise Federal assault quickly captured Maj. Gen. Edward Johnson and Brig. Gen. George Steuart in this area. Union troops reported three thousand Confederate prisoners

Scene of capture of Johnson's division, morning of May 12, 1864. Library of Congress.

and twenty pieces of artillery seized. Seeing prisoners being taken to the rear, Gibbon moved his reserve troops forward to support Barlow's troops and take possession of the captured works. At this point, Gibbon also reported confusion among the Union troops. Later, when Confederate reinforcements had pushed his men back, "the men [were] fighting from the opposite side of the same breastwork" in part of his line.[197]

After meeting Hancock at his headquarters, Generals Johnson and Steuart were escorted to Grant's and Meade's headquarters at the Brown House. (See the "A Mass Capture" interpretive marker on the east side of the salient.) While they were there, Grant received Hancock's 7:15 a.m. report stating that he had "finished Johnson's division and commenced on Early's." "General Grant passed this dispatch around, but did not read it aloud, as usual, out of consideration for Johnson's feelings."[198]

Following Barlow's initial breakthrough, his second line of Smyth's and Brown's men was ordered to advance through the captured works and press the enemy back. The advance had happened so quickly that some of the Confederates surrendered without a fight. Soon, the advancing troops struck a second line of Confederate works located along the present-day Gordon Drive, and the Federal advance ground to a halt. The Confederates attacked five times in ten hours, but Barlow had enough strength to hold.[199]

> ### Report of Brig. Gen. John R. Brooke, USA, Commanding Fourth Brigade
>
> The enemy was apprised of the attack by cheers of some new troops in the division as we swept over and down the last descent, and opened a terrific fire of artillery and musketry upon us, notwithstanding which our brave men marched on, and dragging away the abatis to effect a passage poured in one irresistible mass upon them, and after a sharp, short fight, killed and captured nearly all who occupied the works. . . . Never during the war have I seen such desperate fighting. . . . The right of my brigade struck the works about 40 yards to the right of the Angle, thus giving us a great advantage, in sweeping down the line to our left of the Angle. After crossing the first line I pushed forward in pursuit of the flying enemy. After proceeding about 500 yards, I encountered a second line of works with a marsh in its front. Owing to the disorganization of my command I could not make a determined attack on this line. The enemy came out in strong force, when I retired, fighting to the line already captured . . .Up to this time many prisoners were taken, among them Major-General Johnson and Brigadier-General Steuart, of the rebel service. . . .

Walk back to the apex of the salient, and face south to look along the west face of the salient.

Birney's First Brigade, now commanded by Col. Thomas Egan on Barlow's right, hit the salient where York's Louisiana brigade manned the Confederate works. Birney's Second Brigade, commanded by Col. John Crocker on Egan's right, hit closer to Walker's troops farther down the west side. Gibbon's Second Division followed Barlow's troops on the east side, while Mott's Fourth Division followed Birney's troops on the west side. Hand-to-hand fighting ensued that "defie[d] description." The weight of this attack pushed down the west side of the salient and hit the right of Walker's Stonewall Brigade, temporarily stopping the advance. Unable to fire into their own men to stop the Federal assault, the guns of Carrington's battery that had been left in Johnson's front were captured. Walker's brigade was "annihilated," and the Stonewall Brigade ceased to exist.[200]

Walker stated, "[Had the] muskets of our men been serviceable they never would have gotten within three hundred yards of our line." Col. Thomas

Carter, Long's second-in-command of the artillery, felt that although the infantry's damp ammunition might have prevented a successful resistance to the assault, "the chief cause of the capture of the division was the absence of artillery from the line."[201]

Walk back down the path toward the parking lot, pausing at the various interpretive markers that represent fighting after the initial charge, and stop at the bend in the road where Grant Drive becomes Anderson Drive. Looking south toward the Dole's Salient markers, you can see the approximate location where the Union advance progressed on this side of the salient.

After pushing through Walker's position, Birney's troops ran into resistance from Daniel's brigade and three regiments of Evans's brigade. Daniels had his two rightmost regiments, the Second and Forty-Fourth North Carolina, refuse their flank, and with the help of artillery firing over their heads, they enfiladed the attackers. This fighting on the west side of the salient would have been near Dole's Salient. Next, Ramseur's brigade of four North Carolina regiments who had been located on Daniel's left flank, were pulled out of the line. They swung back, came up past the McCoull House, and formed on Daniels's right with orders to "check the enemy's advance and to drive him back." Ramseur's troops hit just below where Walker had been and retook the works with a charge of "unsurpassed gallantry."[202]

On this side of the salient, the temporary standoff that Daniels, who was killed here, and Evans had made against Birney's troops was broken. Battle's brigade of Rodes's division, which had shifted to Ramseur's previous position, came up and finally pushed Birney's men back outside the works. As the Federals were pressed back, they created their own defenses on the opposite side of the Confederate works. The adversaries were then so close together that at times they were firing and bayoneting through the works. Evans's three brigades that had been sent to this area were moved to the east side of the salient, and they rejoined the other three brigades on Hoffman's right.

Just before 6:00 a.m., Hancock called for reinforcements from Wright, who was immediately ordered to support the attack. Wright sent his second division, commanded by Brig. Gen. Thomas Neill, forward. Neill's Third Brigade, commanded by Col. Daniel Bidwell, moved to this area. Bidwell had previously commanded the Forty-Ninth New York Regiment, and its presence in the battle is marked by the "49th NY" regimental monument that you passed on the walk here. Brig. Gen. David Russell's First Division, Fifth Brigade, now under Col. Oliver Edwards, went in. Edwards's troops would remain in this vicinity for the remainder of the day.[203]

Appendix I

In the late morning, Wright's only division that had not been used was Brig. Gen. James Ricketts's Third Division, and it was sent in for support:[204]

> ### Report of Col. J. Warren Keifer, 110th Ohio Infantry, Commanding Second Brigade
>
> On the 12th the brigade, with the division, was formed 1 mile to the left, about 11 a.m., in support of the First and Second Divisions, Sixth Army Corps, but was not heavily engaged. The One hundred and twenty-sixth Ohio was detached about 12 m., and went to the assistance of Brigadier-General Wheaton's brigade, Second Division, Sixth Corps. It was marched to the front line and engaged the enemy.

The "126th Ohio" marker commemorates their effort here.

The overall contour of the land here on the west side of the salient tended to funnel troops to this area, "the Bloody Angle," where the three Federal monuments are located. From a timing perspective, the monuments to the "49th N.Y. Inf'y" of Bidwell's brigade and the "15th Reg't N.J. Vol's." of Col. Henry Brown's First Brigade of Wright's First Division mark the first two units to arrive. The 126th Ohio detached to Wheaton, who in turn was supporting Edwards here at the salient, and it would come up later in the morning. The 15th New Jersey troops' actual movement would have put them closer to Dole's Salient.

View looking northeast from the Bloody Angle with the west side of the salient on the right. Courtesy of the author.

As the fighting continued, Neill's troops were reinforced by Brooke, who had returned to the Landrum House after running out of ammunition. Brooke now moved to attack more from the west to assist Edwards, who was pinned down on the outside of the Confederate fieldworks.

Take the small marked trail about ten yards up the main trail (refer to map for this tour stop) through the woods to the McCoull House location. Once you have reviewed the interpretive markers, face them.

Note on the "Mayhem in the Mule Shoe" marker that the area looking toward the parking lot was open field, and the terrain looking to the north and east was wooded. Here in the center of the salient, Hancock's men advanced to and then past the McCoull House (previously Johnson's divisional headquarters), but they were becoming increasingly disorganized. No one seemed to know what to do next. As the Federals moved south, word of the pending disaster reached Gordon.

Turn to face south. The unpaved road to your right leads to Gordon Drive, where Gordon's troops were located. There, a trail leads to the Harrison House, where Lee's headquarters were.

Ewell had ordered Gordon to support "either Johnson's or Rhodes' division, or both, as circumstances should require." He had placed Col. Clement Evans, now commanding Gordon's old brigade, along the modern-day Gordon

Harrison House, location of Lee's headquarters.
Library of Congress.

Appendix I

Road, and his other two brigades, led by Col. John Hoffman and Brig. Gen. Robert Johnston, were stationed closer to the Harrison House. With the warnings in the night, these units were moved up out of concern for the enemy's movement. Gordon sent Johnson's brigade forward, and the screaming Confederates quickly ran into the enemy. Surprised, the Federals gave way. A temporary reprieve achieved, Gordon sent Hoffman's brigade and Evans's three remaining regiments to the front. At this critical time, Lee arrived at the front of the formation. As in The Wilderness, he appeared ready to lead the assault, and his troops once again turned him back. Hoffman's troops pushed Brooks back, while on Hoffman's right, Evans's three remaining regiments drove Miles's brigade back to the outer edge of the works. "Gordon, carrying the colors, led them forward in a headlong, resistless charge, which carried everything before it, recapturing the trenches on the right salient, and a portion of those on the left."[205]

With the crisis unfolding in the Mule Shoe, Lee made the difficult decision to trade casualties for time. Lee's engineers, along with the remnants of Johnson's division, began constructing a defensive line across the base of the salient (see the additional driving stop at the end of the tour). To buy time, Lee would commit additional troops from other parts of his defenses to temporarily regain what had been lost and keep the Federals at bay. Grant's overall plan for coordinated assaults on all fronts had failed to materialize.

The Confederate troops entering the Mule Shoe to retake the lost works while a new line was formed across the base of the salient would have to fight from traverse to traverse as they moved to retake their former position. The traverses were earthworks perpendicular to the main earthworks, positioned at intervals to prevent attacking forces from easily moving along a trench line and taking the defenders in flank—what the Confederates were now attempting. It was dark during the early morning attack, and the surprise and size of the Federal assault overtook these works before the defenders knew what had happened. Those circumstances were not present when the works were retaken. Many personal accounts describe soldiers' horror during the hand-to-hand fighting in close quarters and in rainy, muddy conditions.

Early's Corps manned the Confederate defenses near Spotsylvania Court House that formed the southeastern portion of Lee's overall line. When Hancock threatened Lee's left flank near the Po River on the evening of May 10, Mahone's division of Early's Corps and two brigades of Wilcox's division, those commanded by Scales and Thomas, were dispatched to counter the threat. Wilcox's remaining two brigades, Lane's and McGowan's, were left on the eastern flank facing Burnside's troops, as was Heth's division. With these reinforcements, the Confederate position became too strong for Hancock's planned maneuver across the Po River. Hancock's Corps was then pulled back on the night of May 11, and it moved to make Grant's frontal assault on the salient.

With the threat shifted from the far left flank, Mahone's Corps and Wilcox's brigades were recalled back to the Harrison House (its ruins are just over a quarter mile south of here), four miles from where they were starting from and arrived at about 7:00 a.m. Brig. Gen. Abner Perrin's and Brig. Gen. Nathanial Harris's brigades of Mahone's division were sent to this area to meet the threat from Hancock's breakthrough in the salient, while Wilcox's soldiers went back to Early's front.

The "Wood Road" that Upton had used on May 10 extended through the Confederate works, past the Harrison House, and on to the Brock Road. Looking to the west from this location, you can see the markers of Dole's Salient where the road entered the Confederate works and turned south. Mahone's returning Confederates used this road to come up into this critical portion of the salient. In addition, Brig. Gen. Samuel McGowan's brigade of five South Carolina regiments that had stayed near the courthouse was now moving up as reinforcements, following Harris on the same road.[206]

From the markers at the McCoull House, take the trail cut in the grass to the Ramseur Monument to the west.

This is the area where Ramseur's men had moved earlier, and subsequent Confederate troops moved to retake the west side of the salient. From this slight depression, Dole's Salient is not visible, but it is located just over the ridge to the west.

Gordon directed Perrin's returning brigade to Ramseur's right flank. They passed the McCoull House and pressed on to the front, at which time Perrin was killed and replaced by Col. John Sanders. Harris's brigade went in on the right of Perrin's brigade, extending the attack farther up the west side of the salient. These men were going against Bidwell's and Edwards's brigades of Neill's division, and against Brewster's brigade of Mott's division. Birney's two brigades commanded by Ward/Egan and Crooker supported Brewster's men.

By 8:00 a.m. the Rebels had pushed the Federals back on the wings of the salient, but a gap remained near the apex of the original Confederate defensive works. This opening was between the works retaken on the west side by Ramseur's, Perrin's/Saunders's, and Harris's brigades, and those recovered on the east side by Hoffman's and Johnson's/Toon's brigades. McGowan's brigade of 1,300 men was tasked with plugging it. They arrived in the

Appendix I

salient about 9:00 a.m. in the knee-deep mud on the right of Harris's men. McGowan was wounded in the arm, his second-in-command was killed, and Col. Joseph Brown assumed command of the brigade. The South Carolina monument along the west side of the salient where you previously were commemorates this movement. At the apex, Confederates drove Brewster's Excelsior Brigade from the salient and then encountered and pushed out the 86th and 124th New York of Birney's First Brigade, now commanded by Col. Thomas Egan. Hardaway's Confederate artillery was able to fire along the western edge, pounding the Union troops in the Bloody Angle, making it a killing field. The Confederates were finally able to retake their original works.[207]

On their return, Wilcox's brigades under Thomas and Scales were sent back to strengthen Lane's position on the east side of the Mule Shoe. Potter's Second Division of Burnside's Corps was positioned on the right side of the salient across from Lane's brigade. It threatened but did not advance. These returning brigades eased the menace Barlow's troops posed. Later in the day, Mahone's returned troops would make an attack against Burnside that continued the Federals' lack of success on this flank.

From the Ramseur's Brigade Monument, proceed about a quarter of a mile back to the parking lot by heading north through the gap in the trees and then turning to your left. Please ensure that you do not damage any of the remains of the earthworks in the area.

With Hancock's Corps moved to the middle of the Federal line, Warren's Corps once again became the main force on the Confederate left flank. To relieve pressure on the Federal troops fighting at the salient and prevent Lee from moving troops from Laurel Hill, Meade ordered Warren at 8:00 a.m. to attack "immediately." But Warren delayed. At 9:15 a.m. he received a "peremptory" order to attack Laurel Hill, and this attack was quickly repulsed. This was the fourth or fifth unsuccessful assault, and experience had taught the men that to attack Laurel Hill "was certain death on *that* front." This movement against the well-defended position only resulted in more Union casualties. Although Warren correctly assessed the attack's futility, Grant was ready to relieve him if he did not attack. But Humphreys, who was at the front with Warren, was satisfied that the attack could not succeed, and he stopped the continuation of the assault and saved Warren from dismissal. Meade and Grant decided the best way to use the Fifth Corps was to send troops to support the fighting at the salient. To that end, Crawford's and Cutler's divisions were sent to Wright, and Griffin's men were sent to

Hancock, leaving Warren only in charge of Robinson's division on the Federal right flank.[208]

Rather than using the troops sent to him to prepare for another assault as Grant had intended, Wright positioned the men only to support his corps. He began to rationalize not attacking, and at 5:10 p.m., he sent word to Meade that he had decided not to launch the offensive. Meade approved.[209]

HEADQUARTERS SIXTH CORPS,
May 12, 1864—5.10 p.m.

Major-General HUMPHREYS, Chief of Staff:

I have decided not to make the assault for the reasons given in previous dispatch—those reasons being, not that it might not succeed, but in view of the disaster which would possibly follow a failure; also the want of a sufficient available and suitable force to insure a reasonable prospect of success. I shall abandon entirely my old position of yesterday, and extend my present right to the morass in front of it. This, with Griffin's column and some relieved troops, who have been fighting all day, are all the reserve I have. General Hancock desires me to say that he fully concurs in the views I have expressed. I shall make a return on my right with what force I can collect.

H. G. WRIGHT,
Brigadier-General, Commanding.

HEADQUARTERS ARMY OF THE POTOMAC,
May 12, 1864—6.15 p.m.

General WRIGHT:

Your dispatch abandoning attack received and approved. Rectify your lines and connect with Warren, keeping the Fifth Corps as far as practicable together and on your right.

MEADE.

Communicate to Warren and Humphreys and prepare to rearrange the right flank.

At 10:20 a.m., Grant ordered Burnside to move one of his divisions to connect with Hancock and to push an attack with the rest. As a final prod to the slow-moving Burnside, Grant ended by stating, "See that your orders are executed."[210]

About 2:00 p.m., Burnside decided to move his Third Division under Brig. Gen. Orlando Willcox against the Confederate position known as Heth's Salient. Brig. Gen. Joseph Davis's and Col. Robert Mayo's brigades were in position there. In hopes of relieving pressure on Ewell in the salient, Lee planned to use two brigades of Early's Corps, Weisenger's and Lane's, to attack Burnside's exposed left flank. Lane's troops got into some intense fighting with parts of Willcox's troops. As the Confederates were advancing on one side, another part of Willcox's men started their movement on Heth's Salient and were hit by Weisenger's troops. The intense fighting ended after a half hour, and the troops returned to their sides. This engagement occurred at the seventh stop of the Park Service's Spotsylvania Tour. which you well pass on your way out of the park. Although unable to damage the Ninth Corps as he had hoped, Lee still foiled Grant's plans on this front.[211]

Although the major attacks were over for the day, pockets of fighting continued into the night. At 3:00 a.m. on May 13, the Confederate troops quietly withdrew from the salient to their new, stronger line across its base. Grant would test this new line later that day, and he made a major attack on May 18. The new, more defensible line with artillery in place proved strong, and Grant failed to achieve any tactical advantage here.[212]

Additional nearby sites of interest

It is highly recommended that you visit Lee's new defensive line at the base of the Salient.

To travel to Lee's final line from the parking lot, proceed south for three-quarters of a mile on Anderson Drive to its end.

This optional stop is where the Confederates constructed their final line while the fighting was ongoing in the Mule Shoe. Rebel troops were withdrawn to this area at 3:00 a.m. on May 13. With Lee strongly entrenched here, Grant sought to maneuver for a better opportunity. Warren's and Wright's Corps were moved east of the town (outside of the park boundary) on the night of May 13–14, with Lee moving Field's division on the evening of May 14 to counter. Fighting in that area proved to be inconsequential, and Grant once again moved troops. On the night of May 17–18, Wright's Corps was moved from the far left Union flank to a position a little south and west of his position on May 12. Hancock's Corps was moved from near the Ni River to just north of the McCoull House. Grant ordered Wright and Hancock to attack this position at dawn on May 18. The artillery that was lacking in the salient was present here now. Its effectiveness demonstrated how fortunate the Federals had been when Lee made the critical decision to move artillery out of the salient on the night of May 11. The attack was effectively repulsed.

At this time, the Confederate defenses ran from this area to well south of Spotsylvania Court House. Grant would pull his troops back toward the east. Warren would be positioned about a mile east of the McCoull House in a north–south line running to the east of the town, with Wright below Warren, and Burnside angling back from Wright's left flank. Hancock was once again positioned east of the Ni River and south of the Fredericksburg Road.

Seeking an opportunity, Lee would send Ewell's force on a reconnaissance from this area on May 19 to probe the changing Federal right. Ewell barely avoided a disaster, and Hancock's planned move south was delayed by one day.

Conclusion of the Tour

From the parking lot at the Bloody Angle head south, or from Lee's last line head north, to the intersection of Anderson Drive and Gordon Drive. There are interpretive markers about the Harrison House at the intersection. Proceed east on Gordon Drive, where you can see some of the remains of Gordon's defensive line. Once you pass the gravel road on your left to the McCoull House, the road becomes one way. As you exit the park, you will pass the stop 6 of the Park Service's Spotsylvania Battlefield Tour, "East Face of Salient," which you saw from the walking part of the tour. You will then move on to the Park Service's previously mentioned stop 7, "Heth's Salient," which covers Burnside's efforts on this flank. This one-way road will exit onto Route 208, Courthouse Road. This concludes the tour.

APPENDIX II

UNION ORDER OF BATTLE

COMMANDER US ARMIES
    Lieut. Gen. Ulysses S. Grant (USMA 1843)

ARMY OF THE POTOMAC
    Maj. Gen. George G. Meade (USMA 1835)
    Chief of Staff Maj. Gen. Andrew Humphreys (USMA 1831)
    Asst. Adjutant General Brig. Gen. Seth Williams

Provost Guard
    Brig. Gen. Marsena R. Patrick
    1st Massachusetts Cavalry, Companies C and D
    80th New York Infantry
    3rd Pennsylvania Cavalry
    68th Pennsylvania Infantry
    114th Pennsylvania Infantry

Artillery
    Brig. Gen. Henry J. Hunt

Artillery Reserve
    Col. Henry S. Burton

First Brigade
    Col. J. Howard Kitching

Appendix II

    6th New York Heavy
    15th New York Heavy

SECOND BRIGADE
    Maj. John A. Tompkins
    Maine Light, 5th Battery E
    1st New Jersey Light, Battery A
    1st New Jersey Light, Battery B
    New York Light, 5th Battery
    New York Light, 12th Battery
    1st New York Light, Battery B

THIRD BRIGADE
    Maj. Robert H. Fitzhugh
    Massachusetts Light, 9th Battery
    New York Light, 15th Battery
    1st New York Light, Battery C
    New York Light, 11th Battery
    1st Ohio Light, Battery H
    5th United States, Battery E

VOLUNTEER ENGINEER BRIGADE
    Brig. Gen. Henry W. Benham
    15th New York Engineers
    50th New York Engineers

BATTALION US ENGINEERS
    Capt. George H. Mendell

HORSE ARTILLERY

FIRST BRIGADE
    Capt. James M. Robertson
    New York Light, 6th Battery
    2nd United States, Batteries B and L
    2nd United States, Battery D
    2nd United States, Battery M
    4th United States, Battery A
    4th United States, Batteries C and E

SECOND BRIGADE
    Capt. Dunbar R. Ransom

1st United States, Batteries E and G
1st United States, Batteries H and I
1st United States, Battery K
2nd United States, Battery A
2nd United States, Battery G
3rd United States, Batteries C, F, and K

SECOND ARMY CORPS
Maj. Gen. Winfield S. Hancock (USMA 1844)

FIRST DIVISION
Brig. Gen. Francis C. Barlow (USMA 1861)

First Brigade
Col. Nelson A. Miles
26th Michigan
61st New York
81st Pennsylvania
140th Pennsylvania
183rd Pennsylvania

Second Brigade
Col. Thomas A. Smyth
28th Massachusetts
63d New York
69th New York
88th New York
116th Pennsylvania

Third Brigade
Col. Paul Frank
(Col. Hiram L. Brown) (C-S)
39th New York
52nd New York
57th New York
111th New York
125th New York
126th New York

Fourth Brigade
Col. John R. Brooke
2nd Delaware

64th New York
66th New York
53rd Pennsylvania
145th Pennsylvania
148th Pennsylvania

SECOND DIVISION
Brig. Gen. John Gibbon (USMA 1847)

FIRST BRIGADE
Brig. Gen. Alexander S. Webb (W-S)
(Col. Boyd McKeen)
19th Maine
1st Company Sharpshooters
15th Sharpshooters
19th Massachusetts
20th Massachusetts
7th Michigan
42nd New York
59th New York
82nd New York

SECOND BRIGADE
Brig. Gen. Joshua T. Owen
152nd New York
69th Pennsylvania
71st Pennsylvania
72nd Pennsylvania
106th Pennsylvania

THIRD BRIGADE
Col. Samuel S. Carroll (W-W, S)
(Col. Theodore Ellis)
14th Connecticut
1st Delaware
14th Indiana
12th New Jersey
10th New York Battalion
108th New York
4th Ohio
8th Ohio
7th West Virginia

THIRD DIVISION
Maj. Gen. David B. Birney

First Brigade
Brig. Gen. J. H. Hobart Ward (R-S)
(Col. Thomas Egan)
20th Indiana
3rd Maine
40th New York
86th New York
124th New York
99th Pennsylvania
141st Pennsylvania
2nd US Sharpshooters

Second Brigade
Brig. Gen. Alexander Hays (K-W)
(Col. John Crocker)
4th Maine
17th Maine
3rd Michigan
5th Michigan
93d New York
57th Pennsylvania
63rd Pennsylvania
105th Pennsylvania
1st US Sharpshooters

FOURTH DIVISION[1]
Brig. Gen. Gershom Mott

First Brigade
Col. Robert Mcallister (W-W)
(Col. Napoleon McLaughlen)
1st Massachusetts
16th Massachusetts
5th New Jersey
6th New Jersey
7th New Jersey
8th New Jersey
11th New Jersey

26th Pennsylvania
115th Pennsylvania

SECOND BRIGADE (Excelsior Brigade)
Col. William R. Brewster
11th Massachusetts
70th New York
71st New York
72nd New York
73rd New York
74th New York
120th New York
84th Pennsylvania

ARTILLERY BRIGADE
Col. John C. Tidball
6th Maine, Battery (F)
10th Massachusetts Battery
1st New Hampshire Battery
1st New York, Battery G
4th New York Heavy, 3rd Battalion
1st Pennsylvania, Battery F
1st Rhode Island, Battery A
1st Rhode Island, Battery B
4th US, Battery K
5th US, Batteries C and I

FIFTH ARMY CORPS
Maj. Gen. Gouverneur K. Warren (USMA 1850)

PROVOST GUARD
12th New York Battalion, Maj. Henry W. Rider
Aide-de-Camp Maj. Washington Roebling
Asst. Adjutant General Col. Frederick Locke

FIRST DIVISION
Brig. Gen. Charles Griffin (USMA 1847)

FIRST BRIGADE
Brig. Gen. Romeyn B. Ayres
140th New York
146th New York

91st Pennsylvania
155th Pennsylvania
2nd US, Companies B, C, F, H, I, and K
11th US, Companies B, C, D, E, F, and G, 1st Battalion
12th US, Companies A, B, C, D, and G, 1st Battalion
12th US, Companies A, C, D, F, and H, 2nd Battalion
14th US, 1st Battalion
17th US, Companies A, C, D, G, and H, 1st Battalion
17th US, Companies A, B, and C, 2d Battalion

SECOND BRIGADE
   Col. Jacob B. Sweitzer
   9th Massachusetts
   22nd Massachusetts
   32nd Massachusetts
   4th Michigan
   62nd Pennsylvania

THIRD BRIGADE
   Brig. Gen. Joseph J. Bartlett
   20th Maine
   18th Massachusetts
   1st Michigan
   16th Michigan
   44th New York
   83rd Pennsylvania
   118th Pennsylvania

SECOND DIVISION[2]
   Brig. Gen. John C. Robinson (W-S)
   (USMA—did not graduate)

FIRST BRIGADE
   Col. Samuel H. Leonard (Ill)
   (Col. Peter Lyle)
   16th Maine
   13th Massachusetts
   39th Massachusetts
   104th New York

SECOND BRIGADE
   Brig. Gen. Henry Baxter (W-W)
   (Col. Richard Coulter) (W-S)

12th Massachusetts
83rd New York (9th Militia)
97th New York
11th Pennsylvania
88th Pennsylvania
90th Pennsylvania

THIRD BRIGADE
Col. Andrew W. Denison (W-S)
(Col. Charles Phelps) (C-S)
(Col. Richard Bowerman)
1st Maryland
4th Maryland
7th Maryland
8th Maryland

THIRD DIVISION
Brig. Gen. Samuel W. Crawford (W-S)

FIRST BRIGADE
Col. William McCandless (W-W)
(Col. William Talley) (C-S)
(Col. Wellington Ent)
(Col. Samuel Jackson)
1st Pennsylvania Reserves
2nd Pennsylvania Reserves
6th Pennsylvania Reserves
7th Pennsylvania Reserves
11th Pennsylvania Reserves
13th Pennsylvania Reserves (1st Rifles)

THIRD BRIGADE
Col. Joseph W. Fisher
5th Pennsylvania Reserves
8th Pennsylvania Reserves
10th Pennsylvania Reserves
12th Pennsylvania Reserves

FOURTH DIVISION
Brig. Gen. James S. Wadsworth (MW-W)
(Brig. Gen. Lysander Cutler)

First Brigade (Iron Brigade)
    Brig. Gen. Lysander Cutler
    (Col. William Robinson)
    7th Indiana
    19th Indiana
    24th Michigan
    1st New York Battalion Sharpshooters
    2nd Wisconsin
    6th Wisconsin
    7th Wisconsin

Second Brigade
    Brig. Gen. James C. Rice (K-S)
    (Col. Edward Fowler)
    76th New York
    84th New York (14th Militia)
    95th New York
    147th New York
    56th Pennsylvania

Third Brigade
    Col. Roy Stone (W-W)
    (Col. Edward Bragg)
    121st Pennsylvania
    142nd Pennsylvania
    143rd Pennsylvania
    149th Pennsylvania
    150th Pennsylvania

Artillery Brigade
    Col. Charles S. Wainwright
    Massachusetts Light, Battery C
    Massachusetts Light, Battery E
    1st New York Light, Battery D
    1st New York Light, Batteries E and L
    1st New York Light, Battery H
    4th New York Heavy, 2nd Battalion
    1st Pennsylvania Light, Battery B
    4th United States, Battery B
    5th United States, Battery D

SIXTH ARMY CORPS
Maj. Gen. John Sedgwick (K-S) (USMA 1837)
(Brig. Gen. Horatio G. Wright) (USMA 1841)

FIRST DIVISION
Brig. Gen. Horatio G. Wright (USMA 1841)
(Brig. Gen. David A. Russell) (USMA 1845)

First Brigade
    Col. Henry W. Brown
    1st New Jersey
    2nd New Jersey
    3d New Jersey
    4th New Jersey
    10th New Jersey
    15th New Jersey

Second Brigade
    96th Pennsylvania

Third Brigade
    Brig. Gen. David A. Russell
    (Brig. Gen. Henry L. Eustis)
    6th Maine
    49th Pennsylvania
    119th Pennsylvania
    5th Wisconsin

Fourth Brigade
    Brig. Gen. Alexander Shaler (C-W)
    (Col. Nelson Cross)
    65th New York
    67th New York
    122nd New York
    82nd Pennsylvania (detachment)

SECOND DIVISION
Brig. Gen. George W. Getty (W-W) (USMA 1840)
(Brig. Gen. Thomas H. Neill) (USMA 1847)

First Brigade
    Brig. Gen. Frank Wheaton
    62nd New York

93rd Pennsylvania
98th Pennsylvania
102nd Pennsylvania
139th Pennsylvania

SECOND BRIGADE
Col. Lewis A. Grant.
2nd Vermont
3d Vermont
4th Vermont
5th Vermont
6th Vermont

THIRD BRIGADE
Brig. Gen. Thomas H. Neill
(Col. Daniel Bidwell)
7th Maine
43rd New York
49th New York
77th New York
61st Pennsylvania

FOURTH BRIGADE
Brig. Gen. Henry L. Eustis (to 1st Division)
(Col. Oliver Edwards)
7th Massachusetts
10th Massachusetts
37th Massachusetts
2nd Rhode Island

THIRD DIVISION
Brig. Gen. James B. Ricketts (USMA 1839)

FIRST BRIGADE
Brig. Gen. William H. Morris (W-S)
(Col. John Schall)
14th New Jersey
106th New York
151st New York
87th Pennsylvania
10th Vermont

SECOND BRIGADE
Brig. Gen. Truman Seymour (C-W)

(Col. Benjamin Smith)
6th Maryland
110th Ohio
122nd Ohio
126th Ohio
67th Pennsylvania (detachment)
138th Pennsylvania

Artillery Brigade
　Col. Charles H. Tompkins
　Maine Light, 4th Battery (D)
　Massachusetts Light, 1st Battery (A)
　New York Light, 1st Battery
　New York Light, 3rd Battery
　4th New York Heavy, 1st Battalion
　1st Rhode Island Light, Battery C
　1st Rhode Island Light, Battery E
　1st Rhode Island Light, Battery G
　5th United States, Battery M

NINTH ARMY CORPS
　Maj. Gen. Ambrose E. Burnside (USMA 1847)

FIRST DIVISION
　Brig. Gen. Thomas G. Stevenson (K-S)
　(Maj. Gen. Thomas Crittenden)

First Brigade
　Col. Sumner Carruth
　35th Massachusetts
　56th Massachusetts
　57th Massachusetts
　59th Massachusetts
　4th United States
　10th United States

Second Brigade
　Col. Daniel Leasure (W-S)
　(Col. Gilbert Robinson)
　3rd Maryland
　21st Massachusetts
　100th Pennsylvania

ARTILLERY
 Maine Light, 2nd Battery (B)
 Massachusetts Light, 14th Battery

SECOND DIVISION
 Brig. Gen. Robert B. Potter

FIRST BRIGADE
 Col. Zenas R. Bliss (W-S)
 (Col. John Curtin)
 36th Massachusetts
 58th Massachusetts
 51st New York
 45th Pennsylvania
 48th Pennsylvania
 7th Rhode Island

SECOND BRIGADE
 Col. Simon G. Griffin
 31st Maine
 82nd Maine
 6th New Hampshire
 9th New Hampshire
 11th New Hampshire
 17th Vermont

ARTILLERY
 Massachusetts Light, 11th Battery
 New York Light, 19th Battery

THIRD DIVISION
 Brig. Gen. Orlando B. Willcox (USMA 1847)

FIRST BRIGADE
 Col. John F. Hartranft
 2nd Michigan
 8th Michigan
 17th Michigan
 27th Michigan
 109th New York
 51st Pennsylvania

SECOND BRIGADE
    Col. Benjamin C. Christ
    1st Michigan Sharpshooters
    20th Michigan
    70th New York
    60th Ohio
    50th Pennsylvania
ARTILLERY
    Maine Light, 7th Battery (G)
    New York Light, 34th Battery

FOURTH DIVISION
    Brig Gen. Edward Ferrero

FIRST BRIGADE
    Col. Joshua K. Sigfried
    27th US Colored Troops
    30th US Colored Troops
    39th US Colored Troops
    43rd US Colored Troops
SECOND BRIGADE
    Col. Henry G. Thomas
    30th Connecticut (Colored)
    19th US Colored Troops
    23rd US Colored Troops
ARTILLERY
    Pennsylvania Light, Battery D
    Vermont Light, 3rd Battery
CAVALRY
    3rd New Jersey
    22nd New York
    2nd Ohio
    13th Pennsylvania
RESERVE ARTILLERY
    Capt. John Edwards Jr.
    New York Light. 27th Battery
    1st Rhode Island Light, Battery D
    1st Rhode Island Light, Battery H
    2nd United States, Battery E

3rd United States. Battery G
3rd United States, Batteries L and M

PROVISIONAL BRIGADE
   Col. Elisha G. Marshall.
   24th New York Cavalry (dismounted)
   14th New York Heavy Artillery
   2nd Pennsylvania Provisional Heavy Artillery

CAVALRY CORPS
   Maj Gen. Philip H. Sheridan (USMA 1853)

FIRST DIVISION
   Brig. Gen. Alfred T. A. Torbert
   (Brig. Gen. Wesley Merritt) (USMA 1860)

FIRST BRIGADE
   Brig. Gen. George A. Custer
   1st Michigan
   5th Michigan
   6th Michigan
   7th Michigan

SECOND BRIGADE
   Col. Thomas C. Devin
   4th New York
   6th New York
   9th New York
   17th Pennsylvania

RESERVE BRIGADE
   Brig. Gen. Wesley Merritt
   (Col. Alfred Gibbs)
   19th New York (1st Dragoons)
   6th Pennsylvania
   1st United States
   2nd United States
   5th United States

SECOND DIVISION
   Brig. Gen. David Mcm. Gregg (USMA 1855)

First Brigade
    Brig. Gen. Henry E. Davies Jr.
    1st Massachusetts
    1st New Jersey
    6th Ohio
    1st Pennsylvania

Second Brigade
    Col. J. Irvin Gregg
    1st Maine
    10th New York
    2nd Pennsylvania
    4th Pennsylvania
    8th Pennsylvania
    16th Pennsylvania

THIRD DIVISION
    Brig. Gen. James H. Wilson (USMA 1860)

First Brigade
    Col. Timothy M. Bryan Jr.
    Col. John B. Mcintosh
    1st Connecticut
    2nd New York
    5th New York
    18th Pennsylvania

Second Brigade
    Col. George H. Chapman
    3rd Indiana
    8th New York
    1st Vermont

APPENDIX III

CONFEDERATE ORDER OF BATTLE

ARMY OF NORTHERN VIRGINIA
    Gen. Robert E. Lee (USMA 1829)

FIRST ARMY CORPS
    Lieut. Gen. James Longstreet (W-W) (USMA 1842)
    (Maj. Gen. Richard H. Anderson) (USMA 1842)
    Chief of Staff Lieut. Col. Moxley Sorrel

KERSHAW'S DIVISION
    Brig. Gen. Joseph B. Kershaw

KERSHAW'S BRIGADE
    Col. John W. Henagan
    2nd South Carolina
    3rd South Carolina
    7th South Carolina
    8th South Carolina
    15th South Carolina
    3rd South Carolina Battalion

WOFFORD'S BRIGADE
    Brig. Gen. William T. Wofford (W-W)
    16th Georgia

18th Georgia
24th Georgia
Cobb's (Georgia) Legion
Phillip's (Georgia) Legion
3rd Georgia Battalion Sharpshooters

Humphreys's Brigade
Brig. Gen. Benjamin G. Humphreys
13th Mississippi
17th Mississippi
18th Mississippi
21st Mississippi

Bryan's Brigade
Brig. Gen. Goode Bryan
10th Georgia
50th Georgia
51st Georgia
53rd Georgia

FIELD'S DIVISION
Maj. Gen. Charles W. Field (W-W) (USMA 1849)

Jenkins's Brigade
Brig. Gen. Micah Jenkins (K-W)
(Col. John Bratton)
1st South Carolina
2nd South Carolina (Rifles)
5th South Carolina
6th South Carolina
Palmetto (South Carolina) Sharpshooters

Law's Brigade
Brig. Gen. E. McIver Law
4th Alabama
15th Alabama
44th Alabama
47th Alabama
48th Alabama

Anderson's Brigade
Brig. Gen. George T. Anderson
7th Georgia

8th Georgia
9th Georgia
11th Georgia
59th Georgia

GREGG'S BRIGADE
Brig. Gen. John Gregg
3rd Arkansas
1st Texas
4th Texas
5th Texas

BENNING'S BRIGADE
Brig. Gen. Henry L. Benning (W-W)
(Col. Dudley DuBose)
2nd Georgia
15th Georgia
17th Georgia
20th Georgia

ARTILLERY
Brig. Gen. E. Porter Alexander

HUGER'S BATTALION
Lieut. Col. Frank Huger
Fickling's (South Carolina) Battery
Moody's (Louisiana) Battery
Parker's (Virginia) Battery
Smith's, J. D. (Virginia), Battery
Taylor's (Virginia) Battery
Woolfolk's (Virginia) Battery

HASKELL'S BATTALION
Maj. John C. Haskell
Flanner's (North Carolina) Battery
Garden's (South Carolina) Battery
Lamkin's (Virginia) Battery
Ramsay's (North Carolina) Battery

CABELL'S BATTALION
Col. Henry C. Cabell
Callaway's (Georgia) Battery
Carlton's (Georgia) Battery

McCarthy's (Virginia) Battery
Manly's (North Carolina) Battery

SECOND ARMY CORPS
Lieut. Gen. Richard S. Ewell (USMA 1840)

EARLY'S DIVISION
Maj. Gen. Jubal A. Early (To Hill-S) (USMA 1837)
(Brig. Gen. John B. Gordon)

Hays's Brigade[1]
Brig. Gen. Harry T. Hays (W-S)
(Col. William Monaghan)
5th Louisiana
6th Louisiana
7th Louisiana
8th Louisiana
9th Louisiana

Pegram's Brigade
Brig. Gen. John Pegram (W-W)
(Col. John Hoffman)
13th Virginia
31st Virginia
49th Virginia
52nd Virginia
58th Virginia

Gordon's Brigade
Brig. Gen. John B. Gordon.
(Col. Clement Evans)
13th Georgia
26th Georgia
31st Georgia
38th Georgia
60th Georgia
61st Georgia

JOHNSON'S DIVISION
Maj. Gen. Edward Johnson (C-S) (USMA 1838)

STONEWALL BRIGADE
   Brig. Gen. James A. Walker (W-S)
   2nd Virginia
   4th Virginia
   5th Virginia
   27th Virginia
   33rd Virginia
JONES'S BRIGADE
   Brig. Gen. John M. Jones (K-W)
   (Col. William Witcher)
   21st Virginia
   25th Virginia
   42nd Virginia
   44th Virginia
   48th Virginia
   50th Virginia
STEUART'S BRIGADE
   Brig. Gen. George H. Steuart (C-S)
   1st North Carolina
   3rd North Carolina
   10th Virginia
   23rd Virginia
   37th Virginia
STAFFORD'S BRIGADE[2]
   Brig. Gen. Leroy A. Stafford (Mw-W)
   (Col. Zebulon York)
   1st Louisiana
   2nd Louisiana
   10th Louisiana
   14th Louisiana
   15th Louisiana

RODES'S DIVISION
   Maj. Gen. Robert E. Rodes (VMI 1848)

DANIEL'S BRIGADE
   Brig. Gen. Junius Daniel (K-S)
   (Lieut. Col. James Morehead)
   32nd North Carolina

43rd North Carolina
45th North Carolina
53rd North Carolina
2nd North Carolina Battalion

Doles's Brigade
Brig. Gen. George Doles
4th Georgia
12th Georgia
44th Georgia

Ramseur's Brigade
Brig. Gen. Stephen D. Ramseur (W-S)
(Col. Bryan Grimes)
2nd North Carolina
4th North Carolina
14th North Carolina
30th North Carolina

Battle's Brigade
Brig. Gen. Cullen A. Battle (W-S)
3rd Alabama
5th Alabama
6th Alabama
12th Alabama
26th Alabama

Johnston's Brigade[3]
Brig. Gen. Robert D. Johnston (W-S)
(Col. Thomas Toon)
5th North Carolina
12th North Carolina
20th North Carolina
23rd North Carolina

ARTILLERY
Brig. Gen. Armistead L. Long

Hardaway's Battalion
Lieut. Col. Robert A. Hardaway
Dance's (Virginia) Battery
Graham's (Virginia) Battery
Griffin's, C. B. (Virginia) Battery

Jones's (Virginia) Battery
Smith's, B. H. (Virginia), Battery

BRAXTON'S BATTALION
  Lieut. Col. Carter M. Braxton
  Carpenter's (Virginia) Battery
  Cooper's (Virginia) Battery
  Hardwicke's (Virginia) Battery

NELSON'S BATTALION
  Lieut. Col. William Nelson
  Kirkpatrick's (Virginia) Battery
  Massie's (Virginia) Battery
  Milledge's (Georgia) Battery

CUTSHAW'S BATTALION
  Maj. Wilfred E. Cutshaw
  Carrington's (Virginia) Battery
  Garber's, A. W. (Virginia), Battery
  Tanner's (Virginia) Battery

PAGE'S BATTALION
  Maj. Richard C. M. Page
  Carter's, W. P. (Virginia), Battery
  Fry's (Virginia) Battery
  Page's (Virginia) Battery
  Reese's (Alabama) Battery

THIRD ARMY CORPS
  Lieut. Gen. Ambrose P. Hill (USMA 1847)
  (Maj. Gen. Jubal A. Early, May 8 1864) (USMA 1837)

ANDERSON'S DIVISION
  Maj. Gen. Richard H. Anderson (assigned corps commander after wounding of Longstreet-) (USMA 1842)
  (Brig. Gen. William Mahone) (VMI 1847)

PERRIN'S BRIGADE
  Brig. Gen. Abner Perrin (K-S)
  (Col. John Sanders)
  8th Alabama
  9th Alabama
  10th Alabama

11th Alabama
14th Alabama

HARRIS'S BRIGADE
Brig. Gen. Nathaniel H. Harris
12th Mississippi
16th Mississippi
19th Mississippi
48th Mississippi

MAHONE'S BRIGADE
Brig. Gen. William Mahone
(Col. David Weiseger)
6th Virginia
12th Virginia
16th Virginia
41st Virginia
61st Virginia

WRIGHT'S BRIGADE
Brig. Gen. Ambrose R. Wright
3rd Georgia
22nd Georgia
48th Georgia
2nd Georgia Battalion

PERRY'S BRIGADE
Brig. Gen. Edward A. Perry (W-W)
2nd Florida
5th Florida
8th Florida

HETH'S DIVISION
Maj. Gen. Henry Heth (USMA 1847)

DAVIS'S BRIGADE
Brig. Gen. Joseph R. Davis
2nd Mississippi
11th Mississippi
42nd Mississippi
55th North Carolina

COOKE'S BRIGADE
Brig. Gen. John R. Cooke (W-S)

15th North Carolina
27th North Carolina
46th North Carolina
48th North Carolina

KIRKLAND'S BRIGADE
Brig. Gen. William W. Kirkland
11th North Carolina
26th North Carolina
44th North Carolina
47th North Carolina
52nd North Carolina

WALKER'S BRIGADE
Brig. Gen. Henry H. Walker (W-S)
(Col. Robert Mayo)
40th Virginia
47th Virginia
55th Virginia
22nd Virginia Battalion

ARCHER'S BRIGADE
Brig. Gen. James J. Archer.
13th Alabama
1st Tennessee (Provisional Army)
7th Tennessee
14th Tennessee

WILCOX'S DIVISION
Maj. Gen. Cadmus M. Wilcox (USMA 1846)

LANE'S BRIGADE
Brig. Gen. James H. Lane
7th North Carolina
18th North Carolina
28th North Carolina
33rd North Carolina
37th North Carolina

SCALES'S BRIGADE
Brig. Gen. Alfred M. Scales
13th North Carolina
16th North Carolina

22nd North Carolina
34th North Carolina
38th North Carolina

McGowan's Brigade
   Brig. Gen. Samuel McGowan (W-S)
   (Col. Joseph Brown)
   1st South Carolina (Provisional Army)
   12th South Carolina
   13th South Carolina
   14th South Carolina
   1st South Carolina (Orr's Rifles)

Thomas's Brigade
   Brig. Gen. Edward R. Thomas
   14th Georgia
   35th Georgia
   45th Georgia
   49th Georgia

ARTILLERY
   Col. R. Lindsay Walker

Poague's Battalion
   Lieut. Col. William T. Poague
   Richards's (Mississippi) Battery
   Utterback's (Virginia) Battery
   Williams' (North Carolina) Battery
   Wyatt's (Virginia) Battery

Pegram's Battalion
   Lieut. Col. William J. Pegram
   Brander's (Virginia) Battery
   Cayce's (Virginia) Battery
   Ellett's (Virginia) Battery
   Marye's (Virginia) Battery
   Zimmerman's (South Carolina) Battery

McIntosh's Battalion
   Lieut. Col. David G. Mcintosh
   Clutter's (Virginia) Battery
   Donald's (Virginia) Battery
   Hurt's (Alabama) Battery
   Price's (Virginia) Battery

CUTTS' BATTALION
   Col. Allen S. Cutts
   Patterson's (Georgia) Battery
   Ross's (Georgia) Battery
   Wingfield's (Georgia) Battery

RICHARDSON'S BATTALION
   Lieut. Col. Charles Richardson
   Grandy's (Virginia) Battery
   Landry's (Louisiana) Battery
   Moore's (Virginia) Battery
   Penick's (Virginia) Battery

CAVALRY CORPS
Maj. Gen. James E. B. Stuart (USMA 1854)

HAMPTON'S DIVISION
   Maj. Gen. Wade Hampton

YOUNG'S BRIGADE
   Brig. Gen. Pierce M. B. Young
   7th Georgia
   Cobb's (Georgia) Legion
   Phillips (Georgia) Legion
   20th Georgia Battalion
   Jeff. Davis (Mississippi) Legion

ROSSER'S BRIGADE
   Brig. Gen. Thomas L. Rosser
   7th Virginia
   11th Virginia
   12th Virginia
   35th Virginia Battalion

BUTLER'S BRIGADE
   Brig. Gen. Matthew C. Butler
   4th South Carolina
   5th South Carolina
   6th South Carolina

FITZHUGH LEE'S DIVISION
   Maj. Gen. Fitzhugh Lee (USMA 1856)

LOMAX'S BRIGADE
　Brig. Gen. Lunsford L. Lomax
　5th Virginia
　6th Virginia
　15th Virginia

WICKHAM'S BRIGADE
　Brig. Gen. Williams C. Wlckham
　1st Virginia
　2nd Virginia
　3rd Virginia
　4th Virginia

WILLIAM H. F. LEE'S DIVISION
　Maj. Gen. William H. F. Lee

CHAMBLISS'S BRIGADE
　Brig. Gen. John R. Chambliss Jr.
　9th Virginia
　10th Virginia
　13th Virginia

GORDON'S BRIGADE
　Brig. Gen. James B. Gordon
　1st North Carolina
　2nd North Carolina
　5th North Carolina

HORSE ARTILLERY
　Maj. R. Preston Chew

BREATHED'S BATTALION
　Maj. James Breathed
　Hart's (South Carolina) Battery
　Johnston's (Virginia) Battery
　McGregor's (Virginia) Battery
　Shoemaker's (Virginia) Battery
　Thomson's (Virginia) Battery

NOTES

Preface

1. U. S. Grant to Maj. Gen. Henry Halleck, May 11, 1864, in US War Department, *The War of the Rebellion: A Compilation of the Official Records of the Union and Confederate Armies* (Washington DC: 1880–1901), series 1, vol. 36, pt. 2, p. 627. Hereafter cited as *OR*.
2. E. P. Alexander, *Military Memoirs of a Confederate: A Critical Narrative* (New York: Charles Scribner's Sons, 1907), 504; *OR*, vol. 36, pt. 1, pp. 437,787.
3. *OR*, vol. 36, pt. 1, p. 13; U. S. Grant, *Personal Memoirs of U. S. Grant in Two Volumes* (New York: Charles L. Webster, 1885), 2:193; *OR*, vol. 36, pt. 2, p. 403.
4. William Blair, "Grant's Second Civil War," in, *Spotsylvania Campaign*, ed. Gary W. Gallagher (Chapel Hill: University of North Carolina Press, 1998), 247, 248. Note, this chapter provides an in-depth perspective on Grant's *Memoirs*.
5. "Response to Serenade, Washington DC, November 10, 1864", in *Collected Works of Abraham Lincoln*, ed. Roy B. Basler (New Brunswick, NJ: Rutgers University Press, 1953), 8:101.

Introduction

1. Horace Porter, *Campaigning with Grant* (New York: Century, 1906), 40; Charles Lawrence Peirson, "Operations of the Army of the Potomac, May 7-11, 1864," in, vol. 4 of *Papers of the Military Historical Society of Massachusetts*, (Boston: Military Historical Society of Massachusetts, 1905), 409.
2. *OR*, vol. 36, pt. 1, p. 13.
3. Grant, *Memoirs*, 2:125–26.
4. Ibid., 12, 127.
5. Capt. Robert E. Lee, *Collections and Letters of General Robert E. Lee* (New York: Doubleday, Park, 1905), 118; Maj. Gen. E. M. Law, From The Wilderness to Cold Harbor, *The Way to Appomattox*, vol. *4 of Battles and Leaders of the Civil War* (New York: Castle Books, 1956), 134.
6. Army Wagon Transportation, *Journal of the Military Service Institution of the United States 3*, no. 9 (1882): 106.
7. Lee, *Collections and Letters of General Robert E. Lee*, 24; Rev. J. William Jones, *Personal Reminisces, Anecdotes, and Letter of Gen. Robert E. Lee* (New York: D. Appleton, 1875), 142; Alan T. Nolan, *Lee Considered: General Robert E. Lee and Civil War History* (Chapel Hill: University of North Carolina Press, 1991), 178–79.
8. *OR*, vol. 11, pt. 3, p. 569; Horace Porter, *Campaigning with Grant* (New York: Century, 1906), 40; Lieut. Gen. Charles Venable, *General Lee In The Wilderness, Battles and Leaders of the Civil War*, vol. *4, 242*.
9. J. P. Clark, *Preparing for War: The Emergence of the Modern U.S. Army, 1815–1917* (Cambridge, MA: Harvard University Press, 2017), 86; Porter, *Campaigning with Grant*, 20.
10. *OR*, vol. 36, pt. 1, p. 67; Clark, *Preparing for War*, 88–89; Edwin P. Rutan II, *If I Have to Go and Fight, I Am Willing: A Union Regiment Forged in the Petersburg Campaign* (Park City, UT: RTD Publications, 2015), 9; The American Presidency Project (APP) website accessed November 27, 2018, https://www.presidency.ucsb.edu/documents/executive-order-drafting-500000-men, https://www.presidency.ucsb.edu/documents/executive-order-calling-for-200000-men.
11. Casualties: According to the American Battlefield Trust (formerly the Civil War Trust) website (https://www.battlefields.org), casualties at The Wilderness numbered 29,800, those at Spotsylvania numbered 30,000, and those at First Bull Run numbered 4,878. The D Day

Museum website (www.ddaymuseum.co.uk) asserts that casualties at D Day totaled 6,603."; *OR*, vol. 36, pt. 2, p. 841; "Abraham Lincoln's First Inaugural Address, March 4, 1861," accessed November 18, 2017, http://avalon.law.yale.edu/19th_century/lincoln1.asp. The statistic for 17,000 casualties comes from Peter Maugle, historian at Fredericksburg & Spotsylvania National Military Park, email to author, August 9, 2017.

12. Henry W. Thomas, *History of the Doles-Cook Brigade, Army of Northern Virginia, C.S.A.* (Atlanta: Franklin Printing, 1903), 479.

13. Porter, *Campaigning with Grant*, 118–19.

14. Andrew A. Humphreys, *The Virginia Campaign: 1864 and 1865.* (1883; repr., New York: Da Capo, 1995), 117; Edward Porter Alexander, *Fighting for the Confederacy: The Personal Recollections of General Edward Porter Alexander*, ed. Gary W. Gallagher (Chapel Hill: University of North Carolina Press, 1989), 381; Alexander, *Military Memoirs*, 516.

15. Martin McMahon, "The Death of General John Sedgwick," in *Battles and Leaders of the Civil War*, 4:175; Gordon Rhea, *The Battles for Spotsylvania Court House and the Road to Yellow Tavern, May 7–12, 1864* (Baton Rouge: Louisiana State University Press, 1997), 140; Hyland Kirk, *Heavy Guns and Light: A History of the 4th New York Heavy Artillery* (New York: C. T. Dillingham, 1890), 201; Alexander, *Military Memoirs*, 516; W. C. King and W. R. Derby, *Campfire Sketches and Battlefield Echoes* (Springfield, MA: King, Richardson, 1886), 272.

16. Ibid., 40, 122; Peirson, "Operations of the Army of the Potomac", 4:77; *OR*, vol. 36, pt. 2, pp. 406, 580; William Matter, *If It Takes All Summer* (Chapel Hill: University of North Carolina Press, 1988), 110; *OR*, vol. 36, pt. 2, p. 583.

Chapter 1

1. *New York Herald*, March 18, 1864.

2. Grant, *Memoirs*, 2:125–26; Brooks Simpson, "Great Expectations" in *The Wilderness Campaign*, ed. Gary W. Gallagher, (Chapel Hill: University of North Carolina Press, 1997), 2; 1860 US Census data https://www2.census.gov/library/publications/decennial/1860/population/1860a-02.pdf; Abraham Lincoln to Agenor-Etienne de Gasparin, August 4, 1862, in *Collected Works of Abraham Lincoln*, 5:355–56.

3. Capt. Robert E. Lee, *Collections and Letters of General Robert E. Lee* (New York: Doubleday, Page, 1905), 122.

4. James M. McPherson, *Battle Cry of Freedom* (New York: Oxford University Press, 1988), 274, 275. Formally abolishing slavery in the United States, the Thirteenth Amendment was passed by Congress in January 1865 and ratified by the states at the end of the year.
5. Ibid., 488, 525; H. W. Halleck to Major General McClellan, October 6, 1862, in *OR*, series 1, vol. 19, pt. 1, p. 72; *OR*, series 3, vol. 2, pp. 702–4.
6. McPherson, *Battle Cry of Freedom*, 584, 652; Abraham Lincoln to Joseph Hooker, June 10, 1863, in *Collected Works of Abraham Lincoln*, 6:257.
7. Abraham Lincoln to George Meade, July 14, 1863, (not sent) in *Collected Works of Abraham Lincoln*, 6:327–28; Report of the Joint Committee on the Conduct of the War, March 4, 1864, p. XIX, accessed April 25, 2015, https://archive.org/details/reportofjointcomm01unit/page/n27; *OR*, vol. 27, pt. 3, p. 519; William Thayer, ed., "Lincoln and Some of his Generals," *Harper's Magazine*, Christmas 1914, 99.
8. Jonathan P. Dolliver, "Lincoln's Legacy" in *The Abraham Lincoln Companion*, ed. Helene Henderson (Detroit: Omnigraphics, 2008), 297; Geoffrey Perret, *Lincoln's War: The Untold Story of America's Greatest President as Commander in Chief* (New York: Random House, 2004), 356.
9. Statues at Large: 38th Congress, Chapter 14. Library of Congress accessed December 12, 2015. https://www.loc.gov/law/help/statutes-at-large/38th-congress/session-1/c38s1ch14.pdf, pp. 11,12.11, 12; OR, vol. 36, pt. 3, p. 332.
10. Perret, *Lincoln's War*, 356.
11. Noah Brooks, *Washington in Lincoln's Time* (New York: Century, 1896), 143, 139, 141; *New York Daily Tribune*, February 2, 1864, p. 4 / col. 1.
12. U. S. Grant to Maj. B. Burns, Esq., December 17, 1863, in *The Papers of Ulysses S. Grant*, ed. John Simon (Carbondale: Southern Illinois University Press, 1967/1982), 9:217.
13. Brooks, *Washington in Lincoln's Time*, 136; U. S. Grant to Barnabas Burns, December 17, 1863, and U. S. Grant to E. B. Washburn, December 12, 1863, Simon, *Papers of Ulysses S. Grant*, 9:541, 523; *New York Daily Tribune*, February 10, 1864, p. 4 / col. 5; Perret, *Lincoln's War*, 356.
14. Ida M. Tarbell, *The Life of Abraham Lincoln* (New York: Doubleday and McClure, 1900), 2:188; U. S. Grant to Barnabas Burns, December 17, 1863, Simon, *Papers of Ulysses S. Grant*, 9:xv, 543. Note: There is no record of a Jones visiting Lincoln in this time frame in "The Lincoln Log," (http://www.thelincolnlog.org/Home.aspx), nor is a Jones visit mentioned in John Hay's diary *Inside Lincoln's White House*; Perret, *Lincoln's War*, 356.

15. Brooks D. Simpson, *Ulysses S. Grant: Triumph Over Adversity* (Boston: Houghton Mifflin, 2000), 272; John Y. Simon, "Grant, Lincoln, and Unconditional Surrender" in, , *Lincoln's Generals* ed. Gabor S. Boritt (New York: Oxford University Press, 1994), 176.
16. Bruce Catton, *Grant Takes Command* (Boston: Little, Brown, 1968), 122, 175–76; William O. Stoddard, *Lincoln's Third Secretary* (New York: Exposition, 1955), 198, 199; Military superpowers through history, *Business Insider*. Accessed December 12, 2018, https://www.business insider.com/this-ambitious-graphic-shows-the-size-of-standing -armies-from-antiquity-to-the-present-2014-11; Burns, Ken, dir. *The Civil War*, Florentine Films, 1990; Episode 6: 1864- Valley of the Shadow of Death
17. Grant, *Memoirs*, 9:115.
18. Ulysses S. Grant, "Preparing for the Campaign of 64," in *Battles and Leaders of the Civil War* (New York: Castle Books, 1956), 4:110, 112.
19. Catton, *Grant Takes Command*, 128. Army sizes are as follows: Army of the Potomac, 99,438 (Gordon Rhea, *The Battle of the Wilderness, May 5–6, 1864* [Baton Rouge: Louisiana State University Press, 1994], 34); Army of Ohio, 13,559; Army of the Cumberland, 66,773; Army of Tennessee, 24,465 (*Battles and Leaders of the Civil War*, 4:252); Army of the James, 21,500 (Mark Grimsley, *And Keep Moving On: The Virginia Campaign, May–June 1864* [Lincoln: University of Nebraska Press, 2002], 6); Geoffrey C. Ward, *The Civil War: An Illustrated History* (New York: Alfred A .Knopf, 1990), 282.
20. Catton, *Grant Takes Command*, 128. When Grant discovered that Major General Sigel had directly contacted a congressman to get troops reassigned to him, General Order Number 129 (*OR* vol. 33, pp. 769–70) was issued stating that those who did not follow proper military protocol were subject to arrest.
21. Shelby Foote, *The Civil War: A Narrative* (New York: Random House, 1974), 4:10; Catton, *Grant Takes Command*, 128.
22. Catton, *Grant Takes Command*, 129. Hooker felt he didn't get the respect he deserved, and he wasn't even listed on the official organization chart for the battle. James Lee McDonough, *Chattanooga—Death Grip on the Confederacy* (Knoxville: University of Tennessee Press, 1984), 254.
23. Grant, *Memoirs*, 2:117.
24. Peter Cozzens, *The Shipwreck of Their Hopes* (Urbana: University of Illinois Press, 1994), 41.

25. McDonough, *Chattanooga*, 55; *The Diary of Cyrus B. Comstock*, ed. Merlin E. Summer (Dayton, OH: Morningside, 1987) March 10, 1864, p.260.
26. *OR*, vol. 27, pt. 1, p. 92.
27. Grant, *Memoirs*, 2:117; Adam Badeau, *Military History of Ulysses S. Grant* (New York: D. Appleton, 1881), 2:16, 307; Comstock, *Diary of Cyrus B.*, 261.
28. George Meade, *The Life and Letters of George Gordon Meade, Major General United States Army* (New York: Charles Scribner's Sons, 1913), 2:178.
29. Catton, *Grant Takes Command*, 129; Grant, *Memoirs*, 2:117, 118.
30. Grant, *Memoirs*, 2:133.
31. Ibid., 2:539.
32. In the Western Theater, where Grant was much more familiar with the personnel, he had made organizational changes because of his promotion. Ibid., 2:116. The intent of the raid was to attack Richmond and free Union prisoners of war. Kilpatrick was stopped outside of the city and never posed a threat. Colonel Dahlgren commanded one of the supporting columns, and when he was killed, he was found carrying papers detailing plans to burn Richmond and assassinate Confederate president Jefferson Davis and members of his cabinet. These papers were published in various newspapers, and they created a political firestorm. *Battles and Leaders of the Civil War*, 4:95–96; Elon J. Farnsworth web site. Accessed December 12, 2018, https://en.wikipedia.org/wiki/Elon_J._Farnsworth.
33. Grant, *Memoirs*, 2:133.
34. Following his excellent performance at Brandy Station, Merritt had been promoted from the rank of captain to brigadier general in June 1863.
35. Sheridan had been Halleck's quartermaster general at one time, and Halleck had approved Sheridan's promotion to brigadier general. Grant, *Memoirs*, 2:133.
36. Ibid., 2:459.
37. Philip Sheridan, *Civil War Memoirs* (1888; repr. New York: Bantam Books, 1991), 138; Brooks D. Simpson, email to author March 20, 2015.
38. Rhea, *Battle of the Wilderness*, 93, 73–74, 434, 110.
39. Albert D. Richardson, *Personal History of Ulysses S. Grant* (Boston: D. L. Guernsey, 1885), 389; Catton, *Grant Takes Command*, 132; *OR*, vol. 32, pt. 3, p. 18; U. S. Grant to Hon. E. B. Washburn, August 30, 1863, in

Papers of Ulysses S. Grant, 217; Richardson, *Personal History of Ulysses S. Grant*, 387; *OR*, vol. 32, pt. 3, p. 49.

40. Grant, *Personal Memoirs*, 2:116; In the 1863 report where the Committee was investigating Major General John Freemont's conduct while he commanded out west the report stated that the committee could not call the general from so far away because his "services were constantly required in the field" and that committee members could not travel out west to take Fremont's testimony because they were "compelled to remain in attendance upon Congress." Basically, no action was taken; Meade, *Life and Letters*, 2:186.
41. Humphreys, *Virginia Campaign*, 5.
42. *OR*, vol. 32, pt. 3, p. 58.
43. F. C. Grugan "The Use, Development and Influence of the Electrical Telegraph in Warfare" in *Journal of the Military Institution of the United States* 3, no. 9 (1882): 380; James Rusling, *Men and Things I Saw in Civil War Days* (New York: Eaton and Mains; Cincinnati: Curts and Jennings; 1899), 136; William Sherman, *Memoirs of General William T. Sherman* (1889; repr., Bloomington: Indiana University Press, 1957), Two Volumes in One, 398.
44. *OR*, vol. 36, pt. 2, p. 407. Grant further illustrated his dislike for Washington in the following incident. After his visiting Meade in the field and returning to Washington, Grant met with the president and briefed him on his overall plan and his intention to leave immediately for the West. Lincoln then informed Grant that Mrs. Lincoln had set up a dinner in his honor that evening. Even so, Grant insisted that he needed to go and added, "Besides, I have had enough of this show business, Mr. President." (Noah Brooks, "The Life of Lincoln" in *The Writings of Abraham Lincoln*, ed. Arthur Brooks Lapsley [New York: Putman's, 1888] 8:386).
45. *OR*, vol. 33, p. 394; Grant, *Memoirs*, 2:133; *Diary of Cyrus B. Comstock*, April 1, 1864, 262.
46. Catton, *Grant Takes Command*, 138.
47. Meade, *Life and Letters*, 2:187, 182; Catton, *Grant Takes Command*, 138; Grant, *Memoirs*, 2:118, 131.
48. Grant, *Memoirs*, 2:118; *OR* vol. 36, pt. 2, p. 481; Meade, *Life and Letters*, 2:184–85.
49. *OR*, vol. 36, pt. 2, p. 481.
50. Ibid., vol. 36, pt. 2, p. 695.
51. Meade, *Life and Letters*, 2:162.

52. Grant, *Memoirs*, 1:303–4; Mary A. Livermore, *My Story of The War: A Woman's Narrative of Four Years Personal Experience* (Hartford, CT: A. D. Worthington, 1889), 558.
53. U. S. Grant to William Sherman, April 4, 1864, in *Papers of Ulysses S. Grant*, 10:251.
54. Grant, *Memoirs*, 2:127.
55. Humphreys, *Virginia Campaign*, 5; Foote, *Civil War*, 3:8, 13; Catton, *Grant Takes Command*, 146, 147.
56. Foote, *Civil War*, 3:14.
57. *OR*, vol. 34, pt. 1, pp. 11–13.
58. General-In-Chief Halleck had previously authorized Maj. Gen. Bank's Red River Campaign which was not part of Grant's plan.
59. U. S. Grant to George Meade, April 9, 1864, in *Papers of Ulysses S. Grant*, 10:273–75; Rhea, *Battle of the Wilderness*, 57.
60. U. S. Grant to George Meade, April 9, 1864, in *Papers of Ulysses S. Grant*, 10:274.
61. *New York Herald*, July 24, 1878, page 4, column 2; *OR* 36, 1, 15.
62. Gabor S. Boritt, ed., *Lincoln's Generals* (New York, NY, 1994), 165; Brooks D. Simpson, *Triumph over Adversity 1822–1865* (Boston New York 2000), 288
63. *OR*, vol. 33, p. 394.
64. Grant, *Memoirs*, 2:123, 141.
65. Humphreys, *Virginia Campaign*, 8.
66. Ibid., 9.
67. *OR*, vol. 33, p. 828.
68. Ibid.; Humphreys, *Virginia Campaign*, 9
69. Humphreys, *Virginia Campaign*, 10.
70. Horace Porter, *Campaigning with Grant*, 37.
71. Rhea, *Battle of the Wilderness*, 33–34, 55; Gordon Rhea, *The Battles of the Wilderness & Spotsylvania*, National Parks Civil War Series (Eastern National, 2007), 3; Humphreys, *Virginia Campaign*, 20.
72. Porter, *Campaigning with Grant*, 43–44; numbers from American Battlefield Trust (formerly The Civil War Trust) website. Accessed December 12, 2018, https://www.battlefields.org/about/civil-war-trust.
73. *New York Herald*, July 24, 1878; Grant, *Memoirs*, 2:137.

Chapter 2

1. Rhea, *Battle of the Wilderness*, 64.
2. Humphreys, *Virginia Campaign*, 12
3. Ibid., 11; Rhea, *Battle of the Wilderness*, 64; *OR,* vol. 36, pt. 2, pp. 331–34.
4. *OR,* vol. 33, pp. 1113–14; R. E. Lee to Jefferson Davis, April 15, 1864, in *The Wartime Papers of R. E. Lee,* ed. Clifford Dowdey (New York: Bramhall House, 1961), 698; Rhea, *Battle of the Wilderness,* 81; R. E. Lee to General James Longstreet, March 28, 1864, in *Wartime Papers of R. E. Lee,* 685.
5. Charles Venable, "General Lee in The Wilderness Campaign," in *Battles and Leaders of the Civil War,* 4:240; A. L. Long, *Memoirs of Robert E. Lee: His Military and Personal History* (Secaucus, NJ: Blue and Gray Press, 1983), 323, 326; Rhea, *Battle of the Wilderness,* 56.
6. *OR,* vol. 36, pt. 1, p. 18; Long, *Memoirs of Robert E. Lee,* 325.
7. *OR,* vol. 36, pt. 2, p. 358; Washington A. Roebling, entry for May 4, 1864, in *Major Washington A. Roebling's journal of the Overland Campaign,* bound volume #408, Fredericksburg and Spotsylvania National Military Park, Fredericksburg, VA; Humphreys, *Virginia Campaign,* 20; Rhea, *Battle of the Wilderness,* 55; *Battles and Leaders of the Civil War,* 4:153.
8. Rhea, *Battle of the Wilderness,* 53.
9. Ibid.
10. "General Ewell arrived in camp with his wife—a new acquisition—and with one leg less than when I saw him last. From a military point of view the addition of the wife did not compensate for the loss of the leg. We were of the opinion that Ewell was not the same soldier he had been when he was a whole man—and a single one." Randolph H. McKim, *A Soldier's Recollections* (London: Longmans, Green, 1910), 134.
11. Rhea, *Battle of the Wilderness,* 53.
12. Humphreys, *Virginia Campaign,* 10.
13. Rhea, *Battle of the Wilderness,* 22.
14. Terry L. Jones, ed., *Campbell Brown's Civil War: With Ewell and the Army of Northern Virginia* (Baton Rouge: Louisiana State University Press, 2001), 246; Rhea, *Battle of the Wilderness,* 52; Humphreys, *Virginia Campaign,* 22; Alexander, *Military Memoirs,* 498.
15. Alexander, *Military Memoirs,* 499; Campbell Brown papers, July 1, 1864, in *Campbell Brown's Civil War,* ed. Terry L. Jones (Baton Rouge: Louisiana State University, 2001), 247.

16. Roebling, journal entry for May 5, 1864, 4.
17. Rhea, *Battle of the Wilderness*, 94, 104, 128, 132; *OR*, vol. 36, pt. 2, p. 403; *Annals of the War Written by Leading Participants North and South. Originally Published in the Philadelphia Weekly Times* (Philadelphia: The Times Publishing Company, 1879), 491; National Parks Civil War Series, *The Battles of Wilderness & Spotsylvania*, 6.
18. Rhea, *Battle of the Wilderness*, 108, 109, 119, 131, 133, 134; *OR*, vol. 36, pt. 2, pp. 407, 418; *OR*, vol. 36, pt. 1, p. 676.
19. Porter, *Campaigning with Grant*, 47.
20. William Lawrence Royall, *Some Reminiscences* (New York: Neale, 1909), 31; Rhea, *Battle of the Wilderness*, 240, 264; Gregory A. Mertz, "No Turning Back," *Blue & Gray Magazine*, June 1995, 10.
21. Rhea, *Battle of the Wilderness*, 208, 240; *OR*, vol. 36, pt. 2, p. 411; Humphreys, *Virginia Campaign*, 33.
22. Rhea, *Battle of the Wilderness*, 238, 241.
23. Ibid., *Battle of the Wilderness*, 272, 276; James Longstreet, *From Manassas to Appomattox: Memoirs of the Civil War in America* (Philadelphia: J. P. Lippincott, 1896), 557.
24. *Annals of The War*, 274, 494; Major W. S. Dunlop, *Lee's Sharpshooters* (Little Rock, AK: Tunnam and Pittard, 1899), 32; Longstreet, *From Manassas to Appomattox*, 560.
25. Royall, *Some Reminiscences*, 30; Rhea, *Battle of the Wilderness*, 277; *Annals of the War*, 494.
26. Alexander, *Military Memoirs*, 502; Humphreys, *Virginia Campaign*, 41.
27. Rhea, *Battle of the Wilderness*, 269; Douglas Southall Freeman, *Lee's Lieutenants* (New York: Charles Scribner's Sons, 1944), 3:442–43.
28. Royall, *Some Reminiscences*, 30; *Annals of The War*, 494; Longstreet, *From Manassas to Appomattox*, 559.
29. Roebling, journal entry for May 5, 1864, 4.
30. *OR*, vol. 36, pt. 2, pp. 404–5; Henry Heth, *The Memoirs of Henry Heth*, ed. James Morrison, Jr. (Westport, CT: Greenwood, 1974), 184; Rhea, *Battle of the Wilderness*, 270; *Battles and Leaders of the Civil War*, 4:241; Grant, *Memoirs*, 2:197.
31. Royall, *Some Reminiscences*, 31–32; Rhea, *Battle of the Wilderness*, 295; Col. C. S. Venable "Campaign from the Wilderness to Petersburg," *Southern Historical Society Papers*, Rev. J. William Jones, D.D., Secre-

tary of the Society, vol. 14, *January to December, 1886* (Richmond, VA: Wm. Ellis Jones, 1886), 525.

32. Rhea, *Battle of the Wilderness*, 300; C. S. Venable, "Campaign from Wilderness to Petersburg," in *Southern Historical Society Papers*, 14:526.

33. Moxley Sorrel, *Recollections of a Confederate Staff Officer* (New York: Neale, 1905), 240–41; Rhea, *Battle of the Wilderness*, 291.

34. Heth, *Memoirs of Henry Heth*, 185; Mertz, "No Turning Back," 11.

35. "Newly Found Record of Private Talks Shows . . . Lee Blamed Ewell and Longstreet For His Failure in the Wilderness," *Civil War Times Illustrated* 5, no. 1 (April 1996): 5.

36. Mertz, "No Turning Back," 14.

37. Note that the typically more aggressive mind-set Confederate commanders displayed was what Grant wanted to instill in the Army of the Potomac. Gallagher, *Wilderness Campaign*, 238, 240; Mertz, "No Turning Back," 14.

38. Gallagher, *Wilderness Campaign*, 240; Sorrel, *Recollections*, 242; Gallagher, *Wilderness Campaign*, 242.

39. Mertz, "No Turning Back," 15; After Wadsworth's wounding, Confederate surgeons cared for him until he died. Although a map and other personal items had been taken from him after he fell, most of them were eventually returned to his family. Sorrel, *Recollections*, 243, 248.

40. Gallagher, *Wilderness Campaign*, 248; Mertz, "No Turning Back," 18.

41. Gallagher, *Fighting for the Confederacy*, 360.

42. Ibid., 390; Gallagher, *Wilderness Campaign*, 249.

43. Sorrel, *Recollections*, 247.

44. Ibid., 244; Rhea, *Battle of the Wilderness*, 373; *Military Quantitative Physiology: Problems and Concepts in Military Operational Medicine* (Fort Detrick, MD: Office of the Surgeon General / Borden Institute, 2012), 169.

45. Rhea, *Battle of the Wilderness*, 374.

46. Edward Steere, *The Wilderness Campaign: The Meeting of Grant and Lee* (Mechanicsburg, PA: Stackpole Books, 1960), 427; *OR*, vol. 36, pt. 2, pp. 444–45; C. W. Field, "Campaign of 1864–'65," in *Southern Historical Society Papers*, 14:545.

47. Mertz, "No Turning Back," 20; Rhea, *Battle of the Wilderness*, 390; Steere, *Wilderness Campaign*, 425.

48. Charles Page, *Letters of a War Correspondent* (Boston: L. C. Page, 1899), 55; *OR*, vol. 36, pt. 1, p. 354; *OR*, vol. 36, pt. 2, p. 446.
49. *OR* vol. 36, pt. 1, p. 948.
50. Gallagher, *Fighting for the Confederacy*, 363; Alexander, *Military Memoirs*, 507.
51. Richardson, *Personal History of Ulysses S. Grant*, 400.
52. Gallagher, *Wilderness Campaign*, 256.
53. Note: Seymour had only taken over command of the brigade that morning, and the troops were raw. Rhea, *Battle of the Wilderness*, 244; *OR*, vol. 36, pt. 1, p. 1077.
54. *OR*, pt. 36, pt. 1, pp. 927, 1071, 1077; *OR*, vol. 36, pt. 1, p. 1071. In his 1912 autobiographical sketch, Ewell stated that it was Gordon who reported the situation to him and suggested the flank attack. Ralph Eckert, *John Brown Gordon: Soldier, Southerner, American* (Baton Rouge: Louisiana State University Press, 1989), 67; "Battle of The Wilderness," 6 map set, map 3, Frank O'Reilly Historical Researcher (Eastern National, 2003); Mertz, "No Turning Back," 14.
55. Brown, *Campbell Brown's Civil War*, 248; *OR*, vol. 36, pt. 2, p. 953.
56. *OR*, vol. 33, pt. 1, p. 1124; *OR*, vol. 51, pt. 2, p. 890.
57. William Seymore, , *The Civil War Memoirs of Captain William J. Seymour: Reminiscence of a Louisiana Tiger*, ed. Terry L. Jones (Baton Rouge: Louisiana State University Press, 1991), 114.
58. At the time of the Battle of Chancellorsville, Ewell was recovering from a wound and thus not involved in Jackson's dramatic flank attack. Early did participate in the assault, and he knew of its effectiveness.
59. *OR*, vol. 36, pt. 1, p. 1071; *OR*, vol. 36, pt. 2, p. 451.
60. Ibid., vol. 36, pt. 1, p. 1071.
61. Jubal Early, *Lieutenant General Jubal Early C.S.A.: Autobiographical Sketch and Narrative of the War Between the States* (Philadelphia: J. B. Lippincott, 1912), 348.
62. *OR*, vol. 36, pt. 1, p. 1077.
63. John B. Gordon, *Reminiscences of the Civil War* (New York: Charles Scribner's Sons, 1903), 258.
64. Alexander, *Military Memoirs*, 508.
65. Rhea, *Battle of the Wilderness*, 412; Transcript of conversation between William Allan and R. E. Lee, March 3, 1868, William Allen Papers

#2764, Southern Historical Collection, Wilson Library, University of North Carolina at Chapel Hill.

66. *OR*, vol. 36, pt. 2, p. 970. This correspondence, written after twelve o'clock, states Lee is going to visit Ewell's lines; Taylor, *Campaigns in Virginia*, 238; George Quintus Peyton, "*A Civil War record for 1864–1865*"), transcribed by Robert Hodge. In 1929, Family History Library, Salt Lake City, UT. United States Collection, Film 6087252.

67. "Cover Letter for 1864 report (6 February 1868): J. B. Gordon to Lee," MSS 3 L515a, Folder 31—Reports and Reminiscences, 1866 June–1876, Item 603, Robert E. Lee Headquarters Papers, 1850–1876, Virginia Historical Society, Richmond.

68. R. E. Lee to J. B. Gordon, February 22, 1868, Hargrett Rare Book and Manuscript Library, University of Georgia Libraries, Athens.

69. Transcript of conversation between William Allan and R. E. Lee, March 3, 1868.

70. Porter, *Campaigning with Grant*, 68-69.

71. Designated Culpeper Mine Road on the Gordon Trail walk, this route is also referred to as Flat Run Road on O'Reilly Wilderness map set, and as Spotswood Road in Bradley M. Gottfried, *The Maps of the Wilderness: An Atlas of the Wilderness Campaign, Including All Cavalry Operations, May 2–6, 1864* (El Dorado Hills, CA: Savas Beatie, 2016).

72. Early, *Autobiographical Sketch and Narrative*, 350; Grant, *Memoirs*, 2:202.

73. *OR*, vol. 36, pt. 1, pp. 1077, 1078.

74. Ibid., vol. 36, pt. 2, p. 480.

75. Porter, *Campaigning with Grant*, 68, 70.

76. Gallagher, *Wilderness Campaign*, xi; Morris Schaff, *The Battle Of The Wilderness* (Boston: Houghton Mifflin, 1910), 320; Freeman, *Lee's Lieutenants*, 3:442; *Civil War Times Illustrated*, 5, no. 1 (April 1966): 5.

Chapter 3

1. Humphreys, *Virginia Campaign*, 118.
2. American Battlefield Trust (formerly the Civil War Trust), https://www.battlefields.org; *OR*, vol. 36, pt. 2, p. 480.
3. American Battlefield Trust web site.
4. Rhea, *Battle of the Wilderness*, 432; Alexander Webb, "Through The Wilderness," in *Battles and Leaders of the Civil War*, 4:163.

5. Rhea, *Battle of the Wilderness*, 432–34; Alexander Webb, "Through The Wilderness," in *Battles and Leaders of the Civil War*, 4:163.
6. Rhea, *Battle of the Wilderness*, 437.
7. *OR*, vol. 36, pt. 2, p. 480; Porter, *Campaigning with Grant*, 65–66.
8. Rhea, *Battle of the Wilderness*, 426.
9. Ibid., 431, 437; Steere, *Wilderness Campaign*, 457; Agassiz, *Meade's Headquarters*, 102n1.
10. Louis M. Starr, *Bohemian Brigade: Civil War Newsmen in Action* (New York: Alfred A. Knopf, 1954), 299; F. B. Carpenter, *Six Months at the White House with Abraham Lincoln* (New York: Hurd and Houghton, 1866), 283; Henry E. Wing, *When Lincoln Kissed Me: The Story of the Wilderness Campaign* (New York/Cincinnati: Abingdon, 1913) 38–39.
11. Rhea, *Battle of the Wilderness*, 437–38; Porter, *Campaigning with Grant*, 68; *OR*, vol. 36, pt. 2, p. 481.
12. *OR*, vol. 36, pt. 2, pp.481, 482.
13. *OR*, vol. 36, pt. 1, p. 1041.
14. *OR* vol. 36, pt. 2, p. 484; Humphreys, *Virginia Campaign*, 58, 69; Rhea, *Battles for Spotsylvania Court House*, 16, 44; Mark De Wolf Howe, ed., *Touched With Fire: Civil War Letters of Oliver Wendell Holmes Jr., 1861–1864* (Cambridge, MA: Harvard University Press, 1946), May 7, 1864 entry, 108.
15. Agassiz, *Meade's Headquarters*, 102; *OR* vol. 36, pt. 1, pp. 1041, 974.
16. *OR* vol. 36, pt. 1, pp. 969–70; Rhea, *Battles for Spotsylvania Court House*, 22; Gallagher, *Fighting for the Confederacy*, 395.
17. *OR*, vol. 36, pt. 2, p. 969; *OR*, vol. 36, pt. 1, p. 1041.
18. Sorrel, *Recollections*, 248; *OR*, vol. 36, pt. 2, pp. 955, 966.
19. *OR*, vol. 36, pt. 2, pp. 968, 1041; Richard Heron Anderson Manuscript, 1864, David M. Rubenstein Rare Book and Manuscript Library, Duke University, Durham, NC (hereafter cited as Anderson Manuscript).
20. *OR*, vol. 36, pt. 1, p.540.
21. Ibid., vol. 36, pt. 1, p.354.
22. Anderson Manuscript; Ibid., vol. 36, pt. 1, p. 1041.
23. Anderson Manuscript; Rhea, *Battles for Spotsylvania*, 29.
24. Anderson Manuscript; Gallagher, *Fighting for the Confederacy*, 395; Rhea, *Battles for Spotsylvania*, 52; *OR*, vol. 36, pt. 1, p. 1056; Peirson, "Operations of the Army of the Potomac," in *Papers of the Military His-*

torical Society of Massachusetts, 4:230; D. Augustus Dickert, *History of Kershaw's Brigade, with Complete Roll of Companies, Biographical Sketches, Incidents, Anecdotes, Etc.* (Newberry, SC: Elbert H. Aull, 1899), 357; Sorrel, *Recollections*, 254.

25. Rhea, *Battles for Spotsylvania*, 50, 53, 60; Gallagher, *Fighting for the Confederacy*, 395.
26. Rhea, *Battles for Spotsylvania*, 30.
27. *OR*, vol. 36, pt. 2, pp. 484, 970; Rhea, *Battles for Spotsylvania*, 30, 31.
28. *OR*, vol. 36, pt. 2, p. 970.
29. Ibid., vol. 36, pt. 1, p. 816.
30. Ibid., vol. 36, pt. 1, p. 833; Ibid., vol. 36, pt. 2, p. 513; Ibid., vol. 36, pt. 1, p. 833.
31. Rhea, *Battles for Spotsylvania*, 32, 33. R. H. Peck, *Reminisces of a Confederate Soldier of Co. C. 2nd Va. Cavalry* (Fincastle, VA: 1913) Rufus Peck puts Wickham first on this day, while the May 11 edition of the *Richmond Examiner* puts Lomax first.
32. *OR*, vol. 36, pt. 2, p. 515; Rhea, *Battles for Spotsylvania*, 34; National Parks Civil War Series, *Battles of the Wilderness & Spotsylvania*, 23.
33. Rhea, *Battles for Spotsylvania*, 40; *OR*, vol. 36, pt. 2, pp.552, 538–39; Journal entry in Roebling's report for May 8, 1864; Charles S. Wainwright, *A Diary of Battle: The Personal Journals of Colonel Charles S. Wainwright 1861–1865*, ed. Allan Nevins (Pickle Partners, 2014), diary entry for May 8, 1864. In his diary, Wainwright (head of Warren's artillery) stated that he found Warren having breakfast while waiting for Merritt to open the way. Wainwright felt both Meade and Warren were not pushing hard enough.
34. National Parks Civil War Series, *Battles of the Wilderness & Spotsylvania*, 25; Journal entry in Roebling's report for May 8, 1864; James Breathed's report, dated September 14, 1864, in folder 4 of the Edwin L. Halsey Papers, Louis Round Wilson Special Collections Library, The University of North Carolina at Chapel Hill.; "Battle of Spotsylvania Court House," 24 map set, map 1, Frank O'Reilly Historical Researcher (Eastern National, 2000), Per O'Reilly map, Lomax was placed on the west side of Spindle Field (left flank of infantry), and Fitzhugh Lee went with Wickham to Spotsylvania Court House to drive Wilson out.
35. Journal entry in Roebling's report for May 8, 1864; *OR*, vol. 36, pt. 2, pp. 539, 540; Rhea, *Battles for Spotsylvania*, 37.

36. Gen. Charles E. Phelps, *History of the Battle of The Wilderness and the Civil War experiences of Col./Gen. Charles E. Phelps*, The Grassland Foundation, Annapolis, MD; Humphreys, *Virginia Campaign*, 70; Rhea, *Battles for Spotsylvania*, 64.
37. Humphreys, *Virginia Campaign*, 63; *OR*, vol. 36, pt. 1, p. 878.; Rhea, *Battles for Spotsylvania*, 37.
38. *OR*, vol. 36, pt. 1, p. 787, *OR*, vol. 36, pt. 2, p. 466.
39. Philip Sheridan, *Civil War Memoirs*,146–48; Humphreys, *Virginia Campaign*, 117; Gallagher, *Wilderness Campaign*, 93; Rhea, *Battles for Spotsylvania*, 10; *OR*, vol. 36, pt. 2, pp. 529, 546, 547, 545, 515, 526.
40. Rhea, *Battles for Spotsylvania*, 37; Humphreys, *Virginia Campaign*, 68, 69.
41. *OR*, vol. 36, pt. 2, pp. 551–53; Humphreys, *Virginia Campaign*, 67–69, and maps in back of book; Badeau, *Military History of Ulysses S. Grant*, 2:139.
42. Porter, *Campaigning with Grant*, 24, 83–84.
43. Ibid.
44. John Russell Young, *Around the World with General Grant*, (New York: American News Company, 1879), 2:464.
45. Porter, *Campaigning with Grant*, 84; Catton, *Grant Takes Command*, 137; King, *Campfire Sketches*, 230.
46. Grant, *Memoirs*, 2:108, 109.
47. *OR*, vol. 36, pt. 2, 552; Map 39, in vol. 1 of *The West Point Atlas of American Wars*, ed. Brig. Gen. Vincent J. Esposito (New York: Henry Holt, 1995).
48. *OR*, vol. 36, pt. 2, pp. 540–41; Sheridan, *Civil War Memoirs*, 155–56; Rhea, *Battles for Spotsylvania*, 120.
49. *OR*, vol. 36, pt. 2, 566, 581.
50. Rhea, *Battles for Spotsylvania*, 212.
51. Grant, *Memoirs*, 2:153, 154.
52. Ibid., 2:101, 120; O'Reilly Spotsylvania maps 4, 5, 6.

Chapter 4

1. Thomas Hyde, *Following the Greek Cross, or Memories of the Sixth Army Corps* (Boston: Houghton, Mifflin, 1894), 189.

2. David Bongi, "Offensive Infantry Tactics During the Battle of Chancellorsville, May 1863" (master's thesis, U. S. Army Command and General Staff College, 1993), 51. When Hooker had ordered the withdrawal of troops, Warren had urged Couch to disobey the order. Warren was correct that the Federals would be giving up an excellent position; John Hay, "Lincoln and Some Union Generals," *Harper's Magazine*, Christmas 1914, 99.

3. Rhea, *Battle of the Wilderness*, 139–40, 142; *OR*, vol. 36, pt. 2, p. 403.

4. Charles Peirson, "The Operations of the Army of the Potomac May 7-11, 1864," in *Papers of the Military Historical Society of Massachusetts*, 4:134; Gallagher, *Spotsylvania Campaign*, 67.

5. Journal entry in Roebling's report for May 8, 1864.

6. Block House Road is sometimes referred to as Old Court House Road.

7. *OR*, vol. 36, pt. 2, p. 403; Charles Phelps, *History of the Battle of The Wilderness*, Grassland Foundation, Annapolis, MD, 8.

8. Peirson, "Operations of the Army of the Potomac May 7-11, 1864," 4:214.

9. Ibid., 4:215; William H. Powell, *The Fifth Army Corps (Army of the Potomac)* (New York: G. P. Putnam's Sons, 1896), 634; A. M. Judson, *History of the Eighty-Third Regiment Pennsylvania Volunteers*, (Erie, PA: B. F. Lynn, LOC date 1881), 95; A. R. Small, *The Sixteenth Maine Regiment in the War of the Rebellion, 1861–1865* (Portland, ME: B. Thurston, 1886), 177; Lyman, *Meade's Headquarters*, 100.

10. Greg Mertz, "The Fighting at Laurel Hill and the Spindle Field," *Blue & Gray Magazine*, Summer 2004, maps pp. 57–58.

11. Lyman, *Meade's Headquarters*, 100; John Coxe, "Last Struggles and Successes of Lee," *Confederate Veteran* Vol. 22, (1914) 356–57; L. W. Hopkins, *From Bull Run to Appomattox* (Baltimore: Fleet-McGinley, 1908), 154.

12. Phelps, *History of The Battle of The Wilderness*, 5. Phelps also noted that the reward of rations was quite relevant: "The last was a practical and appealing suggestion, probably more appealing than the preceding high flown exhortations. The general well understood that what the men then wanted was not the Constitution and the Union, but bread and meat." A. R. Small, *The Sixteenth Maine Regiment in the War of the Rebellion, 1861–1865* (Portland, ME: B. Thurston, 1886), 178; Journal entry in Roebling's report for May 8, 1864, 7.

13. Capt. Eugene Arus Nash, *A History of the 44th New York Volunteer Infantry in the Civil War, 1861–1865* (Chicago: R. R. Donnelly and Sons, 1911), 187–88.

14. Mertz, "Fighting at Laurel Hill and the Spindle Field" map p. 58; Phelps, *Battle of the Wilderness*, 12; *OR*, vol. 36, pt. 2, pp. 539–40; Judson, *History of the Eighty-Third*, 96; Powell, *Fifth Army Corps*, 633–34.
15. James Harrison Wilson, *Under the Old Flag* (New York: D. Appleton, 1912), 1:396; *OR*, vol. 36, pt. 2, p. 600; Rhea, *Battles for Spotsylvania*, 143.
16. Horace Porter, *Campaigning with Grant*, 108; *OR*, vol. 36, pt. 2, 654; Gordon Rhea, "The Testing of a Corps Commander," in *The Spotsylvania Campaign*, ed. Gary Gallagher, (Chapel Hill: University of North Carolina Press, 1998), 70; Lyman, *Meade's Headquarters*, 110n1.
17. Alexander, *Military Memoirs*, 523.
18. Judson, *History of the Eighty-Third*, 95; Peirson, "Operations of the Army of the Potomac," in *Papers of the Military Historical Society of Massachusetts*, 4:215–16.
19. "Grant as a Critic," July 24, 1878, *New York Herald*; Catton, *Grant Takes Command*, 137.
20. Rawlins would briefly be secretary of war when Grant was president. Catton, *Grant Takes Command*, 57, 137.
21. Grant, *Memoirs*, 2:216.
22. Meade, *Life and Letters*, 2:197–98.
23. William Swinton, *Campaigns of the Army of The Potomac: A Critical History of Operations in Virginia Maryland and Pennsylvania from the Commencement to the Close of the War 1861–1865* (New York: Charles Scribner's Sons, 1882), 423; Grant, *Memoirs*, 2:216.
24. Rhea, *Battles for Spotsylvania Court House*, 86; McMahon, "Death of General John Sedgwick," in *Battles and Leaders of the Civil War*, 4:175; *OR*, vol. 36, pt. 2, pp. 577, 822.
25. Grant, *Memoirs*, 2:232; *OR*, vol. 36, pt. 2, p. 663.
26. Matter, *If It Takes All Summer*, 2; Lyman, *Meade's Headquarters*, 94; Rhea, *Battles for Spotsylvania*, 5657.
27. Grimsley, *And Keep Moving On*, 80.
28. *OR*, vol. 36, pt. 2, p. 547.
29. A. L. Long, *Memoirs of Robert E. Lee*, 336
30. *OR*, vol. 36, pt. 2, p. 481.
31. Lyman, *Meade's Headquarters*, 100.
32. *OR*, vol. 36, pt. 2, pp. 566, 581; Grant, *Memoirs*, 2:218; Matter, *If It Takes All Summer*, 110; *OR*, vol. 36, pt. 2, p. 566; Rhea, *Battles for Spotsylvania*, 110; *OR*, vol. 36, pt. 2, pp. 567, 568; Rhea, *Battles for Spotsylvania*, 112–13.

33. Per Humphreys, *Virginia Campaign*, 104, the breakdown for the wounded in the Army of the Potomac on May 12 is 53 percent for the Second Corps, 25 percent for the Fifth Corps, and 22 percent for the Sixth Corps. I have assumed that this ratio is representative of all casualties in the Army of the Potomac. *OR*, vol. 36, pt. 2, pp. 629, 635.
34. *OR*, vol. 36, pt. 1, p. 72.
35. Gordon Rhea, *To the North Anna River* (Baton Rouge: Louisiana State University Press, 2000), 156; ibid., vol. 36, pt. 2, pp. 864–65; Rhea, *North Anna River*, 170; ibid., vol. 36, pt. 2, p. 910; Porter, *Campaigning with Grant*, 134.
36. *OR*, vol. 36, pt. 1, p. 71.
37. Ibid., vol. 36, pt. 2, p. 976.
38. Porter, *Campaigning with Grant*, 91.
39. Ibid.
40. *OR*, vol. 36, pt. 22, pp. 609–10; Isaac O. Best, *History of the 121st New York State Infantry*, (Chicago: Lt. Jas. Smith, 1921), 135; *OR*, vol. 36, pt. 2, p. 610.
41. O'Reilly Spotsylvania maps; *OR*, vol. 36, pt. 2, p. 297; Best, *History of the 121st*, 135; Hyde, *Following the Greek Cross*, 196. Note: some sources identify Mackenzie's rank as captain. In his official report, Maj. Nathaniel Michler, a member of the engineering corps to which Mackenzie was assigned, refers to Mackenzie's rank as lieutenant. *OR*, vol. 36, pt. 1, p. 297.
42. Andrew A. Humphreys, *The Army of the Potomac, July, 1863 to April, 1864* (New York: Charles Scribner's Sons, 1883), 44.
43. Rhea, *Battles for Spotsylvania*, 163; Hyde, *Following the Greek Cross*, 196; *OR*, vol. 36, pt. 2, p. 667; Best, *History of the 121st*, 128.
44. *OR*, vol. 36, pt. 2, p. 596.
45. Ibid., 603; Best, *History of the 121st*, 136–37; ibid., vol. 36, pt. 2, p. 67.
46. *OR*, vol. 36, pt. 2, p. 662; Rhea, *Battles for Spotsylvania*, 165–66; *OR*, vol. 36, pt. 2, 602–3.
47. Rhea, *Battles for Spotsylvania*, 167; *OR*, vol. 36, pt. 2, p. 490.
48. *OR*, vol. 36, pt. 1, p. 1072.
49. Thomas, *History of the Doles-Cook Brigade*, 478; *OR*, vol. 36, pt. 2, p. 668; Emory Upton to Theodore Lyman, April 26, 1879, in Massachusetts Military Historical Society Collection, Howard Gotlieb Archival Research Center, Boston University.

50. *OR*, vol. 36, pt. 1, p. 668; Upton to Lyman, April 26, 1879.
51. Upton to Lyman, April 26, 1879.
52. Humphreys, *Virginia Campaign*, 75–76; Captain W. W. Old, "Trees Whittled Down at Horseshoe," in *Southern Historical Society Papers*, ed. R. A. Brock (Richmond, VA: published by the society, 1905), 33:20, 23, 33.
53. Grimsley, *And Keep Moving On*, 82; Old, "Trees Whittled Down at Horseshoe," 21; Henry White, "Lee's Wrestle with Grant in the Wilderness," in *Papers of the Military Historical Society of Massachusetts*, 4:59; Dunlop, *Lee's Sharpshooters*, 444; *OR*, vol. 36, pt. 1, p. 1071.
54. Alexander, *Military Memoirs*, 518; *OR*, vol. 51, pt. 2, pp. 916, 917; Sumner, *Diary of Cyrus B. Comstock*, 266; *OR*, vol. 36, pt. 1, p. 909.
55. Long, *Memoirs of Robert E. Lee*, 339; *OR*, vol. 51, pt. 2, p. 917; *OR*, vol. 36, pt. 1, p. 1072; Old, "Trees Whittled Down at Horseshoe," 22; Gallagher, *Spotsylvania Campaign*, 50;Not as relevant as initially thought so deleted; Thomas Galwey, *The Valiant Hours* (Harrisburg, PA: Telegraph, 1961), 208.
56. Old, "Trees Whittled Down at Horseshoe," 21.
57. *OR* vol. 36, pt. 1, pp. 335, 1072; Long, *Memoirs of Robert E. Lee*, 339.
58. Graham T. Dozier, ed., *A Gunner in Lee's Army: The Civil War Letters of Thomas Henry Carter* (Chapel Hill: University of North Carolina Press, 2014), 236; Alexander, *Military Memoirs*, 518; Donald C. Pfanz, *Richard S. Ewell: A Soldier's Life* (Chapel Hill: University of North Carolina Press, 1998), 382; Old, "Trees Whittled Down at Horseshoe," 24.
59. Alexander, *Military Memoirs*, 518.
60. *OR*, vol. 36, pt. 1, p. 1072 circular; Dozier, *A Gunner in Lee's Army*, 236; Major Robert Hunter, "Major-General Johnson at Spotsylvania," in *Southern Historical Society Papers*, 33:337.
61. Dozier, *A Gunner in Lee's Army*, 236; Long, *Memoirs of Robert E. Lee*, 339; *OR* 36, 1, 1044-45, 1087.
62. Old, "Trees Whittled Down at Horseshoe," 23; Dozier, *A Gunner in Lee's Army*, 236.
63. Alexander, *Military Memoirs*, 519; John D. Black, "Reminiscences of the Bloody Angle" in *Glimpses of the National Struggle: Fourth Series; Papers Read Before the Minnesota Commandery of the Military Order of the Loyal Legion of the United States, 1892–1897* (St. Paul, MN: H. L. Collins, 1898), 423.
64. Hunter, "Major-General Johnson at Spotsylvania," 338; *OR*, vol. 36, pt. 1, p. 335; Gallagher, *Spotsylvania Campaign*, 50; Black, "*Reminiscences of*

the Bloody Angle, 425; William Driver, "The Capture of The Salient At Spottsyl-Vania [sic], May 12, 1864," In *Papers of the Military Historical Society of Massachusetts*, 4:281–82.

65. Cyrus Watson, "Forty-Fifth Regiment," in *Histories of the Several Regiments and Battalions from North Carolina*, ed. Walter Clark (Goldsboro, NC: Nash Brothers Books and Job Printers, 1901), 3:51.

66. *OR*, vol. 36, pt. 2, pp. 626–28; Porter, *Campaigning with Grant*, 89, 99; Grant, "Grant on The Wilderness Campaign," in *Battles and Leaders of the Civil War*, 4:146.

67. Lyman, *Meade's Headquarters*, 109–10; *OR*, vol. 36, pt. 2, p. 603; Civil War Trust website.

68. *OR*, vol. 36, pt. 2, p. 628; Porter, *Campaigning with Grant*, 99, 100; Wm. P. Haines, *History of the Men of Co. F, with Descriptions of the Marches and Battles of the 12th New Jersey Vols.* (Mickleton, NJ, 1897), 59–60; *OR*, vol. 36, pt. 1, p. 1072.

69. *OR*, vol. 36, pt. 2, p. 636.

70. Ibid., vol. 36, pt. 2, pp. 629–30; Porter, *Campaigning with Grant*, 100; Humphreys, *Virginia Campaign*, 90; American Battlefield Trust web site; Rhea, *Battles for Spotsylvania*, 225; McHenry Howard, *Recollections of a Maryland Confederate Soldier and Staff Officer Under Johnston, Jackson, and Lee* (Baltimore: Williams and Wilkins, 1914), 292.

71. *OR*, vol. 36, pt. 2, pp. 626, 630.

72. Ibid., vol. 36, pt. 1, pp. 334, 335; *ibid.*, vol. 36, pt. 2, p. 370; Hunter, "Major-General Johnson at Spotsylvania," 338. In *Fighting for the Confederacy*, edited by Gallagher, E. Porter Alexander noted that the abatis did not delay the Federals a minute, which strongly suggested the abatis were incomplete.

73. Maj. Gen. Francis Barlow, "The Capture of the Salient," in *Papers of the Military Historical Society of Massachusetts*, 4:254.

74. Upton to Lyman, April 26, 1879.

75. Alexander, *Military Memoirs*, 54, 522; Humphreys, *Virginia Campaign*, 93; *OR*, vol. 36, pt. 1, p. 336; Colonel C. S. Venable, "Campaign from Wilderness to Petersburg," in *Southern Historical Society Papers*, 14:529; Chris Mackowski and Kristopher D. White, "The Battle of the Bloody Angle, or 'Mule Shoe,' Spotsylvania Court House May 12, 1864," in *Blue & Gray Magazine*, July 2009, maps pp. 16–17; Old, "Trees Whittled Down at Horse shoe," in *Southern Historical Society Papers*, 33:19; Rhea, *Battles for Spotsylvania*, 316.

76. Humphreys, *Virginia Campaign*, 105.
77. Grant, *Memoirs*, 2:2, 232; Alexander, *Military Memoirs*, 521; *OR*, vol. 36, pt. 2, pp. 662–63, 673; Humphreys, *Virginia Campaign*, 97; Venable, "Campaign from Wilderness to Petersburg," 532; Maj. Gen. Lewis Grant, "Review of General Barlow's Paper On The Capture Of The Salient," in *Papers of the Military Historical Society of Massachusetts*, 4:270.
78. Barlow, " Capture of the Salient, May 12, 1864" 4:259; Humphreys, *Virginia Campaign*, 93, 104, 105; *OR*, vol. 36, pt. 2, pp. 652, 998; Alexander, *Military Memoirs*, 525; Rhea, *Battles for Spotsylvania*, 312; *OR*, vol. 36, pt. 2, pp. 657, 658; Peter Maugle, park ranger / historian at Fredericksburg & Spotsylvania National Military Park, e-mail to author, August 9, 2017.
79. Barlow, "Capture of the Salient, May 12, 1864," 256.
80. U. S. Grant to Julia Dent Grant, May 13, 1864, *Papers of Ulysses S. Grant*, 10:444; Meade, *Life and Letters*, 2:195; *Chicago Tribune*, May 14, 1864.
81. *OR*, vol. 36, pt. 2, p. 993; R. E. Lee to James A. Seddon, and R. E. Lee to Jefferson Davis, *Wartime Papers of R. E. Lee*, 727,729.
82. *OR*, vol. 36, pt. 1, p. 335; Barlow, "Capture of the Salient, May 12, 1864," 4:246, 247.
83. Rhea, *Battles for Spotsylvania*, 312; Alexander, *Military Memoirs*, 525.
84. U. S. Grant to Julia Dent Grant, May 13, 1864, *Papers Of Ulysses S. Grant*, 443.
85. *OR*, vol. 36, pt. 2, pp. 698, 724, 726; *OR*, vol. 36, pt. 1, p. 69; Charles A. Dana, *Recollections of the Civil War with the Leaders at Washington and in the Fields in the Sixties* (New York: D. Appleton, 1902), 197; Humphreys, *Virginia Campaign*, 104, 106.
86. Porter, *Campaigning with Grant*, 111; Dana, *Recollections of the Civil War*, 196, 197.
87. *OR*, vol. 36, pt. 2, pp. 700, 757; Summer, *Diary of Cyrus B. Comstock*, 267.
88. *OR*, vol. 51, pt. 2, pp. 929; Gallagher, *Fighting for the Confederacy*, 381.
89. *OR*, vol. 36, pt. 2, pp. 72, 781, 844, 849; Humphreys, *Virginia Campaign*, 110.
90. *OR*, vol. 36, pt. 2, pp. 866, 867, 869, 1019; *OR*, vol. 36, pt. 1, p. 1046; W. E. Cutshaw, "The Battle near Spotsylvania C. H., on May 18th, 1864," in *Southern Historical Society Papers*(1905), 33:332–33; St. Clair A. Mulholland, *The Story of the 116th Regiment, Pennsylvania Volunteers* (n.p., 1903), 222.
91. *OR*, vol. 36, pt. 2, p. 627.
92. Grant, *Memoirs*, 2:238; *ibid.*, vol. 36, pt. 2, p. 1020.

93. Grant, *Memoirs*, 2:238.
94. *OR*, vol. 36, pt. 1, p. 73; ibid., 2:239; *OR*, vol. 36, pt. 2, p. 864; Frank Wilkeson, *Recollections of a Private Soldier in the Army of the Potomac* (New York: G. P. Putnam's Sons, 1887), 100.
95. Walter Taylor, *General Lee: His Campaigns in Virginia, 1861–1865* (Norfolk: Nusbaum, 1906), 243; Rhea, *North Anna River*, 165; *OR*, vol. 36, pt. 1, pp. 1058, 1073; Alexander, *Military Memoirs of a Confederate*, 528.
96. *OR*, vol. 36, pt. 2, pp. 910, 911, 915, 919, 921; Porter, *Campaigning with Grant*, 126; Rhea, *North Anna River*, 183; *OR*. vol. 36, pt. 1, p. 1073; R. E. Lee, March 3, 1868, William Allan Papers, 1802–1937, William Allan Papers #2764, Southern Historical Collection, Wilson Library, University of North Carolina at Chapel Hill; Journal entry in Roebling's report for May 19, 1864,16; Porter, *Campaigning with Grant*, 128.
97. *OR*, vol. 36, pt. 2, p. 921; *OR*, vol. 36, pt. 3, p. 4.
98. *OR*, vol. 36, pt. 3, p. 8; *OR*, vol. 36, pt. 2, p. 840.
99. *OR*, vol. 36, pt. 3, p. 4.
100. Ibid., vol. 36, pt. 3, pp. 800, 801, 812; ibid., vol. 36, pt. 2, pp. 1024–25; Rhea, *North Anna River*, 195, 221.

Conclusion

1. Humphreys, *Virginia Campaign*, 118.
2. James Ford Rhodes, *History of the United States* (London: MacMillan, 1904), 4:417.
3. *OR*, vol. 36, pt. 2, p. 405; John Ropes, "Grant's Campaign in Virginia in 1864," in *Papers of the Military Historical Society of Massachusetts* 4:404-5; Rhea, *Battles for Spotsylvania*, 313.
4. *OR*, vol. 36, pt. 2, p. 627; Porter, *Campaigning with Grant*, 98. Officers were surprised by the publicity Grant's remark had created.
5. Lyman, *Meade's Headquarters*, 102n1.
6. Lt. Col. William Swan, "Battle of The Wilderness," in *Papers of the Military Historical Society of Massachusetts*, 4:134; Porter, *Campaigning with Grant*, 114; Rhea, *Spotsylvania*, 316-16.
7. Humphreys, *Virginia Campaign*, 117.
8. "Address of Hamilton W. Mabie, LL.D." Proceedings of The Eighteenth Annual Lincoln Dinner of the Republican Club of the City of New York, February 12, 1904, 27; Foote, *Civil War*, 2:217; "Lincoln

and Some Union Generals" in *Harper's Magazine*, Christmas 1914, 100.

9. Clark, *Preparing for War*, 86.
10. R. E. Lee to Gen. James Longstreet, Dowdey, *Wartime Papers of R. E. Lee*, 685; Gallagher, *Fighting for the Confederacy*, 121; Nolan, *Lee Considered*, 78n46.
11. Maj. Gen James Wilson, "The Cavalry of the Army of the Potomac" in *Papers of the Military Historical Society of Massachusetts*, 13:49.
12. John J. Hennesscy, *"The Army of the Potomac," in The Wilderness Campaign*, ed. Gary W. Gallagher (Chapel Hill: University of North Carolina Press, 1997), 70.
13. Humphreys, *Virginia Campaign*, 56.
14. According to the American Battlefield Trust website, https://www.battlefields.org (formerly The Civil War Trust), Union casualties at The Wilderness were 18,400, and Union casualties at Shiloh were 13,047.
15. Rhea, *Battles for Spotsylvania*, 319.
16. Ibid.
17. Grant, *Memoirs*, 2:154.
18. *OR*, vol. 36, pt. 3, p. 169; Porter, *Campaigning with Grant*, 144–45.
19. Rhea, *North Anna River*, 259.
20. Venable, "Lee in The Wilderness," in *Battles and Leaders of the Civil War*, 4:244.
21. *OR*, vol. 36, pt. 3, pp. 94–95; Rhea, *North Anna River*, 257.
22. Lyman, *Meade's Headquarters*,148n1: Civil War Trust website; Grant, *Memoirs*, 2:276: *OR*, vol. 36, pt. 3, pp. 600, 603, 639; Ernest B. Furgurson, *Not War But Murder* (New York: Random House, 2000), 206; Gordon Rhea, *Cold Harbor* (Baton Rouge: Louisiana State University Press, 2002), 362; Map 136 text, in vol. 1 in *The West Point Atlas of American Wars*, ed. Brig. Gen. Vincent J. Esposito (New York: Henry Holt and Company, 1995).
23. Meade, *Life and Letters*, 2:201.

Appendix I

1. Henry Heth, *Memoirs of Henry Heth*, ed. James Morrison (Westport, Conn: Greenwood Press, 1974), 184.
2. *Battles and Leaders of the Civil War*, 4:97.

3. Porter, *Campaigning with Grant*, 49; Mertz, "No Turning Back," 56.

4. *OR*, vol. 36, pt. 2, 331–32, 358–59.

5. Humphreys, *Virginia Campaign*, 20.

6. *OR*, vol. 36, pt. 2, pp. 331, 374, 390.

7. Lt. Col. William Swan, "Battle of The Wilderness," in *Papers of the Military Historical Society of Massachusetts*, 4:124.

8. Humphreys, *Virginia Campaign*, 18.

9. Ibid., 19; *OR*, vol. 36, pt. 2, p. 370.

10. *OR*, vol. 36, pt. 2, p. 356; Wilkeson, *Recollections*, 50.

11. *OR*, vol. 36, pt. 2, p. 372.

12. Ibid., vol. 36, pt. 2, pp. 371, 390; Humphreys, *Virginia Campaign*, 20, 21.

13. Porter, *Campaigning with Grant*, 24; Schaff, *Battle of The Wilderness*, 213.

14. *OR*, vol. 36, pt. 2, p. 403.

15. Ibid., vol. 36, pt. 2, p. 403; Porter, *Campaigning with Grant*, 48, 49.

16. Humphreys, *Virginia Campaign*, 5; Porter, *Campaigning with Grant*, 39, 22, 24; Badeau, *Military History of Ulysses S. Grant*, 2:105.

17. Porter, *Campaigning with Grant*, 50.

18. Grant, *Memoirs*, 2:118; Hyde, *Following the Greek Cross*, 183–84; *OR*, vol. 36, pt. 2, p. 405.

19. Verified in author's personal conversation with National Park Service personnel, September 8, 2017.

20. Rhea, *Battle of the Wilderness*, 97.

21. *OR*, vol. 36, pt. 1, p. 580.

22. Rhea, *Battle of the Wilderness*, 98; ibid., vol. 36, pt. 2, p. 413.

23. *OR*, vol. 36, pt. 2, p. 416.

24. Ibid., vol. 36, pt. 2, p. 403.

25. Ibid., vol. 36, pt. 2, p. 408.

26. Swinton, *Campaigns Army of the Potomac*, 421n; ibid., vol. 36, pt. 2, p. 404.

27. *OR*, vol. 36, pt. 2, p. 418.; Rhea, *Battle of the Wilderness*, 108.

28. *OR*, vol. 36, pt. 2, 418–20; Rhea, *Battle of the Wilderness*, 140–41; Lt. Col. William Swan, *"Battle of The Wilderness," in Papers of the Military Historical Society of Massachusetts* (Boston: Military Historical Society of Massachusetts, 1905), 4:129–30.

29. Powell, *Fifth Army Corps*, 609.

30. Judson, *History of the Eighty-Third*, 94.

31. *OR*, vol. 36, pt. 1, p. 540.
32. Agassiz, *Meade's Headquarters*, 90–91n1.
33. *OR*, vol. 36, pt. 1, p. 660.
34. Ibid., vol. 36, pt. 1, p. 665.
35. Rhea, *Battle of the Wilderness*, 178.
36. Gottfried, *Maps of the Wilderness*, 71, 75, 83; Peirson, "Operations of the Army of the Potomac," 4:235.
37. *OR*, vol. 36, pt. 2, pp. 425, 449, 450; Gottfried, *Maps of the Wilderness*, 140; Page, *Letters of a War Correspondent*, 52.
38. *OR*, vol. 36, pt. 2, p. 450.
39. Ibid., vol. 36, pt. 2, p. 451.
40. John B. Gordon, *Reminiscences*, 243; Ibid., vol. 36, pt. 2, p. 962.
41. Gordon, *Reminiscences*, 244; *OR*, vol. 36, pt. 1, p. 1071.
42. Rhea, *Battle of the Wilderness*, 410; Gottfried, *Maps of the Wilderness*, 230.
43. *OR*, vol. 36, pt. 1, pp. 737, 1071.
44. Gottfried, *Maps of the Wilderness*, 230.
45. Humphreys, *Virginia Campaign*, 50–51n2.
46. Gottfried, *Maps of the Wilderness*, 234.
47. Rhea, *Battle of the Wilderness*, 424; *OR*, vol. 36, pt. 1, p. 745.
48. *OR*, vol. 36, pt. 1, pp. 1077–78.
49. Frank Vandiver, ed., *War Memoirs: Autobiographical Sketch and Narrative of the War Between the States*, by Jubal Anderson Early (1912; repr., Bloomington: Indiana University Press, 1960), 350.
50. *OR*, vol. 36, pt. 2, p. 480.
51. Schaff, *Battle of the Wilderness*, 320.
52. *Blue & Gray Magazine*, June 1995, 62; NPS website, https://parkplanning.nps.gov/projectHome.cfm?projectID=43393 accessed October 29, 2017.
53. Thomas, *History of the Doles-Cook Brigade*, 477.
54. *Blue & Gray Magazine*, April 1995, 61.
55. Ibid.
56. Gottfried, *Maps of the Wilderness*, 43.
57. Ibid., 18
58. *OR*, vol. 36, pt. 1, p. 573; Rhea, *Battle of the Wilderness*, 125.

59. Ibid., vol. 36, pt. 1, 610–11. Note: Cutler replaced Wadsworth, who was killed the next day. Although Cutler is the author of the report, he refers to himself as "Cutler".

60. *OR*, vol. 36, pt. 1, pp. 1070, 1076–77; Thomas, *History of the Doles-Cook Brigade*, 14. The exact point of Gordon's attack is uncertain; see Rhea, *Battle of the Wilderness*, 161, 26n.

61. *OR* vol. 36, pt. 1, p. 610; Gottfried, *Maps of the Wilderness*, 53.

62. Rhea, *Battle of the Wilderness*, 183, 184; Brown, *Campbell Brown's Civil War*, 248.

63. Roebling report. After the war, Roebling went on the supervise the construction of the completion of the Brooklyn Bridge that his father had designed.

64. *OR*, vol. 326, pt. 2, p. 418.

65. Ibid., vol. 36, pt. 2, pp. 418, 420.

66. Ibid. vol. 36, pt. 2, p. 419.

67. *Annals of The War*, 492.

68. Lyman, *Meade Letter*, 93; *OR*, vol. 36, pt. 2, p. 441.

69. *History of the Thirty-Sixth Regiment, Massachusetts Volunteers, 1862–1865* (Boston: Press of Rockwell and Churchhill,1884), 150; Royall, *Some Reminiscences*, 33–34.

70. Lyman Jackman and Amos Hadley, *History of the Sixth New Hampshire Regiment in the War of the Union* (Concord, NH; Republican Press Association, 1891), 226–27; Royall, *Some Reminiscences*, 34.

71. *OR*, vol. 36, pt. 1, p. 928; *OR*, vol. 36, pt. 2, p. 460.

72. Ibid., vol. 36, pt. 1, p. 1081.

73. C. S. Venable, "Campaign from the Wilderness to Petersburg," in *Southern Historical Society Papers*, 14:524.

74. *Blue & Gray Magazine*, June 1995, 61; Gottfried, *Maps Of The Wilderness*, 145.

75. Charles Venable, "General Lee in the Wilderness," in *Battles and Leaders of the Civil War* (New York: Castle Books, 1956), 4:241. According to *Annals of the War*, 492, Brigadier General Wilcox said that Heth was present rather than Stuart, implying a slightly different time. I have used the version of Lt. Col. Charles Venable, a member of Lee's staff. Venable was more likely to have been with Lee than one of Hill's division commanders.

76. *Annals of War*, 492.

77. *OR*, vol. 36, pt. 2, p. 953
78. Ibid., vol. 36, pt. 2, p. 951.
79. *Annals of War*, 495; Morrison, *Memoirs of Henry Heth*, 184.
80. Swinton, *Campaigns of the Army of the Potomac*, 431n.
81. *Annals of War*, 496; Walker, *History Second Army Corps*, 422; *OR*, vol. 36, pt. 1, p. 476.
82. Francis A. Walker, *History of the Second Army Corps in the Army Of the Potomac* (New York: Charles Scribner's Sons, 1886), 422; Swinton, *Campaigns of the Army of the Potomac*, 431.
83. *Annals of War*, 496; Royall, *Some Reminiscences*, 31–32.
84. Venable, "Campaign from the Wilderness to Petersburg," in, *Southern Historical Society Papers*, 14:525. Per Royall, *Some Reminiscences*, 31, Longstreet arrived alone. According to *Annals of War*, 496, Longstreet arrived after a brigade had arrived. William Dame, *From the Rapidan to Richmond and the Spotsylvania Campaign* (Baltimore: Green-Lucas, 1920), 85.
85. *Annals of War*, 496.
86. Rhea, *Battle of the Wilderness*, 304; Venable, "Campaign from the Wilderness to Petersburg," in, *Southern Historical Society Papers*, *14*:525–26, 544, 546–47.
87. Rhea, *Battle of the Wilderness*, 313; Royall, *Some Reminiscences*, 33.
88. *OR*, vol. 36, pt. 1, pp. 477, 1061; Rhea, *Battle of the Wilderness*, 310.
89. *OR*, vol. 36, pt. 1, pp. 611, 596.
90. Rhea, *Battle of the Wilderness*, 315; C. W. Field "Campaign of 1864-65" in *Southern Historical Society Papers*, 14:544; Sorrel, *Recollections*, 242.
91. Francis Dawson, *Reminiscences of Confederate Service, 1861–1865* (Charleston, SC: News and Courier Book Presses, 1881), 116; Longstreet, *From Manassas to Appomattox*, 564; Field, "Campaign of 1864-65," in *Southern Historical Society Papers*, , 14:545.
92. Frank Mixson, *Reminiscences of a Private* (Columbia, SC: State Company, 1910), 70.
93. *OR*, vol. 36, pt. 2, p. 960.
94. *Blue & Gray Magazine*, April 1995, 57.
95. Rhea, *Battle of the Wilderness*, 194; *OR*, vol. 36, pt. 2, p. 418, *OR*, vol. 36, pt. 1, p. 676.
96. *OR*, vol. 36, pt. 2, pp. 407, 419; *OR*, vol. 36, pt. 1, p. 677.

97. *OR*, vol. 36, pt. 1, p. 677.
98. O'Reilly Wilderness map 2 of 6.
99. Rhea, *Battle of the Wilderness*, 194; *Annals of War*, 492.
100. *OR*, vol. 36, pt. 1, pp. 473, 488, 697.
101. Ibid., vol. 36, pt. 2, p. 967.
102. Ibid., vol. 36, pt. 1, pp. 681–82.
103. *Annals of War*, 429; Ibid., vol. 36, pt. 1, pp. 410–11.
104. *OR*, vol. 36, pt. 1, p. 320; *OR*, vol. 36, pt. 2, p. 411.
105. *OR*, vol. 36, pt. 2, p. 953.
106. Morrison, *Memoirs of Henry Heth*, 184; Swan, "Battle of The Wilderness'" in , *Papers of the Military Historical Society of Massachusetts*, 4:141; W. A. Graham, "Nineteenth Regiment," Clark ed., in *Histories of the Several Regiments and Battalions*, 2:94.
107. Humphreys, *Virginia Campaign*, 33.
108. In Roebling's report, he states that Grant wanted the Ninth Corps to take the heights at Tuning's (presumably Chewning Farm) and move against Hill's rear. Porter, *Campaigning with Grant*, 54; *OR*, vol. 36, pt.2, pp. 415, 425.
109. *OR*, vol. 36, pt.2, p. 439.
110. Lyman, *Meade's Headquarters*, 64; *OR*, vol. 36, pt. 1, p. 437.
111. *OR*, vol. 36, pt. 1, pp. 321–22.
112. Ibid., vol. 36, pt. 1, p. 322.
113. Ibid., vol. 36, pt. 2, pp. 352, 442, 451–52.
114. Ibid., vol. 36, pt. 1, p. 489.
115. Ibid., vol. 36, pt. 1, p. 323; Longstreet, *From Manassas to Appomattox*, 568.
116. *OR*, vol. 36, pt. 1, pp. 323, 624; Alexander Webb, "Through The Wilderness," in *Battles and Leaders of the Civil War*, 4:161.
117. *OR*, vol. 36, pt. 2, pp. 444–45; Schaff, *Battle of the Wilderness*, 290.
118. *OR*, vol. 36, pt. 1, pp. 324, 1068; Field, "Campaign of 1864-65," in *Southern Historical Society Papers*, 14:546.
119. *OR*, vol. 36, pt. 1, p. 447.
120. Ibid., vol. 36, pt. 1, p. 514.
121. Ibid., vol. 36, pt. 2, pp. 445–46.
122. Ibid., vol. 36, pt. 2, p. 447.
123. Longstreet, *From Manassas to Appomattox*, 556; Ibid., vol. 36, pt. 1, p. 1054.

124. *OR*, vol. 36, pt. 2, pp. 376, 406, 407.
125. Ibid. vol. 36, pt. 2, p. 407.
126. Ibid., vol. 36, pt. 2, p. 427; Rhea, *Battle of the Wilderness*, 22, 349; *ibid.*, vol. 36, pt. 1, p. 372.
127. Rhea, *Battle of the Wilderness*, 378; *OR*, vol. 36, pt. 2, p. 470.
128. Rhea, *Battle of the Wilderness*, 379.
129. *OR*, vol. 36, pt. 2, p. 483.
130. Ibid., vol. 36, pt. 1, p. 833; *ibid.*, vol. 36, pt. 2, pp. 513, 514.
131. Rhea, *Battles for Spotsylvania*, 354n80.
132. *OR*, vol. 36, pt. 2, p. 867; *OR*, vol. 36, pt. 1, pp. 833, 846.
133. *OR*, vol. 36, pt. 2, p. 515.
134. Ibid., vol. 36, pt. 2, pp. 515–16.
135. Porter, *Campaigning with Grant*, 81; Humphreys, *Virginia Campaign*, 58–59; ibid., vol. 36, pt. 2, p. 552.
136. *OR*, vol. 36, pt. 2, p. 553.
137. Ibid., vol. 36, pt. 2, p. 539.
138. Martin McMahon, "Death of General John Sedgwick," in *Battles and Leaders of the Civil War*, 4:175.
139. *OR*, vol. 36, pt. 2, pp. 538–39.
140. Journal entry in Roebling's report for May 5, 1864; Rhea, *Battles for Spotsylvania*, 49; ibid., vol. 36, pt. 2, p. 539.
141. Gallagher, *Fighting for the Confederacy*, 368; Woodford Hackley, *The Little Fork Rangers: A Sketch of Company "D," Fourth Virginia Cavalry* (Richmond, VA: Dietz, 1927), 89; Dickert, *History of Kershaw's Brigade*, 358.
142. Gallagher, *Fighting for the Confederacy*, 368.
143. Dickert, *History of Kershaw's Brigade*, 357–58.
144. Rhea, *Battles for Spotsylvania*, 53; Journal entry in Roebling's report for May 8, 1864; Peirson, "Operations of the Army of the Potomac May 7-11, 1864," in *Papers of the Military Historical Society of Massachusetts*, 4:215, 216.
145. Phelps, "History of the Battle of the Wilderness," p. 8; *OR*, vol. 36, pt. 1, p. 594.
146. Coxe, "Last Struggles and Success of Lee," 2; ibid. p. 8.
147. Judson, *Eight-Third Regiment*, 95
148. Rhea, *Spotsylvania*, 57; *OR* 36, 1, 557.

149. Judson, *Eighty-Third Regiment*, 96.
150. *OR*, vol. 36, pt. 2, pp. 539–40.
151. Ibid., vol. 36, pt. 2, p. 545.
152. Mertz, "Fighting at Laurel Hill," *Blue & Gray Magazine*, Summer 2004, 22; Gallagher, *Fighting for The Confederacy*, 368.
153. *OR*, vol. 36, pt. 2, p. 542.
154. Mertz, "Upton's Attack and the Defense of Dole's Salient," *Blue & Gray Magazine*, Summer 2001, 8; *ibid.*, vol. 36, pt. 2, pp. 601–2.
155. *OR*, vol. 36, pt. 1, p. 331.
156. Ibid., vol. 36, pt. 2, pp. 607, 596, 600, 604, 609–10; *ibid.*, vol. 36, pt. 1, p. 561.
157. Gibbon, *Recollections*, 218–19; *ibid.*, vol. 36, pt. 2, p. 600.
158. Roebling, journal entry for May 10, 1864.
159. *OR*, vol. 36, pt. 2, p. 603.
160. Ibid., vol. 36, pt. 2, p. 603.
161. Upton to Lyman, April 26, 1879; Mertz, "Upton's Attack and the Defense of Dole's Salient," *Blue & Gray Magazine*, Summer 2001, 46–47; map, p. 57; O'Rielly maps.
162. Mackenzie's commander, Major Michler, stated his rank as lieutenant. *OR*, vol. 36, pt. 1, p. 297. Upton's report states his rank as captain. *OR*, vol. 36, pt. 2, p. 667.
163. Best, *History of the 121st*, 136.
164. Ibid., 128, 136–37.
165. *OR*, vol. 36, pt. 1, p. 667.
166. Ibid., vol. 36, pt. 1, pp. 667–68; Peter Michie, Life there is no specific letter to cite and *Letters of Emory Upton* (New York: D. Appleton, 1885), 97.
167. *OR*, vol. 36, pt. 1, p. 668.
168. William White, *Contributions to a History of the Richmond Howitzer Battalion, Pamphlet No. 2* (Richmond, VA: Carlton McCarty, 1883), 244.
169. Upton to Lyman, April 26, 1879; Clark, *Histories of the Several Regiments and Battalions*, 48.
170. *OR*, vol. 36, pt. 1, p. 668.
171. Ibid.
172. *Ibid.*

173. Cyrus Watson, "Forty-Fifth Regiment," in , *Histories of the Several Regiments and Battalions*, 3:47; *ibid.*, vol. 36, pt. 1, p. 1072.
174. Rhea, *Battles for Spotsylvania*, 171; Taylor, *Four Years with General Lee*, 130.
175. Taylor, *Four Years with General Lee*, 172; Mertz, "Upton's Attack," in *Blue & Gray Magazine*, Summer 2001, 47; Rhea, *Battles for Spotsylvania*, 172; S. D. Thurston, "Report of the Conduct of General George H. Steuart's Brigade from the 5th to the 12th on May 1864, inclusive," in *Southern Historical Society Papers*, 14:151.
176. Michie, *Upton Letters*, 98–99; *OR*, vol. 36, pt. 1, p. 668; Upton to Lyman, April 26, 1879.
177. Rhea, *Battles for Spotsylvania*, 176.
178. *OR*, vol. 36, pt. 1, p.668.
179. Rhea, *Battles for Spotsylvania*, 291–92.
180. Venable, "Campaign from the Wilderness to Petersburg," in *Southern Historical Society Papers*, vol. 14, 529.
181. *OR*, vol. 36, pt. 2, p. 630.
182. Ibid., vol. 36, pt. 2, pp. 640–41.
183. Ibid., vol. 36, pt. 2, pp. 635, 637, 643–44.
184. Barlow, "Operations of the Army of the Potomac," in *Papers of the Military Historical Society of Massachusetts*, 4:245–47. Note: Barlow refers to the location for the attack as the "right flank," when in fact it was the middle with Early's Corps on the right.
185. *OR*, vol. 51, pt. 2, pp. 916–17.
186. Ibid., vol. 51, pt. 2, p. 917.
187. Ibid., vol. 36, pt. 1, p. 1044.
188. Ibid., vol. 36, pt. 1, p. 1086.
189. Alexander, *Military Memoirs*, 518; Mackowski and White, "Battle of the Bloody Angle, or 'Mule Shoe" in *Blue & Gray Magazine*, 2009, 19; Page, "Captured Guns at Spotsylvania Courthouse-Correction of General Ewell's Report," in , *Southern Historical Society Papers*, 7:535.
190. Robert Hunter, "Major General Edward Johnson at Spotsylvania," in *Southern Historical Society Papers, 33*:337; *OR*, vol. 36, pt. 1, p. 1086.
191. Hunter, "Johnson at Spotsylvania," in *Southern Historical Society Papers*, 33:337.
192. *OR*, vol. 36, pt. 1, p. 334; Barlow, "Operations of the Army of the

Potomac," in , *Papers of the Military Historical Society of Massachusetts*, 4:246, 249.

193. Barlow, "Operations of the Army of the Potomac," in *Papers of the Military Historical Society of Massachusetts*, 4:247, 248.

194. *OR*, vol. 36, pt. 1, 334–35.

195. *Glimpses of the National Struggle: Fourth Series*, 425, 426; Hunter, "Johnson at Spotsylvania," in *Southern Historical Society Papers* 33:338; James Walker, "Bloody Angle," in *Southern Historical Society Papers* 21:235.

196. Barlow, "Operations of the Army of the Potomac," in *Papers of the Military Historical Society of Massachusetts*, 4:250; Mackowski and White, "Bloody Angle, or Mule Shoe," *Blue & Gray Magazine*, 2009, 24, 22; Page, "Captured Guns at Spotsylvania," in *Southern Historical Society Papers*, 7:536; Walker, "Bloody Angle," in *Southern Historical Society Papers*, 11:240.

197. *Glimpses of the Nation's Struggles: Fourth Series*, 426; Mackowski and White, "Bloody Angle of Mule Shoe," *Blue & Gray Magazine*, 2009, 23; *OR*, vol. 36, pt. 2, p. 658; Gibbon, *Recollections*, 220.

198. *OR*, vol. 36, pt. 2, p. 657; Porter, *Campaigning with Grant*, 105.

199. Rhea, *Battles for Spotsylvania*, 246; King, *Campfire Sketches*, 309; *OR*, vol. 36, pt. 2, p. 410.

200. Charles Weygant, *History of the One Hundred and Twenty-Fourth Regiment NYSV* (Newburg, NY: Journal Printing House, 1877), 322; Mackowski and White, "Bloody Angle or Mule Shoe," *Blue & Gray Magazine*, 2009, 24; Walker, "Bloody Angle," in *Southern Historical Society Papers*, 14:235, 237–38.

201. Walker, "Bloody Angle," in *Southern Historical Society Papers*, 14:236, 239.

202. Mackowski and White, "Bloody Angle or Mule Shoe," *Blue & Gray Magazine*, 2009, 24; O'Reilly Spotsylvania map 10 of 24; Rhea, *Battles for Spotsylvania*, 247 map, 257; *OR*, vol. 36, pt. 1, pp. 1082, 1072.

203. *OR*, vol. 36, pt. 2, pp. 656, 657; Mackowski and White, "Bloody Angle or Mule Shoe," *Blue & Gray Magazine*, 2009, 48.

204. *OR*, vol. 36, pt. 1, p. 733.

205. Mackowski and White, "Bloody Angle or Mule Shoe," *Blue & Gray Magazine*, 2009, 25; *OR*, vol. 36, pt. 1, pp. 410, 1078; Venable, "Wilderness to Petersburg," in *Southern Historical Society Papers*, 14:530.

206. Rhea, *Battles for Spotsylvania*, 268.

207. Mackowski and White, "Bloody Angle or Mule Shoe," *Blue & Gray Magazine*, 2009, 45; O'Reilly Spotsylvania maps 12 and 13 of 24; ibid., 275, 277; Mackowski and White, "Bloody Angle or Mule Shoe," *Blue & Gray Magazine*, 2009, 46.
208. *OR*, vol. 36, pt. 2, pp. 654, 655, 662, 663, 668; Journal entry in Roebling's report for May 12, 1864; Humphreys, *Virginia Campaign*, 101n3; Rhea, *Battles for Spotsylvania*, 289.
209. *OR*, vol. 36, pt. 2, p. 675.
210. Ibid., vol. 36, pt. 2, p. 679.
211. Rhea, *Battles for Spotsylvania*, 298–302.
212. Ibid., 306; *OR*, vol. 36, pt. 1, p. 1073.

Appendix II

1. On May 13, the Third and Fourth Divisions were temporarily consolidated under the command of Birney. *OR*, vol. 36, pt. 2, p. 709.
2. Disbanded on May 9, with 1st and 3rd Brigades to 4th Division and 2nd Brigade to 3rd Division.

Appendix III

1. On May 8, brigades consolidated under Johnson's division. *OR*, vol. 36, pt. 2, p. 974.
2. On May 8, brigades consolidated under Johnson's division. *OR*, vol. 36, pt. 2, p. 974.
3. On May 8, brigade transferred to Early's division. *OR*, vol. 36, pt. 22, p. 974.

BIBLIOGRAPHY

1860 US Census. Accessed April 29, 2015. https://www.census.gov/library/publications/1864/dec/1860a.html.

"*Address of Hamilton W. Mabie, LL.D.*" Proceedings of the Eighteenth Annual Lincoln Dinner of the Republican Club of the City of New York, February 12, 1904.

Agassiz, George R., ed. *Meade's Headquarters 1863–1865: Letters Of Colonel Theodore Lyman From The Wilderness to Appomattox*. Boston: Massachusetts Historical Society, 1922.

Alexander, E. P. *Military Memoirs of a Confederate: A Critical Narrative*. New York: Charles Scribner's Sons, 1907.

American Battlefield Trust website (formally Civil War Trust website) Accessed December 2, 2017. https://www.battlefields.org.

Anderson, Richard Heron. Richard Heron Anderson Manuscript, 1864. David M. Rubenstein Rare Book and Manuscript Library, Duke University, Durham, NC.

Annals of The War Written by Leading Participants North and South, The. Originally Published in the Philadelphia Weekly Times. Philadelphia: The Times Publishing Company, 1879.

Badeau, Adam. *Military History of Ulysses S. Grant*. 3 vols. New York: D. Appleton, 1881.

Basler, Roy P., ed. *The Collected Works of Abraham Lincoln*. New Brunswick, NJ: Rutgers University Press, 1953.

"Battle of Spotsylvania Court House," 24 map set, Frank O'Reilly Historical Researcher. Eastern National, 2000.

"Battle of The Wilderness," 6 map set, Frank O'Reilly Historical Researcher. Eastern National, 2003.

Battles and Leaders of the Civil War. 4 vols. New York: Castle Books, 1956.

Battles of the Wilderness & Spotsylvania, The. National Parks Civil War Series: Easter National, 2007.

Best, Isaac O. *History of the 121st New York State Infantry*. Chicago: Lt. Jas. Smith, 1921.

Black, John D. "Reminiscences of the Bloody Angle." In *Glimpses of the National Struggle: Fourth Series; Papers Read Before the Minnesota Commandery of the Military Order of the Loyal Legion of the United States, 1892–1897* pp. 420-426. St. Paul, MN: H. L. Collins, 1898.

Blue & Gray Magazine, June 1995.

Bongi, David, "Offensive Infantry Tactics During the Battle of Chancellorsville, May 1863," Master's thesis, U. S. Army Command and General Staff College, 1993.

Boritt, Gabor S., ed. *Lincoln's Generals*. New York: Oxford University Press, 1994.

Breathed, James report, dated September 14, 1864, in folder 4 of the Edwin L. Halsey Papers, Louis Round Wilson Special Collections Library, The University of North Carolina at Chapel Hill.

Brock, R. A., ed. *Southern Historical Society Papers*. Vol. 7. 1879.

Brock, R. A., ed. *Southern Historical Society Papers*. Vol. 21. 1893. Reprint. Millwood, NY: Krause, 1977.

———, ed. *Southern Historical Society Papers*. Vol. 33. Richmond, VA: published by the society, 1905.

Brooks, Noah. *Washington in Lincoln's Time*. New York: Century, 1896.

Burns, Ken, dir. The Civil War, Florentine Films, 1990; Episode 6: 1864- Valley of the Shadow of Death

Business Insider. "Military superpowers through history." Accessed December 12, 2018, https://www.businessinsider.com/this-ambitious-graphic-shows-the-size-of-standing-armies-from-antiquity-to-the-present-2014-11

Carpenter, F. B. *Six Months at the White House with Abraham Lincoln*. New York: Hurd and Houghton, 1866.

Catton, Bruce. *Grant Takes Command*. Boston: Little, Brown, 1968.

Chicago Tribune. May 14, 1864

Civil War Times Illustrated, April 1996, vol. 5, issue 1, 5.

Clark, J. P. *Preparing for War: The Emergence of the Modern U.S. Army, 1815–1917*. Cambridge, MA: Harvard University Press, 2017.

Clark, Walter, ed. *Histories of the Several Regiments and Battalions from North Carolina*. Vol. 3. Goldsboro, NC: Nash Brothers Books and Job Printers, 1901.

"Cover Letter for 1864 report (6 February 1868): J. B. Gordon to Lee." MSS 3 L515a, Folder 31—Reports and Reminiscences, 1866 June–1876, Item 603. Robert E. Lee Headquarters Papers, 1850–1876. Virginia Historical Society, Richmond.

Coxe, John. "Last Struggles and Successes of Lee." *Confederate Veteran*, Vol. 22.

Cozzens, Peter. *The Shipwreck of Their Hopes*. Urbana: University of Illinois Press, 1994.

D Day Museum website. August 9, 2017. www.ddaymuseum.co.uk.

Dame, William. *From the Rapidan to Richmond and the Spotsylvania Campaign*. Baltimore: Green-Lucas, 1920.

Dana, Charles A. *Recollections of the Civil War with the Leaders at Washington and in the Field in the Sixties*. New York: D. Appleton, 1902.

Dawson, Francis. *Reminiscences of Confederate Service, 1861–1865*. Charleston, SC: News and Courier Book Presses, 1881.

Dickert, D. Augustus. *History of Kershaw's Brigade, with Complete Roll of Companies, Biographical Sketches, Incidents, Anecdotes, Etc.* Newberry, SC: Elbert H. Aull, 1899.

Dowdey, Clifford, ed. *The Wartime Papers of R. E. Lee*. New York: Bramhall House, 1961.

Dozier, Graham T., ed., *A Gunner in Lee's Army: The Civil War Letters of Thomas Henry Carter*. Chapel Hill: University of North Carolina Press, 2014.

Dunlop, Major W. S. *Lee's Sharpshooters*. Little Rock, AK: Tunnam and Pittard, 1899.

Early, Jubal. *Lieutenant General Jubal Early C.S.A.: Autobiographical Sketch and Narrative of the War Between the States*. Philadelphia: J. B. Lippincott, 1912.

Eckert, Ralph. *John Brown Gordon: Soldier, Southerner, American*. Baton Rouge: Louisiana State University Press, 1989.

Emory Upton to Theodore Lyman, April 26, 1879. Massachusetts Military Historical Society Collection, Howard Gotlieb Archival Research Center, Boston University.

Esposito, Brig. Gen. Vincent J., ed. *The West Point Atlas of American Wars*. 2 vols. New York: Henry Holt, 1995.

Foote, Shelby. *The Civil War: A Narrative*. 4 vols. New York: Random House, 1974.

Freeman, Douglas Southall. *Lee's Lieutenants*. 3 vols. New York: Charles Scribner's Sons, 1944.

Furgurson, Ernest B. *Not War But Murder*. New York: Random House, 2000.

Gallagher, Gary W., ed. *Fighting for the Confederacy: The Personal Recollections of General Edward Porter Alexander*. Chapel Hill: University of North Carolina Press, 1989.

———, ed. *The Spotsylvania Campaign*. Chapel Hill: University of North Carolina Press, 1998.

———, ed. *The Wilderness Campaign*. Chapel Hill: University of North Carolina Press, 1997.

Galwey, Thomas. *The Valiant Hours*. Harrisburg, PA: Telegraph, 1961.

Gibbon, John. *Personal Recollections of the Civil War*. New York: Putman's, 1928. Reprint Dayton, OH: Morningside, 1978.

Gordon, John B. *Reminiscences of the Civil War*. New York: Charles Scribner's Sons, 1903.

Gottfried, Bradley M. *The Maps of the Wilderness: An Atlas of the Wilderness Campaign, Including All Cavalry Operations, May 2–6, 1864*. El Dorado Hills, CA: Savas Beatie, 2016.

Grant, Ulysses S. *The Papers of Ulysses S. Grant*. 32 vols. Edited by John Simon. Carbondale: Southern Illinois University Press, 1967, Digital collection 1982

Grant, U. S. *Personal Memoirs of U. S. Grant In Two Volumes*. 2 vols. New York: Charles L. Webster, 1885.

Grimsley, Mark. *And Keep Moving On: The Virginia Campaign, May–June 1864*. Lincoln: University of Nebraska Press, 2002.

Hackley, Woodford. *The Little Fork Rangers: A Sketch of Company "D," Fourth Virginia Cavalry*. Richmond, VA: Dietz, 1927.

Haines, Wm. P. *History of the Men of Co. F, with Descriptions of the Marches and Battles of the 12th New Jersey Vols*. Mickleton, NJ, 1897.

Harper's Magazine, Christmas 1914.

Henderson, Helene, ed. *The Abraham Lincoln Companion*. Detroit: Omnigraphics, 2008.

Heth, Henry, *The Memoirs of Henry Heth*. Edited by James Morrison, Jr. Westport, CT: Greenwood, 1974.

History of the Thirty-Sixth Regiment, Massachusetts Volunteers, 1862–1865. Boston: Press of Rockwell and Churchhill, 1884.

Hopkins, L. W. *From Bull Run to Appomattox*. Baltimore: Fleet-McGinley, 1908.

Howard, McHenry. *Recollections of a Maryland Confederate Soldier and Staff Officer Under Johnston, Jackson, and Lee*. Baltimore: Williams and Wilkins, 1914.

Howe, Mark De Wolf, ed. *Touched With Fire: Civil War Letters of Oliver Wendell Holmes Jr., 1861–1864*. Cambridge, MA: Harvard University Press, 1946.

Humphreys, Andrew A. *The Army of the Potomac, July, 1863 to April, 1864*. New York: Charles Scribner's Sons, 1883.

———. *The Virginia Campaign: 1864 and 1865*. New York: Charles Scribner's Sons, 1883. Reprint. New York: Da Capo, 1995.

Hyde, Thomas W. *Following the Greek Cross, or Memories of the Sixth Army Corps*. Boston: Houghton, Mifflin, 1894.

Jackman, Lyman, and Amos Hadley. *History of the Sixth New Hampshire Regiment in the War of the Union*. Concord, NH: Republican Press Association, 1891.

Jones, Rev. J. William. *Personal Reminiscences, Anecdotes, and Letter of Gen. Robert E. Lee*. New York: D. Appleton, 1875.

Jones, Rev. J. William, D.D., Sec. of the Society. *Southern Historical Society Papers*. Vol. 7, *January to December, 1879*. Richmond, VA: No publication date.

———. *Southern Historical Society Papers*. Vol. 14, *January to December, 1886*. Richmond, VA: Wm. Ellis Jones, 1886.

Jones, Terry L., ed. *Campbell Brown's Civil War: With Ewell and the Army of Northern Virginia*. Baton Rouge: Louisiana State University Press, 2001.

———, ed. *The Civil War Memoirs of Captain William J. Seymour: Reminiscence of a Louisiana Tiger*. Baton Rouge: Louisiana State University Press, 1991.

Journal of the Military Service Institution of the United States. 3, no. 9 (1882).

Judson, A. M. *History of the Eighty-Third Regiment Pennsylvania Volunteers.* Erie, PA: B. F. Lynn, LOC date 1881.

King, W. C., and Derby, W. P., *Campfire Sketches and Battlefield Echoes.* Springfield, MA: King, Richardson, 1886.

Kirk, Hyland. *Heavy Guns and Light: A History of the 4th New York Heavy Artillery.* New York: C. T. Dillingham, 1890.

Lee, Capt. Robert E. *Recollections and Letters of General Robert E. Lee.* New York: Doubleday, Page, 1905.

Lincoln, Abraham. "Abraham Lincoln's First Inaugural Address, March 4, 1861." Accessed February 9, 2015, http://avalon.law.yale.edu/19th_century/lincoln1.asp.

Livermore, Mary A. *My Story of the War: A Woman's Narrative of Four Years Personal Experience.* Hartford, CT: A. D. Worthington, 1889.

Long, A. L. *Memoirs of Robert E. Lee: His Military and Personal History.* Secaucus, NJ: Blue and Gray Press, 1983.

Longstreet, James. *From Manassas to Appomattox: Memoirs of the Civil War in America.* Philadelphia: J. B. Lippincott, 1896.

Mackowski, Chris, and Kristopher D. White. "The Battle of the Bloody Angle, or 'Mule Shoe,' Spotsylvania Court House May 12, 1864." *Blue & Gray Magazine*, July 2009.

Matter, William. *If It Takes All Summer.* Chapel Hill: University of North Carolina Press, 1988.

McDonough, James Lee. *Chattanooga—Death Grip on the Confederacy.* Knoxville: University of Tennessee Press, 1984.

McKim, Randolph H. *A Soldier's Recollections.* London: Longmans, Green, 1910.

McPherson, James M. *Battle Cry of Freedom.* New York: Oxford University Press, 1988.

Meade, George. *The Life and Letters of George Gordon Meade, Major General United States Army.* 2 vols. New York: Charles Scribner's Sons, 1913.

Mertz, Gregory A. "No Turning Back." *Blue & Gray Magazine*, June 1995.

Michie, Peter S. *The Life and Letters of Emory Upton, Colonel of the Fourth Regiment of Artillery, and Brevet Major-General, U. S. Army.* New York: D. Appleton, 1885.southern historical

Military Quantitative Physiology: Problems and Concepts in Military Operational Medicine. Fort Detrick, Md: Office of the Surgeon General / Borden Institute, 2012.

Mixson, Frank. *Reminiscences of a Private*. Columbia, SC: State Company, 1910.

Mulholland, St. Clair A. *The Story of the 116th Regiment, Pennsylvania Volunteers*. n.p., 1903.

Nash, Capt. Eugene Arus. *A History of the Forty-Fourth Regiment New York Volunteer Infantry in the Civil War, 1861–1865*. Chicago: R. R. Donnelly and Sons, 1911.

New York Daily Tribune. February 2, 1864.

New York Daily Tribune. February 10, 1864.

New York Herald. July 24, 1878.

New York Herald. March 18, 1864.

Nolan, Alan T. *Lee Considered: General Robert E. Lee and Civil War History*. Chapel Hill: University of North Carolina Press, 1991.

Page, Charles. *Letters of a War Correspondent*. Boston: L. C. Page, 1899.

Papers of the Military Historical Society of Massachusetts. The Wilderness Campaign, May–June 1864. Vol. 4, . Boston: Military Historical Society of Massachusetts, 1905.

Papers of the Military Historical Society of Massachusetts. Vol. 13, 1913.

Papers of the Military Historical Society of Massachusetts. Vol. 33. 1905.

Peck, R. H. *Reminisces of a Confederate Soldier of Co. C. 2nd Va. Cavalry*. Fincastle, VA: 1913.

Perret, Geoffrey. *Lincoln's War: The Untold Story of America's Greatest President as Commander in Chief*. New York: Random House, 2004.

Peyton, George Quintus. "A Civil War Record for 1864–1865" Transcribed by Robert Hodge in 1929, Family History Library, Salt Lake City, UT. United States Collection, Film 6087252.

Pfanz, Donald C. *Richard S. Ewell: A Soldier's Life*. Chapel Hill: University of North Carolina Press, 1998.

Phelps, Charles, History of the Battle of The Wilderness and the Civil War experiences of Col./Gen. Charles E. Phelps. Grassland Foundation, Annapolis, MD, accessed December 21, 2016, http://www.grasslandfoundation.com/wp-content/uploads/2013/12/General-Phelps.pdf

Porter, Horace. *Campaigning with Grant*. New York: Century, 1906.

Powell, William H. *The Fifth Army Corps (Army of the Potomac)*. New York: G. P. Putnam's Sons, 1896.

R. E. Lee to J. B. Gordon, 22 February 1868. Hargrett Rare Book and Manuscript Library, University of Georgia Libraries, Athens.

Report of the Joint Committee on the Conduct of the War, March 4, 1864. page XIX accessed April 25, 2015, https://archive.org/details/reportofjointcommo1unit/page/n27.

Rhea, Gordon. *The Battle of the Wilderness, May 5–6, 1864*. Baton Rouge: Louisiana State University Press, 1994.

———. *The Battles for Spotsylvania Court House and the Road to Yellow Tavern May 7–12, 1864*. Baton Rouge: Louisiana State University Press, 1997.

———. *To the North Anna River*. Baton Rouge: Louisiana State University Press, 2000

———. *Cold Harbor*. Baton Rouge: Louisiana State University Press, 2002

———. *The Battles of Wilderness & Spotsylvania, National Park Civil War Series*. Published by Eastern National, copyright 2007

Rhodes, James Ford. *History of the United States*. 7 vols. London: MacMillan, 1904.

Richardson, Albert D. *Personal History of Ulysses S. Grant*. Boston: D. L. Guernsey, 1885.

Roebling, W. A. Major Washington A. Roebling's journal of the Overland Campaign. Bound volume #408. Fredericksburg and Spotsylvania National Military Park, Fredericksburg, VA.

Roth, Dave, with Greg Mertz. "The Fighting at Laurel Hill and the Spindle Field." *Blue & Gray Magazine*, Summer 2004.

Royall, William Lawrence. *Some Reminiscences*. New York: Neale, 1909.

Rusling, James. *Men and Things I Saw in Civil War Days*. New York: Eaton and Mains; Cincinnati: Curts and Jennings; 1899.

Rutan, Edwin P., II. *If I Have to Go and Fight, I Am Willing: A Union Regiment Forged in the Petersburg Campaign*. Park City, UT: RTD Publications, 2015.

Schaff, Morris. *The Battle of The Wilderness*. Boston: Houghton Mifflin, 1910.

Sheridan, Philip. *Civil War Memoirs*. New York: Webster, 1888. Reprint. New York: Bantam Books, 1991.

Sherman, William. *Memoirs of General William T. Sherman*. Ney York: Appleton, 1889. Reprint. Bloomington: Indiana University Press, 1957. Two Volumes in One.

Simpson, Brooks D. Email to author, March 20, 2015.

———. *Ulysses S. Grant: Triumph Over Adversity*. Boston: Houghton Mifflin, 2000.

Small, A. R. *The Sixteenth Maine Regiment in the War of the Rebellion, 1861–1865*. Portland, ME: B. Thurston, 1886.

Sorrel, Moxley. *Recollections of a Confederate Staff Officer*. New York: Neale, 1905.

Starr, Louis M. *Bohemian Brigade: Civil War Newsmen in Action*. New York: Alfred A. Knopf, 1954.

Steere, Edward. *The Wilderness Campaign: The Meeting of Grant and Lee*. Mechanicsburg, PA: Stackpole Books, 1960.

Stoddard, William O. *Lincoln's Third Secretary*. New York: Exposition, 1955.

Summer, Merlin E., ed. *The Diary of Cyrus B. Comstock*. Dayton, OH: Morningside, 1987.

Swinton, William. *Campaigns of the Army of the Potomac: A Critical History of Operations in Virginia Maryland and Pennsylvania from the Commencement to the Close of the War 1861–1865*. New York: Charles Scribner's Sons, 1882.

Tarbell, Ida M. *The Life of Abraham Lincoln*. New York: Doubleday and McClure, 1900.

Taylor, Walter. *Four Years with General Lee*. New York: D. Appleton, 1876.

———. *General Lee: His Campaigns in Virginia, 1861–1865*. Norfolk: Nusbaum, 1906.

Thirty-Eighth Congress, Session I, Ch. 13–14, 1864. Library of Congress accessed December12, 2015. https://www.loc.gov/law/help/statutes-at-large/38th-congress/session-1/c38s1ch14.pdf.

Thomas, Henry W. *History of the Doles-Cook Brigade, Army of Northern Virginia, C.S.A.* Atlanta: Franklin, 1903

Transcript of conversation between William Allan and R. E. Lee, March 3, 1868. William Allen Papers, #2764. Southern Historical Collection, Wilson Library, University of North Carolina at Chapel Hill.

US War Department. *The War of the Rebellion: A Compilation of the Official Records of the Union and Confederate Armies*. 128 vols. Washington, DC: 1880–1901.

Vandiver, Frank, ed. *War Memoirs: Autobiographical Sketch and Narrative of the War Between the States*. By Jubal Anderson Early. 1912. Reprint. Bloomington: Indiana University Press, 1960.

Wainwright, Col. Charles.S *A Diary of Battle: The Personal Journals of Colonel Charles S. Wainwright 1861–1865*. Edited by Nevins, Allan. Pickle Partners, 2014. Kindle

Walker, Francis A. *History of the Second Army Corps in The Army Of The Potomac*. New York: Charles Scribner's Sons, 1886.

Ward, Geoffrey C. *The Civil War: An Illustrated History*. New York: Alfred A. Knopf, 1990.

Weygant, Charles. *History of the One Hundred and Twenty-Fourth Regiment NYSV*. Newburg, NY: Journal Printing House, 1877.

White, William. *Contributions to a History of the Richmond Howitzer Battalion*. Pamphlet No. 2. Richmond, VA: Carlton McCarty, 1883.

Wilkeson, Frank. *Recollections of a Private Soldier in the Army of the Potomac*. New York: G. P. Putnam's Sons, 1887.

Wilson, James Harrison. *Under the Old Flag*. 2 vol. New York: D. Appleton, 1912.

Wing, Henry E. *When Lincoln Kissed Me: The Story of the Wilderness Campaign*. New York/Cincinnati: Abingdon, 1913.

Young, John Russell. *Around the World with General Grant*. New York: American News Company, 1879.

INDEX

Abbreviations
ANV = Army of Northern Virginia
AP = Army of the Potomac
Arty = Artillery
Bn = Battalion
Bde = Brigade
Btry = Battery
Cav = Cavalry
Cmdr = Commander
Div = Division

Alexander, E. Porter, Brig. Gen., CSA (Arty Cmdr), 67, 72, 77, 78, 91, 153, 264, 287
Alexandria Railroad, 36, 37, 193
Allan, William, 78
Alrich, 90, 96, 105, 109, 294, 252
Alsop (town), 105
Alsop Farm, 102; maps, 94, 100
Anaconda Plan, 32
Anderson, George T. "Tige," Brig. Gen., CSA (Bde, Cmdr), 66, 245
Anderson, Richard H., Maj. Gen., CSA (Bde & Div Cmdr), 40, 92, 95, 96, 138, 255; critical decision, 91, 93, 94, 165, 180, 259; maps, 89, 94, 159
Anderson Drive, 275, 280, 293, 300, 301
Antietam (Battle of), 10, 12, 31, 83, 84, 152, 279
Appomattox Court House, 5, 10, 83, 104, 161, 171
Army of Northern Virginia, 2, 38, 54; fighting style, 5, 100, 195; last major assault, 70, 179; leadership, 4, 5; logistics/supplies, 3, 10, 28, 46, 145
Army of the James, 2, 29, 31
Army of the Potomac, 3; critical decision, 18, 168; fighting style, 11, 90, 108, 114, 167, 183, 197, 270; leadership, 6, 12, 13, 14, 15, 19, 20, 21, 22, 23, 24, 25, 38, 80, 84, 104, 107, 124, 179; Lee's army as objective, 12, 33, 34, 36, 146, 1161, 173
Army of the Tennessee, 1, 33
Army of Virginia, 84
Atlanta, 164
Ayres, Romeyn B., Brig. Gen., USA (Bde Cmdr), 114, 189, 200, 201, 214

Baldwin, Clark B., Col., USA, 58
Ball, William H., Col., USA, 210
Ball's Bluff, 27
Banks, Nathan P., Maj. Gen., USA, 13, 33
Barlow, Francis C., Brig. Gen., USA (Div Cmdr), 128, 270, 284; salient 144, 149, 150, 152, 289; maps, 126, 148
Bartlett, Joseph J., Brig. Gen., USA (Bde Cmdr), 117, 119, 201; map, 118
Battle of Five Oaks, 171
Battle of Resaca, 83
Battle of the Crater, 171
Beauregard, P. G. T., Gen., CSA, 160
Belle Plain, 160, 285
Bennett Farm, 83
Bermuda Hundred, 124, 145
Big Black River, 173
Birney, David B., Maj. Gen., USA (Div Cmdr), 149, 234, 241; map, 148
Block House, 94, 253
Block House Bridge, 95, 96, 259
Block House Road (a.k.a. Old Court House Road), 91, 105, 120, 121, 130, 260, 261
Bloody Angle, 113, 122, 137, 152, 174, 278, 279, 294, 298, 301
Bragg, Braxton, Gen., CSA, 160, 173
Brandy Station, 4, 18, 19, 20, 32, 37, 193
Breather, James, Maj., CSA (H Arty), 102, 259
Bristol Station, 114
Brock Road, 39, 50, 90, 102, 116, 117, 180, 181, 232, 234, 246; cavalry operations, 96, 97, 98, 101, 156, 253; Orange Plank intersection, 52, 56, 65, 68; maps, 55, 62, 71, 118
Brooke, John R., Brig. Gen., USA (Bde Cmdr), 243, 290, 295
Brown, Campbell, Maj., CSA, 74, 217
Brown, Henry W., Col., USA (Bde Cmdr), 294
Brown, Hiram L., Col., USA (Bde Cmdr), 288
Brown, Joseph, Col., CSA (Bde Cmdr) 298
Brown House, 128, 149, 168, 284, 288, 289, 291

Buford, John, Brig. Gen., USA (Cav), 24
Burnside, Ambrose E., Maj. Gen., USA (Corps Cmdr), 12, 64, 123, 124, 134, 138, 151, 155, 163, 171, 221
Butler, Benjamin F., Maj. Gen., USA, 2, 22, 28, 109, 145
Butterfield, Daniel S., Brig. Gen., USA, 27

Catharpin Road, 58, 92, 96, 97, 99, 101, 246
Champion Hill, 173
Chancellorsville (Battle of), 19, 49, 67, 85
Chase, Salmon P. (USA Sec. of Treasury), 15
Chattanooga, 6, 15, 19, 20, 22, 24, 36, 87, 108, 122
Chewning Farm, 54, 85, 200, 212; map, 218
Chickamauga, 32
Chilesburg, 109, 125
City Point, 33, 130
Cold Harbor, 114, 172
Comstock, Cyrus B., Lt. Col., USA (Grant's Aide-de-camp), 20, 21, 28
Corbin's Bridge, 92, 105, 250, 254
Crawford, Samuel W., Brig. Gen., USA (Div Cmdr), 54, 200, 219
critical decision, xiii, xiv, xvi, 8; criterion xii; hierarchy xii; how presented. xv; results evaluation, 163–170
Culpeper (town of), 9, 190, 193; map, 35
Culpeper Mine Road, 80, 203, 206, 214, 217, 343n71
Custer, George A., Brig. Gen., USA (Cav Cmdr), 2, 98, 248, 249, 250

D-Day, 7
Dahlgren, Ulric, Col., USA (Cav), 23, 336n32
Dana, Charles (USA Asst. Sec. of War), 15, 19, 129, 155
Daniel, Junius, Brig. Gen., CSA (Bde Cmdr), 202, 214, 267, 274, 276, 293
Davidson, William L., Col., CSA, 58
Davies, Henry E. Jr., Brig. Gen., USA (Cav Bde Cmdr), 99, 250 251, 252

Index

Davis, Jefferson F., CSA President, 3, 4, 5, 10, 11, 23, 44, 69, 160
Davis, Joseph R., Brig. Gen., CSA (Bde Cmdr), 300
DeLacey, Patrick, Sgt., USA (Metal of Honor Recipient), 72
Denison, Andrew W., Col., USA (Bde Cmdr), 116, 117, 119, 258, 260, 261, 262
Devin, Thomas C., Col., USA (Cav), 98, 249, 250, 251, 252
Dodge, Grenville M., Brig. Gen., USA, 23, 122
Dole, George P., Brig. Gen., CSA (Bde Cmdr), 133, 145, 214, 215, 217, 267, 273, 275, 276, 281
Dole's Salient, 7, 41, 134, 135, 146, 152, 266, 272, 274, 279, 293, 294, 297
draft riots (New York), 1, 11
Drewry's Bluff, 157
Driver, William, Maj., USA, 144

Early, Jubal A., Maj. Gen., CSA (Div Cmdr), 99, 130, 342n58; Gordon's flank attack, 74–77, 80, 81, 207,211; map, 79
Ely's Ford, 37, 39, 44, 190, 247, 248; maps, 45, 89, 94
Emancipation Proclamation, 4, 11, 14, 10
Ewell, Richard S., Lt. Gen., CSA (Corps Cmdr), xiv, 38, 49, 50, 171, 235, 281, 300, 339n10; critical decision, 73, 77, 78,169, 179, 206, 207; Dole's Salient defense, 135, 276, 277; Ewell-Hill Gap, 50, 58, 69, 70, 213, 221, 236; Hill-Ewell Drive, 211, 212, 182, 280; Orange Turnpike/Saunders Field, 65, 69, 74, 114, 167, 202, 213, 217; salient artillery movement, 141, 142, 287, 288; maps, 45, 51, 159

Field, Charles W., Maj. Gen., CSA (Div Cmdr), 63, 70, 242
First Bull Run. *See* First Manassas
First Manassas, 2, 7, 10, 33
Forrest, Nathan B., Maj. Gen., CSA (Cav), 109

Fort Donelson, 152
Fort Monroe, 2, 28, 32, 34
Fredericksburg (Battle of), 84, 86
Fredericksburg, 35, 91, 130, 138
Fredericksburg Cemetery, 213
Fredericksburg Road, 8, 126, 129, 158, 285
Freeman, Douglas S., xi
Fremont, John C., Maj. Gen., USA, 15
French, William H. Maj. Gen., USA, 23
Furnace Road, 97, 248, 250, 251

Gayle (Gate) House, 8, 109, 111, 254, 284, 285
Germanna Ford, 39, 189, 207, 247; map, 45
Germanna Road, 46, 74, 89, 189, 206
Getty, George W., Brig. Gen., USA (Div Cmdr), 54, 61, 223, 225, 226, 234, 242; map, 55
Gettysburg (Battle of), 11, 12, 19, 21, 166
Gibbon, John Brig. Gen., USA (Div Cmdr), 128, 149, 270, 291; map, 148
Gordon, John B., Brig. Gen., CSA (Bde & Div Cmdr), 40, 73, 216, 268, 357n60; Gordon Flank Attack Trail, 205, 206, 214; May 6th flank attack, 77, 78, 80, 205, 209, 211; salient/Mule Shoe operations, 136, 150, 151, 215, 296; maps, 79, 148
Gordonsville, 10, 35, 39, 46, 48
Grant, Ulysses S., Lt. Gen., USA (Army Cmdr), xiv, 3, 17, 27, 29, 31, 54, 61, 80, 83, 152, 162, 172, 173, 182, 183, 193, 238; AP, 6, 22, 24, 53, 90, 110, 127, 129; critical decisions, 18, 26, 36, 84, 87, 104, 108, 122, 126, 145, 147, 154, 158, 164, 166, 167, 168, 169; rank of Lieut. Gen., 2, 16, 163; relationship with Gregg, David McM., Brig. Gen., USA (Cav Div Cmdr), 24, 105, 248, 249, 253; relationship with Lincoln, 15, 16, 161, 163, 337n44; maps, 35, 89, 99, 126, 148, 159
Gregg, J. Irvin, Col., USA (Cav Bde Cmdr), 99, 101, 251, 252
Gregg, John, Brig. Gen., CSA (Bde Cmdr), 62, 63, 227, 228

377

Index

Griffin, Charles, Brig. Gen., USA (Div Cmdr), 107, 166, 187, 204, 220, 258, 298, 299; Laurel Hill, 117, 119, 261, 262, 263; Saunders Field, 53, 189, 196, 198, 199, 200; maps, 51, 79

Halleck, Henry W., Maj. Gen., USA, 11–15, 336n35; correspondence with Grant, 80, 86, 104, 162, 189, 211; Sheridan recommendation, 24, 166

Hampton, Wade, Maj. Gen., CSA (Cav Div Cmdr), 97, 101, 106, 170, 180, 250, 254, 286

Hancock, Winfield S., Maj. Gen., USA (Corps Cmdr), 38, 123, 167, 171; corps movement 127, 129, 156, 158, 160, 168; Longstreet's flank attack, 66, 69, 241, 250; Mule Shoe, 142, 147, 149, 150, 151, 293, 295; Orange Plank Road Day 5, 6, 39, 40, 56, 57, 61–62, 85, 178, 226, 233, 234, 238; Po River operation, 110, 128, 130, 268; maps, 45, 62, 126, 148, 159

Harris, Nathaniel H., Brig. Gen., CSA (Bde Cmdr), 297, 298

Harris Farm, 129, 156, 158, 159

Harrison House, 126, 131, 272, 295, 297

Hartranft, John F., Col., USA (Bde Cmdr), 72

Haxall's Landing, 109, 110

Hay, John, 12

Hayes, Harry T., Brig. Gen., CSA (Bde Cmdr), 73

Hays, Alexander, Brig. Gen., USA (Bde Cmdr), 236

Henagan, John. W., Col., CSA (Bde Cmdr), xvi, 119, 228, 259, 260, 262

Heth, Henry, Maj. Gen., CSA (Div Cmdr), 64, 300, 301, 357n75; critical decision, 61, 178, 224, 226; Orange Plank Road (May 5th), 56, 57, 58, 225, 232, 234, 236, 238, 239; Orange Plank Road (May 6th), 227, 228; map, 55

Hill, Ambrose P., Lt. Gen., CSA (Corps Cmdr), 38, 46, 50, 61, 64, 178, 238; critical decision, 61; health and performance, 48, 49, 114, 128, 171; Hill-Ewell Drive, 211, 212; Plank Road (May 5th), 52, 54, 56, 57, 58, 225, 247; Plank Road (May 6th), 63, 64, 221, 227; maps, 45, 62, 159

Hoke, Robert F., Maj. Gen., CSA, 153

Holmes, Oliver Wendall, 90

Hooker, Joseph, Maj. Gen., USA, 12, 13, 14, 19 22, 23, 85

Horn, John W., Col., USA, 208

Huger, Frank, Lt. Col., CSA (Bn Cmdr), 120

Humphreys, Andrew A., Maj. Gen., USA (Meade's chief of staff), 7, 27, 37, 57, 83, 102, 170, 189, 209, 238, 240, 298; casualty estimates, 151, 152; plan, 43, 44, 46, 49, 167

Humphreys, Benjamin G. Brig. Gen., CSA (Bde Cmdr), 118, 119, 122

Jackson, Andrew (7th US President), 15

Jackson, Mississippi, 173

Jackson, Thomas J., (Stonewall) Lt. Gen., CSA, 48, 68, 81, 92, 169

James River, 33, 109

Jenkins, Micah, Brig. Gen., CSA (Bde Cmdr), 67, 229, 231, 242

Johnson, Edward (Allegheny), Maj. Gen., CSA (Div Cmdr), 92, 137, 213, 291; capture, 136; salient, 142, 149, 277, 280, 287, 288; maps, 141, 143, 148

Johnston, Joseph (Joe), E., Gen., CSA, 1, 5, 32, 33, 83, 86, 130, 162, 173

Johnston, Robert D., Brig. Gen., CSA (Bde Cmdr), 208, 209, 296, 297, 268

Johnston's Confederate horse artillery, 117

Joint Committee on the Conduct of the War, 12, 19, 22, 27, 334n7, 337n40

Jones, John M., Brig. Gen., CSA (Bde Cmdr), 212, 214; maps, 118, 126

Jones, Joseph R., 15, 16

Jones, Lorraine F., CSA (Arty), 281

378

Index

Kershaw, Joseph B., Brig. Gen., CSA (Div Cmdr), xvi, 67, 117, 228, 230; map, 71
Kershaw's Brigade, xvi, 116, 158, 259
Kilpatrick, Judson, Brig. Gen., USA (Cav Cmdr), 23, 25, 336n32

Lacey House, 185, 188, 195
Landrum House, 129, 138, 279, 288; map, 126
Laurel Hill, 40, 41, 95, 102, 120, 134, 152, 170, 255, 264, 293
Lee, Fitzhugh, Maj. Gen., CSA (Cav Div Cmdr), 40, 258, 260, 285; Brock Road, 101, 102, 103, 251; critical decision, 96, 100, 165; Spindle Field, 95, 115, 117, 259; Todd's Tavern, 97, 98, 246, 250, 255; maps, 99, 100
Lee, Robert E., Gen., CSA (Army Cmdr), 7, 9, 38, 39, 54, 92, 121, 160, 182, 221, 295; Command Style, 4, 5, 63, 164, 228, 276; critical decision, 43, 50, 56, 60, 70, 136, 141, 168, 169, 170; logistics/supplies issues, 3, 10, 44, 46, 74; Spotsylvania, 137, 138, 151, 152, 153; Wilderness, 61, 199, 213, 224, 226, 227, 230, 242; maps, 45, 71
Lincoln, Abraham (USA President), 7, 13, 64; critical decision, 10, 15, 163; 1864 presidential election, xv, 11, 14, 16, 162; relationship with Grant, 6, 17, 18, 31, 33
Logan, John A., Maj. Gen., USA, 23, 122
Lomax, Lundsford L., Brig. Gen., CSA (Cav Bde Cmdr), 101, 250, 251, 252, 345n31
Long, Armistead L., Brig. Gen., CSA (Arty Div Cmdr), 89, 281, 287, 288
Longstreet, James, Lt. Gen., CSA (Corps Cmdr), 32, 38, 74, 85, 92, 171, 358n84; arrival and fighting at The Wilderness, 63–66, 168, 169, 226, 227, 229, 238, 241, 242; initial position, 10, 48, 49; movement to The Wilderness, 50, 58, 61, 246, 247; wounding, 67–70, 231; map, 45
Lyle, Peter, Col., USA (Bde Cmdr), 116, 117, 261; map, 118

Lyman, Theodre, Lt. Col., USA (Meade's aide-de-camp), 57, 91, 120, 124, 234, 237

Mackenzie, Ranald S., Lieut., USA, 130, 131, 132, 272, 349n41, 361n162
Mahone, William (Billy), Brig. Gen., CSA (Bde Cmdr), 62, 66, 110, 128, 229
McAllister, Robert, Col., USA, 234, 235, 241; map, 62
McClellan, George B., Maj. Gen., USA, 9–12, 14, 15, 18, 19, 27, 29, 31, 84
McClellan, Henry B., Maj., CSA, 61
McClernand, John A., Maj. Gen., USA, 13, 16
McCoull House, 142, 280, 287, 295; maps, 142, 143, 148
McLaws, Lafayette, Maj. Gen., CSA, 92, 263
McMahon, Martin T., Col., USA, 131
McPherson, James B., Maj. Gen., USA, 23, 122
Meade, George G., Maj. Gen., USA (AP Cmdr), 23, 33, 52, 61, 120, 163, 171, 194; AP Commander, 13, 19, 20, 21, 22, 38, 84, 85, 123; headquarters, 29, 30, 31, 38, 164; Meade/Sheridan relationship, 90, 101, 104, 106, 252, 253
Merritt, Wesley, Brig. Gen., USA (Cav Bde & Div Cmdr), 2, 24, 101, 105, 253, 254, 257, 258, 261, 336n34; maps, 99, 100
Miles, Nelson A., Col., USA (Bde Cmdr), 137, 258, 290, 296
Milford Station, 129
Military Division of the Mississippi, 1
military draft, 1, 11
Mine Run Campaign, 12, 20, 34
Mine Run Road. See Culpeper Mine Run Road
Missionary Ridge, 108
Mississippi River, 1
Mott, Gresham, Brig. Gen., USA (Div Cmdr), 134, 146, 149, 234, 269; maps, 133, 148
Moulder, Bob, xv

Mud Tavern, 160
Mule Shoe. *See* salient
Myers Hill, 155, 156

National Park Service, xvi, 175, 177, 206
Neill, Thomas H., Brig. Gen., USA (Div & Bde Cmdr), 152, 209, 293, 295
Newton, John, Brig. Gen., USA, 23
Ni River (also spelled Ny River), 8, 127, 129, 138, 156; map, 45
North Anna River, 47, 56, 154, 172
Ny River. *See* Ni River

Orange Court House, 46
Orange Plank Road, 39, 50, 61, 63, 65, 104, 229; maps, 45, 55, 232
Orange Turnpike, 39, 50, 104, 185, 192, 196, 225; maps, 45, 51
Overland Campaign, xii, xiv, 3, 4, 25, 43, 161, 169, 171

Palmer, William H., Col., CSA, 58, 222
Parker's Store, 50, 54, 188, 219; maps, 45, 94
Pegram, William J., Lt. Col., CSA (Arty Bn), 79, 208,231
Pemberton, John C., Lt. Gen., CSA, 91
Pendleton, Alexander Swift (Sandie), Lieut. Col., CSA, 207
Pendleton, William N., Brig. Gen., CSA, 92, 94, 286
Pendleton Road, 92, 93
Peninsula Campaign, 9, 11, 25, 83
Perrin, Abner, Brig. Gen., CSA (Bde Cmdr), 297
Petersburg, 130, 171, 173
Phelps, Charles E., Col., USA (Bde Cmdr), 119, 262
Phillips, John H., Capt., USA, xv
Pickett, George E., Maj. Gen., CSA, 59, 240
Piney Branch Church, 90, 96; maps, 89, 99
Piney Branch Road, 251
Plank Road. *See* Orange Plank Road
Pleasonton, Alfred, Maj. Gen., USA, 23, 24

Po River, 128, 103, 128, 130, 147, 162, 181, 250, 268, 296, 297; map, 45
Poague, William T., Lt. Col., CSA (Bde Cmdr), 63, 64, 224, 227; map, 62
Pope, John, Maj. Gen., USA, 36, 84, 166,
Port Gibson, 173
Port Royal, 160
Porter, Horace, Lieut. Col., USA, 37, 86, 155, 193
Potter, Robert B., Brig. Gen., USA (Div Cmdr), 74, 298; map, 148

Ramseur, Stephen D., Brig. Gen., CSA (Bde Cmdr), 136, 214; monument, 278, 298
Rapidan River, 37, 68, 69; Union crossing of, 39, 163, 178, 186, 190
Rappahannock River, 87, 160, 174
Rappahannock Station, 131, 132
Rawlins, John A., Brig. Gen., USA, 88, 107, 124, 203, 348n20
Raymond (town of), 173
Reagan, John (CSA Postmaster Gen.), 173, 174
Rhodes, Robert E., Maj. Gen., CSA (Div Cmdr), 81, 268, 295
Richard's Shop, 58, 61
Richmond, 2, 10, 12, 34, 35, 46, 69, 91, 158, 161, 173, 232
Ricketts, James B., Brig. Gen., USA (Div Cmdr), 123, 194
Robertson's Tavern, 50, 186, 189, 197
Robinson, John C. Brig. Gen., USA (Div Cmdr), 102, 116, 117, 261, 262; maps, 51, 141
Rodes, Robert E., Maj. Gen., CSA (Div Cmdr), 136, 148
Roebling, Washington, Maj., USA (Warren's aide-de-camp), 218, 220, 261, 271
Rosecrans, William S., Maj. Gen., USA, 20
Russell, David A., Brig. Gen., USA (Bde & Div Cmdr), 130, 131, 133, 277,

salient, 136, 140, 142–145, 150–152, 155, 170; effect of terrain, 137, 267, 268, 294; tour, 267, 268, 278–281, 294; maps, 141, 148

Index

Sanders, John, Col., CSA (Bde Cmdr), 297
Saunders Field, 39, 53, 85, 107, 170, 189, 195; map, 51
Schaff, Morris, 211
Scott, Winfield, Brevet Lt. Gen., USA, 5, 13, 32
Scott House. *See* Sheldon house
Second Manassas, 36, 49, 83, 84
Seddon, James A. (CSA Sec. of war), 44, 91, 153
Sedgwick, John, Maj. Gen., USA (Corps Cmdr), xvi, 120, 200; death, 9, 40, 127, 130, 255, 256; Wilderness, 53, 73, 123, 206, 212, 238, 241; maps, 51, 79, 89, 94
Seven Days' Battles, 83, 84
Seymour, Truman, Brig. Gen., USA (Bde Cmdr), 73, 79, 208, 209, 210, 342n53
Shady Grove Church, 105, 248, 254, 286; maps, 89, 94
Shady Grove Church Road, 91, 118
Shaler, Alexander, Brig. Gen., USA (Bde Cmdr), 79, 208, 210
sharpshooters, 7, 119, 236, 261; effectiveness, 8, 123, 241
Sheldon House, 131, 272
Shenandoah Valley, 31, 33, 125, 130, 157
Sheridan, Philip H., Maj. Gen., USA (Cav Corps Cmdr), 25, 26, 40, 85, 97, 250, 336n35; as cavalry commander, 24, 101, 103, 109, 110; critical decision, 108, 169, 170; relationship with Meade, 104, 105, 106, 107
Sherman, William T., Maj. Gen., USA, 1, 20, 26, 27, 31, 33, 38, 130; AP command possibility, 19, 21
Shiloh (Battle of), 1, 31, 84, 122, 167, 354n14
Sigel, Franz, Mar. Gen., USA, 33, 157, 335n20
Smith, Benjamin H., Capt., CSA, 274, 275, 281
Smith, Martin L., Maj. Gen., CSA, 66, 137, 229
Smith, William (Baldy) F., Maj. Gen., USA 20, 21, 22

Sorrel, G. Moxley, Lieut. Col., CSA, 66, 69, 92, 95, 229, 241
Special Order No. 22 (CSA special order for ANV designation), 5
Spindle Field/House, 95, 102, 115, 119, 181, 225, 257; map, 118
Spotsylvania Court House, 4, 7, 40, 41, 136, 145; Grant's plan, 87, 88, 90; Sheridan, 103, 105; CSA actions, 95, 96, 102; map, 35
Stafford, Leroy A., Brig. Gen., CSA (Bde Cmdr), 201, 214, 231, 281
Stanton, Edwin M. (USA Sec. of War), 15, 34, 129
Steuart, George "Maryland" H., CSA (Bde Cmdr), 214; capture, 149, 292; salient, 137, 277, 280, 290, 291
Stoddard, William, 17
Stuart, James E. B. "Jeb," Maj. Gen., CSA (Cav Corps Cmdr), 38, 40, 91, 97, 108; actions near Laurel Hill, 96, 102, 118, 259, 260, 262; critical decision, 104, 169, 180; death, 170
Switzer, Jacob b., Col., USA (Bde Cmdr), 189, 276
Sykes, George, Maj. Gen., USA, 23

Tapp Farm, May 6th attack of, 64, 221, 228, 241; Lee's headquarters, 62, 176, 213, 227, 230, 232; maps, 55, 62, 71, 246
Taylor, Walter H., Lieut. Col., CSA, 63, 78, 190, 227
telegraph, usage of, 28, 189
Telegraph Road, 138, 157, 158, 160, 172, 285
Todd's Tavern, 52, 54, 90, 91, 96, 246, 247; cavalry action, 97, 98, 101, 105, 165, 180; maps, 89, 94, 99, 100
Torbert, Alfred T. A., Brig. Gen., USA, 25

Upton, Emory, Col., USA (Bde Cmdr), 2, 131, 150, 204, 274, 289; attack on Dole's Salient, 7, 133, 134, 135, 136, 149, 182, 272, 275, 277; critical decision, 129, 132, 165, 266; map, 133
US Military Academy. *See* West Point

Venable, Charles S., Lieut. Col., CSA, 58, 61
Verdiersville (New), 45, 50, 190, 212
Vicksburg, 1, 15, 36, 66, 173
Virginia Central Rail Road, 145

Wadsworth, James S., Brig. Gen., USA (Div Cmdr), 53, 57; death, 7, 66, 230, 242, 341n39; maps, 51, 62
Walker, Henry H., Brig. Gen., CSA (Bde Cmdr), 57, 275, 277; maps, 55, 62
Walker, James A., Brig. Gen., CSA (Bde Cmdr), 201, 214, 275, 281, 292, 293
Wallace, Lew, Maj. Gen., USA, 14
Warren, Gouverneur K., Maj. Gen., USA (Corps Cmdr), 23, 38, 39, 40, 41, 47, 85; command style, 53, 120, 124, 152, 199, 298; critical decision, 113, 114, 115, 116, 170, 181; Laurel Hill 102, 119; as Meade's possible replacement, 120; move to Spotsylvania, 90, 93, 95, 101, 257, 258; maps, 45, 51, 89, 94, 100, 118, 159
Washburn, Elihu B. (USA congressman), 13, 15, 30
Washington, D.C., 12, 14, 16, 22, 177; Grant's view of, 26–29, 337n44; influence of AP, 1, 19, 20, 27, 335n20
Washington, George, 13
Watson, Cyrus, Sgt., CSA, 145

Webb, Alexander S., Brig. Gen., USA (Bde Cmdr), 46, 240, 242; map, 62
West Point, 2, 13, 113, 131
Wheaton, Frank, Brig. Gen., USA (Bde Cmdr), 54, 236; maps, 55, 62
Wickham, William C., Brig. Gen., CSA (Cav Bde Cmdr), 98, 103, 251, 345n31
Widow Tapp. *See* Tapp Farm
Wilcox, Cadmus M., Maj. Gen., CSA (Div Cmdr), 57, 64, 151, 221; night of May 5th, 58, 59, 61, 226; maps, 141, 143, 148
Wilderness, The, 4, 7, 47, 85, 90, 96, 104, 105, 203, 204, 218, 225; terrain, 37, 38, 48, 49, 69, 80
Wilderness Tavern, 38, 46, 48, 52, 74, 185, 186, 190, 200, 218, 220, 221, 222; maps, 45, 71, 94
Willcox, Orlando B., Brig. Gen., USA (Div Cmdr), 110, 127, 239, 300
Williamsport, 173
Wilson, James H., Brig. Gen., USA (Cav Div Cmdr), 2, 25, 54, 166, 188, 190; actions at Todd's Tavern, 95, 103, 105
Wing, Henry, 87
Wofford, William T., Brig. Gen., CSA (Bde Cmdr), 66, 67, 229, 260
Wright, Horatio G., Brig. Gen., USA (Div & Corps Cmdr), 123, 127, 130, 156, 183, 293; map, 15